COLONIALISI

SCIENCI

Colonialism and Science

❧❦

SAINT DOMINGUE IN
THE OLD REGIME

James E. McClellan III

With a New Foreword by Vertus Saint-Louis

The University of Chicago Press
CHICAGO AND LONDON

For Jackie

The University of Chicago Press, Chicago 60637
Copyright © 1992 by The Johns Hopkins University Press
Foreword © 2010 by James E. McClellan III.
All rights reserved.
University of Chicago Press edition 2010

Printed in the United States of America

19 18 17 16 15 14 13 12 11 10   1 2 3 4 5

FRONTISPIECE
The passage outward from Saint Domingue: "Vue du Débouquement de St. Domingue."
Engraving by N. Ponce in Moreau de Saint-Méry (1791).

ISBN: 978-0-226-51467-3 (paper)
ISBN: 0-226-51467-6 (paper)

Library of Congress Cataloging-in-Publication Data

McClellan, James E. (James Edward), 1946–
  Colonialism and science : Saint Domingue in the old regime / James E. McClellan III ;
with a new foreword by Vertus Saint-Louis.
      p. cm.
  Includes bibliographical references and index.
  Originally published: Baltimore : Johns Hopkins University Press, c1992.
  ISBN-13: 978-0-226-51467-3 (pbk. : alk. paper)
  ISBN-10: 0-226-51467-6 (pbk. : alk. paper) 1. Science—Haiti—History—18th century.
2. Haiti—History. I. Saint-Louis, Vertus. II. Title.
  Q127.H2M38 2010
  509.7294'09033—dc22
                    2010016562

♾ The paper used in this publication meets the minimum requirements of the American
National Standard for Information Sciences—Permanence of Paper for Printed Library
Materials, ANSI Z39.48-1992.

CONTENTS

❦

CONTENTS

FOREWORD

It was in 1994 that I first was able to read Professor James McClellan's *Colonialism and Science: Saint Domingue in the Old Regime*. My own personal experience seems to have prepared me to grasp the thrust of this work.

I come from a peasant family in Haiti. My father owned a primitive distillery. The rusty barrel of an old iron machine for pressing sugar cane lay abandoned in the courtyard, whereas all the milling machines round about were made of wood. I asked about this anomaly and learned that when the iron mills broke down, a repairman had to come from Port-au-Prince. In a part of the world without roads, people preferred to discard their iron mills and be content with wooden ones. I have also not lost sight of the fact that poor worker productivity in Haiti, which I could see everywhere around me, is responsible for many problems in a country where, after Independence, politics became a major industry. With iron losing ground to wood on the Central Plain in Haiti in the first half of the twentieth century, when I later came to study history, these insights became associated in my mind with another development: the shift in the course of the Haitian Revolution from the manufacture of semirefined crystallized sugar (called *terré*, or clayed sugar) to the rudimentary distilling and making of a crude sugar (*rapadou*) that retained its molasses. Questions of science and technology thus appeared to me as major problems of Haitian society, problems from which the exercise of political power could not be separated. These concerns, together with my earlier medical studies, may explain why, unlike my countrymen who care only about social and political issues, I became so interested in the works on natural history by the first historians of colonization, and in particular, in the formal distinction José de Acosta made in 1590 between moral history and natural history.

A version of this foreword was originally written in December 2005 for a French translation of *Colonialism and Science*.

A passage from Cassirer on natural history seemed especially insightful. This author cites Diderot, who uses disciplinary requirements proper to natural history as a foil against the inveterate habit of reasoning whereby one imagines uncovering every truth without the intervention of experience. I saw this as a lesson against a form of history that concerns itself with the interpretation of known facts rather than, as Diderot recommends, seeking out new facts. The botanist finding new plants makes one think of the historian finding new sources. Cassirer then juxtaposes Buffon's *Natural History* (1749–1788) with Newton's *Principia* (1687). He admits that Buffon's body of work cannot be compared to Newton's, but he cannot envision Buffon without Newton. I ascribed even more importance to the commentaries cited in the introduction to the 1894 reprint of Acosta that linked contemporary geophysical research to his first works, and out of the third volume of that same reprint I retained Humboldt's remarks on the origin of certain of Halley's discoveries. Despite the obvious naiveté of some of the descriptions of the first writers on natural history, I perceived them as representatives and witnesses of a certain exoticism; I saw them as pioneers of the movement that would lead Darwin to America. However, I did not yet associate modern European science with the global expansion of Europeans. But bearing in mind my predispositions, when I read McClellan's volume, it immediately became clear to me that the scientific enterprise was a feature of French colonization in Saint Domingue, a feature overlooked by historians, but one that can help us understand both the past and the present in Haiti. This book seemed to me to be definitive.

* * *

In the first part of *Colonialism and Science*, James McClellan sets out to familiarize the reader with "the case of Saint Domingue." He reviews material factors, historical development, demography, the economy, and urban life with a fine-grained command of the facts and a keen sense of historical analysis. In the chapter on population, the mention of free people of color, who were descended from Africans but looked like whites and were socially accepted as whites, belies the pretense of pure European blood as the basis for the colony's racist policies and government. McClellan conveys the vitality of daily life in an urban context in a society where the plantation was the economic unit of production. Reading rooms, the press, pornography, Masonic lodges resembled their counterparts in France. But there were no universities.

Contrary to what one might have thought based on the first part alone, McClellan does not limit himself to Saint Domingue in portraying France's scientific efforts in the colonies; in the second part of the book, he highlights what might be called the combined scientific administration of the French colonies under the supervision of the French Ministry of the Navy operating in the context of great power rivalries. Although the concerns of pure science are not totally absent, the reader discovers natural history applied to medicine and economic production and astronomy with the aim of resolving problems of navigation to support commerce. In addition to the Parisian Académie des Sciences, the Jardin du Roi in Paris and the Navy's headquarters in Brest, Rochefort, and Toulon were institutions directly engaged in promoting science in the colonies where botanical gardens proliferated and astronomical observations were made.

A third part is devoted to the Cercle des Philadelphes of Cap Français in Saint Domingue. Blanche Maurel and Pierre Pluchon have presented the colonists' Cercle des Philadelphes as an organization of Freemasons hiding their designs and political activities beneath the veil of scientific research. McClellan insists on the essentially scientific objectives of the Cercle. Here I would emphasize the difference between science as practiced in metropolitan France and as practiced by colonists in Saint Domingue. The science of the colonists was very limited compared to that in France. It focused uniquely on descriptive natural history and gave little thought to mathematics. Science in the colonies was also the handiwork of the Metropolis, which for its own ends used its human and material resources and true scientific institutions as well. The Saint Domingue physician Arthaud defined the scope of the Cercle. He thought the Metropolis should stop shipping botanical specimens to the colony because too many of them died on the crossing. France persisted in this project because it had the means to do so. France did not think of colonization without thinking of science and technology. The planters of Saint Domingue, in their quest for political autonomy, also understood the importance of science for their projects.

At the end of his account of science in Saint Domingue in the Old Regime, McClellan, in the guise of a conclusion, poses a question: can it be doubted that the history of science and medicine in Saint Domingue depended primarily on the demands of French colonization and not on some logical or institutional necessity of science in France? The author leaves little doubt as to his answer. Colonial science reflected the needs of French colonization. This question and its answer, in my view, pertain to discussions about the origins and role of science that, as far as I know, are of special interest in scholarly circles in the developed world. My periodic rereading of McClellan's work leads me to believe that these discussions should be of concern primarily to the people of colonized territories that are, today, nominally independent, but in fact are not at all. These peoples, even more than those of the West, should wonder about the links between colonization and science. This is a theoretical question of great practical importance because of its connections to politics, and this book treats the history of science in a colony whose situation after independence is exemplary of many countries in Africa. The title, *Colonialism and Science*, invites reflections on the deep connections between and among society, science, and the state.

* * *

From the beginning of modern colonization, the great powers of Europe endeavored to link commerce, technology, and science in the interests of their national agendas and political strategies. Portugal's *Casa da Mina e India*, the bureau charged with the African trade in gold and slaves and then with the spice trade with India, accumulated experience in matters of navigation, geography, and cartography. Spain imitated Portugal in creating the *Casa de Contratación* in Seville, which, in association with the Council of the Indies, concerned itself with commerce, cosmography, geography, hydrography, and natural history. Developments of great importance for the future of science were at play here. For the first time, institutions of commerce and institutions of learning were closely conjoined under the aegis of politics. These linkages could have an unhealthy dimension, but they also resulted in an increase in scientific knowledge. New knowledge had to produce tangible results, forcing scientists to turn away from purely theoretical contemplation and abstract philosophy. Valued and valuable knowledge was above all practical. Resolutely turned toward the future, this knowledge leaned on the past to conquer the present.

Greek and Arabic science developed the theoretical tools to address questions concerning latitude and longitude. Portuguese sailors started to apply these theories not on land, but at sea. Portuguese colonization was already launched through the practice of sailors when learned men such as Abraham Zacuto of the University of Salamanca and his student José Vizinho came to offer their services. News of the exploration of West Africa brought Jerome Munzer to Lisbon. A new class of men was in the process of defining itself as a society of "knights of the ocean sea." Science learned to rely on their prows and prowess. Along with the practice of these seafaring knights came a rethinking of space in terms of latitude and longitude. New knowledge spread from the seas to the whole planet. Whereas the Greeks and Arabs had recorded six hundred species of plants, a Spanish physician-naturalist quickly discovered three thousand new plant species in Mexico. Astronomers found in the heavens, and botanists in the diversity of American plants, reasons to free themselves from the tutelage of old books.

The powers of northwest Europe—Holland, England, and France above all others—sought to combine technique and science in embarking on colonization. It seems the laurels went to England in the sixteenth century, perhaps explaining the superiority it quickly exhibited at sea. Having gained its independence, the Netherlands profited from the conflict between Spain and England to achieve an economic breakthrough by replacing Portugal in Asia. The Dutch were very pragmatic, efficient colonizers. They excelled in cartography, for example, but it is difficult to say that they theorized about colonization. In France, Paris was in the interior. With Rouen as a go-between, highly placed individuals could profit from the aggression of the Normans and Bretons who took the initiative against Portugal and Spain. In England, London emerged as the political and economic capital, and in the struggle to wrest the monopoly on the world's riches from the Iberian powers, close connections quickly developed between craftsmen and learned academics ranging from mathematicians and geographers to philosophers. The thought of Francis Bacon that inspired the Royal Society of London was not foreign to the dreams of empire of the Hakluyt brothers or John Dee. When Richelieu and Colbert launched French colonization, they understood the importance of science for the success of their enterprise. In the eighteenth century, France took advantage of much more than age-old experience in tapping science to install in Saint Domingue an economy and a society founded on slavery.

★ ★ ★

The historian who sees only commerce and slavery in the colonial experience is like those who, as Avelino Teixeira da Mota emphasizes in speaking of the great voyages of exploration, concern themselves with economic, social, and political questions associated with these enterprises, and neglect the sum of the technical means that made them possible. Colonization began with technology. By the end of the eighteenth century, colonization could rely on the progress of cartography, geography, mathematics, astronomy, and the physical sciences in the service of navigation. From the case of Saint Domingue we can pass to the case of Haiti and a number of African countries. Their political leaders are engaged in social and political struggles, battling each other without mercy. While they denounce the very real oppression of Western powers, they do not systematically look to science to improve production and commerce. They fail to secure science for themselves—one of the essential elements of the power of the West. People of the Third World have every interest in learning from Saint Domingue's example through the work of James McClellan.

Vertus Saint-Louis
École Normale Supérieure
Port-au-Prince, Haiti

PREFACE AND
ACKNOWLEDGMENTS

This study explores the interpenetration of French science and French colonialism in the eighteenth-century French West Indian colony of Saint
Domingue (modern-day Haiti).

French Saint Domingue was extraordinarily and surprisingly important in its heyday prior to the French and Haitian revolutions both as a
colony and as a center of science in the Americas. The preeminence of
French science among national science traditions of the eighteenth century, the importance of eighteenth-century Saint Domingue in the history of European colonialism, and recent historiographical interest in the
topic of colonialism and science also suggest that a full-length study of
science in old Saint Domingue can enrich our understanding of the historical interactions of European science and colonialism.

As it turns out, one can chronicle Old Regime Saint Domingue and
the story of science there in fine and fascinating detail, but I want the
reader to know that I did not deliberately set out to create this comprehensive account of science and civilization in Saint Domingue. Rather, I
began innocently enough to undertake a short article in English about the
Cercle des Philadelphes, the scientific society founded in Saint Domingue
in 1784. I originally planned this article as a minor follow-up to an earlier
study of eighteenth-century scientific societies and a modest contribution
to the burgeoning scholarly literature of science and colonialism.

When I began research, however, I was chagrined to discover huge
historiographical lacunae and major disjunctures between and among
various scholarly literatures. Old Saint Domingue proved to be a scholarly *terra incognita*. Readers may be familiar with Haitian history or with
the current literature of science and colonialism, but the particular topic
of science in colonial Saint Domingue probably does not ring many bells.
Separated small communities of experts notwithstanding, and despite the

impression left by the accompanying bibliography, colonial Saint Domingue is not well known, certainly to Anglophone audiences, and the full story of science in Saint Domingue has not been told until now in any language. Put another way, the scholarly circumstances under which this study emerged were and are the very opposite of dealing with a more familiar subject, like the French Revolution, say, or the trial of Galileo, where one begins on familiar and collectively well-trodden ground.

Because no general study exists in English, part I of this work provides an initial survey of French colonization in Saint Domingue, and it forms the basis for further investigations into the history of science and medicine in the colony. What did colonization entail in Saint Domingue? What made the colony so successful? What problems did it face for which European science and medicine were seen to provide solutions? Why would a French scientific society arise in such a seemingly out-of-the-way place? Any study of the connections between colonialism and science in Saint Domingue will fall short unless one first addresses the nature of French colonialism there in the eighteenth century. Part II then explores the manifold ways in which seventeenth- and eighteenth-century science and medicine facilitated (or seemed to facilitate) colonial development. Finally, part III presents an institutional history of the Cercle des Philadelphes, the colony's scientific society. Elevated to the status of a royal society by Louis XVI, the Cercle des Philadelphes became the Société Royale des Sciences et des Arts du Cap François in the very year, 1789. A provincial, if tropical French academy thus crowns the larger history of science in Saint Domingue, and the institution merits detailed treatment both for its intrinsic interest and for the additional light it sheds on the dynamics of science and colonialism in the Old Regime.

The following introduction discusses the reasons for Saint Domingue's peculiar "invisibility" and the many interesting things the case has to contribute to the historiography of science and colonialism. Here I may be permitted to note my sense of being a weary traveler back from the frontiers of research laden with a rich and elaborate tale of Old Regime science in the tropics. Indeed, I've come to feel somewhat like the historian's Ancient Mariner, with "grey beard and glittering eye," and too often in my enthusiasm over this project I've accosted even very knowledgeable historians and eighteenth-century experts with tidbits of my tale, only to be greeted with benign incredulity. In any event, I am glad, at last, to have this story off my chest, and I hope any novelties it contains are of interest and use to those who come across it. I alone, of course, take responsibility for its errors.

Many people helped bring this project to fruition. Heartfelt thanks go first and foremost to Jackie McClellan for her continued support while I again tracked ghosts of the eighteenth century. Her contributions are many, and the present dedication represents the most minimal public expression I can muster of her role and my affection. I also thank my friend Paul Kay for his moral support.

My friend and professorial colleague Harold Dorn did not suggest the title for this volume, but he read several parts of the manuscript with his usual keen intelligence, made many useful suggestions, and otherwise exerted a major influence on this book's coming-to-be. Professor Gerald L. Geison of Princeton University also critically reviewed a number of sections and recommended many improvements, but I owe him as much for the friendly counsel that so sustained the author over the life of this project. I was flattered that Professor Charles C. Gillispie of Princeton took the time to read an early version of the entire manuscript, and his acute suggestions substantially strengthened the presentation and interpretation.

I thank my colleagues and friends Professor Salvatore Prisco III and Professor J. D. Trout for reading parts of this work in manuscript. Friends Philip R. Reilly, M.D., and Neil C. Bodick, M.D., saved me from more than one medical gaff. Discussions and correspondence with Professor Dirk J. Struik of M.I.T. and Professor Jacob W. Gruber of Temple University advanced my thinking on a number of issues. I also warmly acknowledge the particular courtesies shown to me by M. Robert Cornevin, permanent secretary of the Académie des Sciences d'Outre Mer in Paris. Similarly, I cannot omit M. Pierre Pluchon of Paris, who so graciously shared his work on Saint Domingue. That we disagree over the origins of the Cercle des Philadelphes does not minimize my debt to M. Pluchon's scholarship or person.

In the course of preparing this volume, I worked in virtually every division of the New York Public Library, and I wish to express my deepest appreciation to all associated with that most remarkable public institution. I must single out the Schomburg Center for Research in Black Culture of the New York Public Library, where I spent many productive days in most pleasant surroundings, and I specifically wish to acknowledge Mr. André Élizée, who not only guided me to and through the treasures of the rare-book and archival collections of the Schomburg Center but who did so much to facilitate my introduction to Haiti and Haitian culture.

I also did considerable research for this study at the American Philosophical Society in Philadelphia, and I am again happy to credit the aid of several friends and acquaintances at that wonderful establishment. I especially thank Mr. Roy E. Goodman, whose expertise, initiative, and goodwill proved invaluable in uncovering many gems held in the APS

Library; Ms. Hildegard Stephans, not the least for her helping me discover the extraordinary maps of old Saint Domingue stored in the Bank Building; Ms. Elizabeth Carroll-Horrocks for so kindly facilitating access to manuscript materials at the APS; and Mr. Frank Margeson for his genuinely expert photographic work.

Frontispiece and Figures 1, 3, 4, 5, 6, 8, 14, and 16 were reproduced with permission from originals in the Manuscripts, Archives and Rare Books Division, Schomburg Center for Research in Black Culture, The New York Public Library, Astor, Lenox and Tilden Foundations. Figure 9 was reproduced with permission from original in the Rare Books and Manuscripts Division, The New York Public Library, Astor, Lenox and Tilden Foundations. For Figures 2, 7, 10, 11, 12, 15, and 18, photographs are courtesy of the Bibliothèque Nationale, Paris. Figure 13 was reproduced with permission from original in the Archives Nationales, Centre des Archives d'Outre-Mer, Aix-en-Provence. Figures 17, 19, 20, 21, 22, 23, and 24, and Maps 2 and 4, were reproduced with permission from originals in the American Philosophical Society Library.

For helping on site and via the mail with questions about materials in their collections, I thank M. Pierre Berthon and Mme Claudine Pouret at the archives of the Académie des Sciences of the Institut de France; Ms. Roberta Zonghi of the Boston Public Library; Messrs. James H. Hutson and Fred Bauman of the Library of Congress; and J. F. Maurel, conservateur général of the Centre des Archives d'Outre-Mer in Aix-en-Provence.

I owe a word of gratitude to many otherwise anonymous reference desk librarians and stack people at several other magnificent research centers, including the libraries of Columbia University, the New York Academy of Medicine, various departments of the Bibliothèque Nationale in Paris, and the Archives Nationales in Paris—both at the Hôtel Soubise and at the Section d'Outre-Mer, formerly on the rue Oudinot.

It would be silly to single out the Haitian people whose ardor in the colonial period and today merits deep admiration, but several individuals showed me exceptional kindnesses during a research trip to Haiti. I wish particularly to acknowledge Frère Constant of the Bibliothèque Haïtienne des Frères de l'Instruction Chrétienne in Port-au-Prince, M. Guy Maximilien of the Institut Français d'Haïti, M. Gérard Fombrun of the Moulin-sur-Mer establishment in Montrouis, and M. Jacques Cauna, then attached to the French Embassy in Haiti. I am indebted to Jacques Cauna not only for his hospitality when I visited Haiti, but also for continuing to share with me his own scholarly research and insights into colonial Saint Domingue.

A book such as this one customarily comes embellished with acknowledgments of the support of distinguished foundations and public agencies. Thinking I would save myself time and effort, I foolishly neglected to apply, but I do acknowledge my own institution, Stevens Institute of Technology, whose sabbatical leave allowed me to do most of the research for this book.

Finally, I extend my warmest thanks to Dr. Robert J. Brugger and his colleagues at the Johns Hopkins University Press, not only for their expert and most professional handling of this manuscript but also for being so decisive in taking it on in the first place.

NOTE ON WEIGHTS, MEASURES, AND CURRENCY

Carreau. Measure of area. Equivalent to 2.79 English acres, based on the metric equivalent of the *carreau* given by Maurel and Taillemite in *DPF*, p. 16. Maurel and Taillemite in *DPF*, pp. 14–15, also refer to a "Carreau de Saint Domingue," a square with a side of 350 "pieds," which converts to over 3 acres to a *carreau*. See also Zupko, pp. 6, 37; Cauna (1987), p. 253.

Lieue. French league. Standard measure of length. One *lieue* equals 2,000 *toises*, 3.89 kilometers, or 2.42 English miles. See Maurel and Taillemite in *DPF*, p. 16; Zupko, pp. 95–96.

Livre. Unit of currency. Separate systems of currency ruled in France and in the colonies. The colonial *livre* equaled two-thirds of the *livre tournois*, the standard in France. As a result, small confusions sometimes arise in the literature concerning figures in livres reported for Saint Domingue. Every effort has been made to report figures correctly either in colonial livres or livres tournois, as appropriate. Unless otherwise indicated, references to unspecified livres intend colonial livres. See Lacombe, pp. 14–17; Stein (1979), p. 138; Maurel and Taillemite in *DPF*, pp. 14, 16; Devèze, p. 282; Cauna (1987), p. 253.

The currency systems of the Old Regime are not comparable with today's, and the value of the livre—colonial, tournois, or otherwise—is difficult to express in terms of francs, dollars, yen, or whatever current benchmark. Still, various estimates of the value of the livre have been made. The range of five to ten U.S. dollars for the livre proves a helpful rule of thumb in interpreting the figures, and one needs to bear in mind that any number of livres or dollars is worth more to someone with nothing than to someone with millions. In making such heuristic conversions, the salary of a skilled worker at the end of the Old Regime (on the order of 500 livres tournois annually) often provides the benchmark for comparison. On these points, see Gillispie (1983), p. xi; Ott, p. 54; Stein (1988), p. 132. Roche (1981), chap. 3, throws additional light on this question.

Livre.   Measure of weight. The *livre de Paris* equals 489 grams or 1.08 English pounds. See Maurel and Taillemite in *DPF,* p. 16; Zupko, pp. 98–99.

Pied.   Measure of length; slightly longer than an English foot. Maurel and Taillemite in *DPF,* p. 16; Zupko, p. 134.

Toise.   Standard measure of length. One *toise* equals 1.949 meters, or somewhat over 2 English yards.

COLONIALISM AND
SCIENCE

Et enfin si ce sort réelement déplorable [de sa perte absolue] était celui qui menace Saint-Domingue, il serait nécessaire encore à l'Histoire des Nations de réunir un chapitre au grand livre de l'expérience, pour montrer ce qu'a été, dans sa courte existence, une Colonie que sa nature, sa splendeur et sa destruction rendraient le premier exemple de ce genre dans les annales du monde. Nous recherchons avec curiosité les ruines des anciens établissemens qui ont fait la gloire et l'admiration des peuples et nous recourons à de pénibles recherches, à de savantes dissertations pour arriver, par elles, à la connaissance imparfaite des moeurs et du gouvernement de ces peuples. La Grèce, l'Italie appellent chaque jour, les observateurs. Eh bien! avec cet Ouvrage, on méditerait sur Saint-Domingue, et sans doute on peut, à quelques égards, retirer autant de fruit de cette contemplation que de celle des débris d'Herculanum.

—M. L. E. Moreau de Saint-Méry,
*Description de la Partie Française de
l'Isle de Saint-Domingue,* 1797

The Case of
Saint Domingue

THE RISE OF MODERN science and the colonial expansion of Europe after 1492 constitute two fundamental and characteristic features of modern world history. Since the Scientific Revolution of the sixteenth and seventeenth centuries, science has become increasingly important to mankind in a variety of ways: intellectually as a body of knowledge and a means of knowing; technically as applied science; and as a social "institution" and enterprise of ever-increasing scale and importance. Parallel to the cultural progress of science, during the period from Columbus through the great colonial empires of the nineteenth and twentieth centuries, the global expansion of the West has been a major historical force, utterly transforming the world's social, political, and economic character.

How the twin enterprises of science and colonialism have interacted in the past and how to account for their interactions are questions of historical interest that have recently attracted a new attention. At its most general, the present study examines connections between the world-historical movements of science and colonialism, at least as seen in the nontrivial case of the eighteenth-century French West Indian colony of Saint Domingue (modern Haiti) (Map 1). Indeed, one can find no better case than that of Saint Domingue for investigating the interactions of science and colonialism at the close of the eighteenth century.

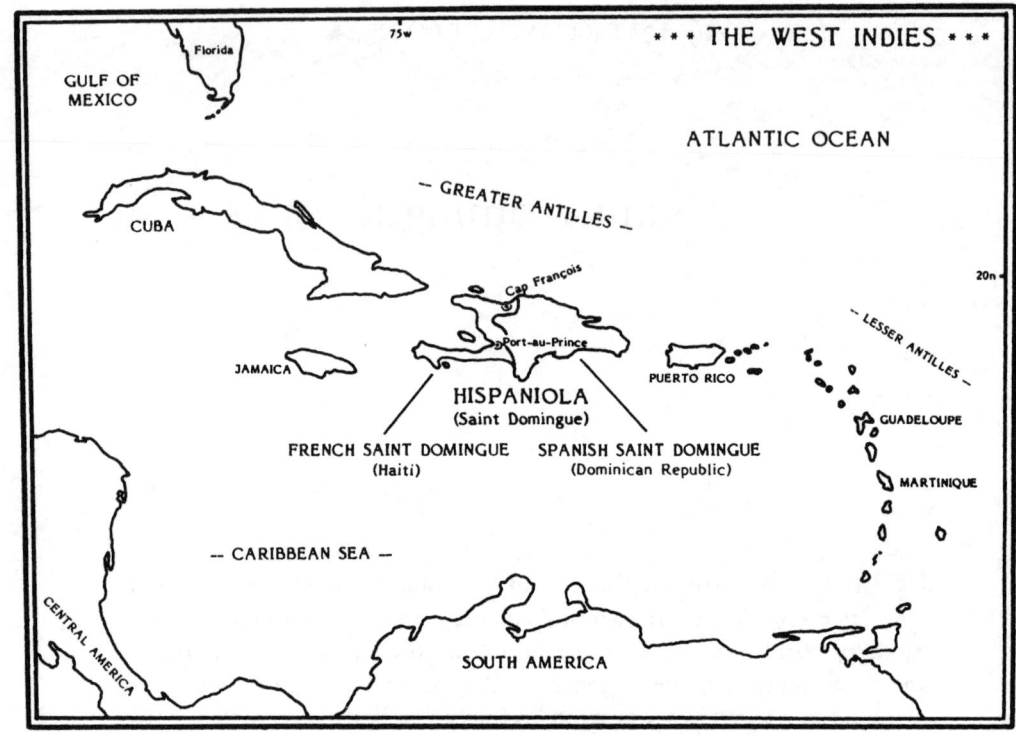

MAP 1. The West Indies.

Old Saint Domingue:
World's Richest Colony and American Center of Science

In the last decades of the eighteenth century, 4,800 sailing miles from France, the French colony of Saint Domingue flourished on the western end of the Caribbean island of Hispaniola where today the Republic of Haiti languishes. Experts agree that at its peak in the 1780s French Saint Domingue was the single richest and most productive colony in the world. The French West Indian possession was then the world's leading producer of both sugar and coffee. Already by 1776 the single colony of Saint Domingue produced more wealth than the whole of the Spanish Empire in the Americas.[1] Contemporary Saint Domingue meant a great deal to France, where one person in eight is said to have lived off colonial trade in one way or another.[2] Eighteenth-century Saint Domingue has rightly been called a "pivot" in the world economy of the day and a key link in the global system of contemporary European colonialism.[3] Other rich and important colonies existed, to be sure, and France was not the only colonial power in the eighteenth century, but Saint Domingue was nonetheless, if one may be permitted the expression, the jewel in the crown of contemporary European colonialism.

The sheer size and density of the colony's eighteenth-century population helped put Saint Domingue at the cutting edge of contemporary European colonial development. In 1790 French Saint Domingue harbored approximately 560,000 inhabitants. That figure about equals the combined populations of contemporary New York and Pennsylvania, then the two most populous states in the new United States. The half a million and more in French Saint Domingue, however, packed into a mountainous area about the size of the state of Maryland, making Saint Domingue one of the most densely populated spots in the New World at the time.[4] Furthermore, slaves constituted nearly 90 percent of the colony's population (500,000), most born in Africa. In dramatic contrast, only 700,000 slaves inhabited the whole of the United States in 1790, and the most heavily slave state, South Carolina, was only 60 percent slave.[5] French Saint Domingue stood at the unhappy vanguard of contemporary race relations and the forging of multiracial towns and societies in the Americas, and it represents what H. Hoetink has labeled "the historical and geographical core of Afro-America."[6] The fact that in 1791 its slaves rose up in history's largest and most successful slave revolt exemplifies the extraordinary character of Saint Domingue as a slave society. To remind the reader, that revolt culminated in the establishment of the Republic of Haiti in 1804, the second independent country in the Western Hemisphere after the United States.

The urban civilization that arose in Saint Domingue added to the colony's contemporary fame. The main town in the colony, Cap François, had a population of 18,850 in 1789.[7] Cap François was thus about the same size as Boston in North America or Dijon in France.[8] (Founded comparatively late in the century, 1749, the capital and second city of Saint Domingue, Port-au-Prince, was only half the size of Cap François.) Cap François was vibrant and lively, much more so than its sleepy Caribbean and South American counterparts. The quality and diversity of life in Cap François, for a privileged minority at least, largely matched anything offered in Havana, Philadelphia, New York, or Charleston.[9] Its busy port and opportunities for profit, its social and institutional complexity, its government and military presence, its theater, its scientific and intellectual life, its decadence—all conspired to make Cap François and the social order that flourished in Saint Domingue at the end of the eighteenth century an advanced and in many ways attractive center of civilization on American shores.

In a like vein colonial Saint Domingue became a major, if not the major center of organized and institutionalized science and medicine in the Western Hemisphere at the end of the eighteenth century. The best evidence to support this perhaps surprising claim comes from the existence and activity of the government-supported scientific society in Saint Domingue, the Société Royale des Sciences et des Arts du Cap François (a.k.a. the Cercle des Philadelphes). The Cercle des Philadelphes began on private initiative in 1784; it received official recognition and government subventions in 1786, and on May 17, 1789, Louis XVI signed royal letters patent—the last such ever issued—for Saint Domingue's scientific society. Fully the equivalent of any French provincial scientific society, the Cercle des Philadelphes maintained an extraordinarily active existence in the short period until its demise in 1792: it published five volumes of scientific memoirs and had others in press before the collapse of the colony; it elected an international membership of over 160, including Benjamin Franklin; it established extraordinary, formal ties with the Académie Royale des Sciences in Paris, in addition to effecting institutional affiliations with the American Philosophical Society in Philadelphia and several French provincial academies and musées. The recognition and support given to the Cercle des Philadelphes testify to its exceptional institutional status as a scientific society in the latter eighteenth century and to the maturation of scientific culture generally in the colonial context of eighteenth-century Saint Domingue.

But the Cercle des Philadelphes did not emerge out of a vacuum. Rather, it culminated a century of scientific involvements in Saint Domingue that entailed many different agencies and powers of French colonialism, all of which variously expressed interest in using science and useful knowledge to improve aspects of the economy and governance of the colony. For example, colonial medicine constituted an important instrumentality of colonization and an essential institutional and scientific base on which science and the Cercle des Philadelphes developed further. This topic will be explored in subsequent chapters, but in the final analysis, it may well be that contemporary European medicine became more firmly established in eighteenth-century Saint Domingue than anywhere else in the Americas at the time. In addition, several formal botanical gardens and experimental agricultural stations enriched the colony's scientific infrastructure, as we will see. The French military—notably the royal navy—was another colonizing institution in Saint Domingue that made considerable use of scientific and technical knowledge, particularly in a series of astronomical and cartographic expeditions sponsored by the navy ministry, the Marine Academy at Brest, and the Royal Academy of Sciences in Paris. In these and like ways it becomes clear that science became firmly institutionalized in old Saint Domingue and a key element in government policies for colonial development.

Few, if any other contemporary colonial settings matched colonial Saint Domingue as an enclave for organized science in the world outside of Europe at the end of the eighteenth century.

In North America only Philadelphia and Boston can compare in importance as ultramarine extensions of European science. In the latter 1780s Cap François and the Cercle des Philadelphes certainly rivaled, if they did not eclipse, Philadelphia with its American Philosophical Society (1768) and Boston with its American Academy of Arts and Sciences (1780), the two major scientific centers in the United States at the time.[10] Elsewhere globally, the Dutch East Indian possession of Batavia with its scientific society, the *Bataviaasch Gnootschap van Kunsten en Wetenschappen* (1778), presents the only other case comparable with Saint Domingue as an extra-European colonial scientific outpost in the next to the last decade of the eighteenth century.[11] The Batavian Society of Arts and Sciences and the Cercle des Philadelphes resembled one another closely, both arising as tropical learned societies at great distances from mother countries overseas. Important differences exist, but the cases of Dutch Batavia and French Saint Domingue suggest a causal connection between eighteenth-century European colonialism and the organization of colonial scientific societies.

Whatever a further comparison of Batavia and Saint Domingue might reveal, if Saint Domingue were not the leading scientific center outside of Europe in the 1780s, it was indubitably among the leading centers. Given its preeminence (or something close to preeminence) both as a colony generally and as an overseas center of organized science, Saint Domingue becomes the ideal choice for investigating colonial science in the late eighteenth century.

Destroyed in the French and Haitian revolutions, French Saint Domingue has more or less disappeared from historical consciousness, but a better understanding of this leading colonial center of science in the Old Regime can help show in detail how (together and separately) the great enterprises of science and colonialism have grown to transform the world.[12]

Historiographical and Conceptual Background

A significant new historiographical trend has recently developed, as historians of science have turned to systematic inquiries into science as an element of Western colonialism and imperialism. Put better, they have returned to the study, as recent work takes up the pioneering and earlier-ignored efforts begun by I. B. Cohen in 1959, Donald Fleming in 1962, and especially George Basalla in 1967. Among the newer literature, Lucille Brockway's *Science and Colonial Expansion* (1979) broke new ground in its exploration of the role of scientific institutions, particularly the botanical garden, in the expanding British empire of the nineteenth century.[13] The relatively recent articles by Weinberg (1978), Struik (1984), and Chambers (1987) confirm this new trend in the study of colonial science in Spanish and Latin American contexts. Headrick's book (1981) opened up the study of technology and colonialism, and Osborne's (1987) investigation of zoology in the Second Empire pioneered the study of French colonial science. Similarly, in an ongoing series of works published throughout the 1980s and into the 1990s Lewis Pyenson has sharpened questions for debate in his studies of German, Dutch, and now French overseas expansion in the nineteenth and twentieth centuries.[14] A conference on science in colonial settings held in Australia, papers from which appeared in 1987 in the volume, *Scientific Colonialism,* evidences this new scholarly trend.[15] An international congress sponsored by the American, Spanish, and Latin American societies for the history of science and held in Madrid in 1991 likewise testifies to the vigor with which scholars presently pursue the topic of colonial science.

The study at hand builds on this evolving body of literature. It seeks to provide a definitive case study of European science in an important New World setting at a key point in time, and to analyze generally the role of science in a colonial context. The present work strives to complement the existing literature in three ways. One is simply to add a new case study for comparative purposes. Experts perceive the historiography of science and colonialism as still not well developed, and they have called for further case studies in order for the inquiry to advance.[16] Second, because recent work has tended to focus on the nineteenth and twentieth centuries, coincident with the colonial and scientific expansion of that period, the present concern with the eighteenth century provides a needed point of chronological contrast at an earlier stage of colonial development. Third, scholars have essentially ignored the colonial experience of the French.[17] Studying French colonial science promises to rectify what may be a fundamental imbalance in the literature to date. The French case is especially relevant for the eighteenth century, in that France was then the leading scientific power of Europe, and at the same time it shared with Britain the role of the world's leading colonial power. The case of science in old Saint Domingue adds a valuable and unique perspective to considerations of science and colonialism raised in this new historiography.

To speak of science and colonialism thus poses and juxtaposes a set of interpretative questions. Two points of usage and definition need to be clarified at the outset. First, for our purposes "science" in the eighteenth century should not be considered a monolithic intellectual or social entity. It needs to be thought of functionally, as a dynamic complex of ideas, institutions, individuals, and social groups embodying and operationally defining natural knowledge across a range of social levels and in extended social connections. Such a broad definition encompasses, for example, Euler's empyrean work on the three-body problem and voodoo mesmerism among slaves as elements of one grand cultural constellation that defined eighteenth-century science. Scientific ideas deserve close consideration when and where they come up, but one must be alert to the whole of the scientific enterprise and anything even vaguely "scientific." Such an inclusive approach seems the best strategy in tracing a way through the uncharted story of science in colonial Saint Domingue.

The second point to be emphasized here concerns a fundamental and largely unrecognized duality inherent in the Janus-like topic of science and colonialism. The story is a dual one. One of its aspects concerns how science and the scientific enterprise formed part of and facilitated colonial development. The other deals with how the colonial experience affected science and the contemporary scientific enterprise. These two major facets of the subject cannot be entirely separated, but they also should not be confused. It is one thing, in other words, to be concerned with the significance of science for colonialism and to inquire into the role of science in the colonial process, as in improving maps or slowing disease. It is another thing to consider the impact of the colonial experience on science and the scientific enterprise—how, for example, the global experience of colonialism affected the science of botany. These two aspects of "colonial science" have not always been clearly distinguished, nor have they received equal treatment. Earlier historical work in this area dealt mostly with the effects of colonialism on science and the scientific enterprise. Only with the more recent historiographical trend just mentioned has attention shifted more to the role and effect of science on colonialism.

However that may be, a major moral of this tale is that science and organized knowledge did not come to Saint Domingue as something separate from the rest of the colonizing process but, rather, formed an inherent part of French colonialism from the beginning. In other words, the French did not colonize Saint Domingue and then import science and medicine as cultural afterthoughts. French science and learning came part and parcel with French colonialism, virtually as a "productive force." Because they were already so institutionalized in the culture and state apparatus of France in the eighteenth century, science and medicine played—or seemed to play—important roles in the development of French West Indian colonial interests.

As for why science became associated with the forces and agencies of colonialism, common wisdom responds with the notion of the perceived utility of science and the practical ends it could serve. It has almost gone without saying that the powers controlling colonization considered science and organized knowledge useful for their ends; according to Lewis Pyenson, "It becomes a relatively straightforward task, if by no means trivial, to identify the practical origins" of colonial science.[18]

This concept of the useful application of science provides a principal point of departure for understanding the connections between science and colonial development in Saint Domingue: the drumbeat of utility sounds a tattoo throughout this volume. Clearly, French science received support because of the many practical benefits it offered or seemed to offer in establishing, maintaining, and enlarging French Saint Domingue. Science could investigate causes of tropical diseases, determine Saint Domingue's precise location, identify useful flora and fauna, and explore potential new avenues for colonial development, particularly promising economic enterprises. The full armamentarium of French science was constantly, relentlessly employed for its real or perceived utility across a broad range of activities.

That science was useful for colonial development stands to reason, but the criterion of science's utility by itself remains of only limited value. The issue is not simply the utility of science in a colonial context, but utility for whom and for what ends and purposes.

Without doubt, French science and medicine bolstered slavery and the slave system in Saint Domingue. We will encounter many telling examples of this unfortunate but unsurprising fact: plantation owners inoculating their slaves against smallpox by the tens of thousands, the Cercle des Philadelphes categorizing Africans according to kinds of work that best suited them, botanists importing the "precious" breadfruit tree to feed slaves, and so on. Yet, it would be misleading to claim that French colonial science and medicine served solely or even primarily to fortify the institution of slavery in Saint Domingue, for a number of other areas of colonial development—colonial cartography or economic botany among them—benefited more immediately and more substantially than did science in the service of slavery in Saint Domingue. But the issue cannot be science supporting slavery versus science supporting other aspects of contemporary colonialism, for slavery absolutely permeated Saint Domingue. The colony's entire character and texture derived from slavery. Slavery was never far from the surface of thought or conversation, and no colonist questioned the institution of slavery or the proposition that colonial development depended fundamentally upon slavery. Directly and indirectly, then, scientific expertise deployed to build the colony constituted support for slavery and the slave system, and the theme of science and slavery in old Saint Domingue provides a sober reminder that in some instances at least science has served "unfreedom" and human oppression.

As a second major theme, this book argues that French colonial science flourished in Saint Domingue not so much from private or strictly colonial interests or initiatives as from metropolitan imperatives. The mercantilist French state turns out to have been the primary agent in control of colonial development in Saint Domingue, and the French state alone possessed the resources to enlist science effectively in the interests of colonial development. Individual colonists and colonial enterprises were essential to the growth of a colony like Saint Domingue, to be sure. Clearly, too, contemporary science ministered to the economic and social interests of planters and others in the colony. Furthermore, in the long run, as we will see, the agroindustrial and capitalist plantation economy established in Saint Domingue ran counter to the mercantilist interests of the French state that promoted colonization earlier in the seventeenth and eighteenth centuries. Nevertheless, all evidence suggests that mainly the paternalistic French state and not private parties promoted science and medicine in Saint Domingue. Almost without exception, support and patronage for science in Saint Domingue came from one or another of the bureaucratic tentacles of prerevolutionary French colonial government.

Given the ties between contemporary science and slavery and the fact that colonial science and medicine primarily served the ends of an economically and politically retrograde French state, the case of colonial Saint Domingue leads one to question received notions of science as inevitably a progressive historical force. Although this theme is too large to be developed in any analytical detail here, the case forces us to think of eighteenth-century science less as a heroic and forward-looking enterprise, dispelling the clouds of benighted ignorance, and more as an agent beholden to reactionary powers and the entrenched interests of the status quo. Along these lines, the story of science and medicine in Old Regime Saint Domingue challenges our usual view of science in a progressive French Enlightenment. At the least, colonial Saint Domingue provides a sensitive measure of the spread of Enlightenment thought, as well as its science.[19]

A further point to be emphasized here at the outset concerns the European perspective in which Saint Domingue and science in Saint Domingue have to be considered. Saint Domingue effectively constituted another French province or administrative region (*généralité*). The territory happened to be in the West Indies, but organizationally it resembled Béarn and Navarre, say, or the Boulonnais. Saint Domingue became a "provincial" scientific center, and the colony needs to be thought of in the larger context of French and Parisian science and culture. In other words, the inquiry at hand turns out to be fully comparable with studying science in Montpellier and the Languedoc or in Dijon and Burgundy in the eighteenth century. To ask, as some might, "Who were the scientists in Saint Domingue?" or "What discoveries did colonists make?" misses the point that, despite the existence of colonial science and scientists, the history of science in Saint Domingue forms part of the larger history of Europe and European science. The story of science in Saint Domingue speaks to the place of science in eighteenth-century French society and contemporary French colonialism. Therein lies its historical significance, not the extent or quality of scientific research done in the West Indies.

The history of science in old Saint Domingue also illuminates the growing levels of social support for science in the provinces and in general since the seventeenth century. In analyzing the nature and agency of support for science in such a seemingly out-of-the-way place like colonial Saint Domingue, we can learn more of the reasons why science received support in major centers like Paris and, in general, why science has increasingly become such a key part of world culture since the seventeenth century.

Finally, our general knowledge of Saint Domingue and the place of the colony in French and Caribbean history becomes enriched by a deeper understanding of the story of science there. Discovering how and to what ends the scientific enterprise came to be employed in Saint Domingue helps to reveal the nature of Saint Domingian society and culture on the eve of the French Revolution. Given the economic importance of Saint Domingue and the issue of slavery in the French and Haitian revolutions, whatever new we can learn about the colony becomes even more valuable historiographically. But more, the state of the colony at its height at the end of the eighteenth century provides a valuable indicator of the nature and extent of European colonization of the Americas at a point three hundred years after Columbus. On Hispaniola in 1492 Columbus established the first European outpost in the new world, Navidad, and at the same spot the Saint Domingian town of Cap François later rose and flourished.[20] Saint Domingue thus provides a notable measure of the "progress" of European colonization in three centuries from 1492. As the world marks the quincentennial of Columbus's discovery of America, focusing on the tricentennial center of Saint Domingue seems especially appropriate.

Patterns of Colonial Development

Two overarching phases characterize modern colonial history: a first, commercial phase running from 1500 to the end of the eighteenth century, and a second, industrial phase unfolding in the nineteenth and twentieth centuries. Reaching its peak in 1789, Saint Domingue thus culminated the first phase of modern colonialism, and the case typifies the state of European colonialism at a critical point just prior to the Industrial Revolution. Situating the colony in the larger context of European colonial history will enhance our appreciation of Saint Domingue and its larger importance.

France and Britain vied for top honors as the leading European colonial powers in the last decades of the Old Regime.[21] France entered late into the colonial game, with the first French colonies established along the Saint Lawrence in Canada in the opening decade of the seventeenth century. The French began colonizing the West Indies in 1624 with the settlement on Saint Christopher (Saint Kitts); the colonies in Guadeloupe and Martinique followed in 1635, with the colony in Saint Domingue not officially getting off the ground until 1665.[22]

France's eighteenth-century colonial empire was not very extensive. In 1789 it consisted of Saint Domingue, Martinique, Guadeloupe and lesser possessions in the West Indies; Cayenne and French Guiana in South America; a toehold in India at Pondicherry; and an Indian ocean outpost in the Mascarene islands off Madagascar. The French empire in fact contracted at the end of the eighteenth century, with the Seven Years' War marking an important turning point. In a famous exchange the French gave up Canada in 1763, they lost in India, and they ceded their claim in Louisiana to Spain.[23] To consider the comparatively small colonial holdings of the French as constituting one of the two most important colonial empires of the later eighteenth century does seem startling at first glance.

On the surface the contemporary British colonial empire appears far more powerful. But the seeming power of the British colonial empire in the eighteenth century is somewhat illusory. The British suffered a major loss to their colonial system with the independence of the American colonies, and governing Canada bought little advantage. Similarly, British colonial control of India did not amount to much before the end of the eighteenth century.[24] In the West Indies British sugar production about equaled the French. However, the main British possession there, Jamaica, was itself only half the size of Saint Domingue in terms of land mass and population, and exports from Saint Domingue doubled those of Jamaica in 1788.[25] Overall, their respective Antilles possessions remained comparatively less important to Britain than to France.[26] No British or any other colonial possession could compare with the single French holding in Saint Domingue. That colony alone thrust France to the forefront of European colonization and international trade in the eighteenth century. As Robert Louis Stein has put it, "Together, the French colonies formed an empire that was the envy of France's rivals. It was a commercial empire whose value far outweighed its limited geographic size."[27]

France and Britain established several different types of colonies.[28] With its tentative enclaves in India, Britain continued the Dutch example of strictly commercial colonies. In North America Britain and France also began new settlement colonies of largely self-sufficient agricultural communities. But in the eighteenth-century the agroindustrial colonies created by the English and French powers in the West Indies became the most important and productive.[29]

Unlike other colonial situations, few native peoples remained by the time the English gained control of Barbados and Jamaica in the middle of the seventeenth century or, contemporaneously, when the French established themselves in Guadeloupe, Martinique, and Saint Domingue, because the Spanish had exterminated the aborigines of these islands a century earlier. Culminating in the second half of the eighteenth century, both the English and the French developed similar plantation systems of production.[30] Two main elements characterize the plantation system: slave labor and single-crop agriculture. Caribbean plantations provided a ready market for slaves and propelled the international slave trade. Owners used slaves as agricultural laborers primarily to tend and process sugar cane, but slaves also worked on plantations producing coffee, indigo, cotton, and cacao. Under the hot sun of the tropics the agroindustrial colonies of the Antilles and the Caribbean produced an abundant harvest and incredible surplus wealth, and these colonies fueled trade and economic activity around the triangle linking Europe, Africa, and the West Indies. Although Indian and East Indian trade was important to the expanding world economy at the end of the eighteenth century, the West Indian colonies, and especially Saint Domingue, became the prizes.

Everyone seems to agree that commercial and maritime activities drove the first major phase of European colonial expansion. According to Marx and Marxist historiography, merchant capital supplied the motive force behind colonization in this first phase.[31] Colonial development in the period *was* commercial, dependent on the activity of trading, and suited to the "manufacturing" economies of England, France, and the United Netherlands. For Marx, through the eighteenth century, surplus capital gleaned from trading and colonial development stimulated industrial and economic growth at home, rather than the other way around.

Several other features characterize colonial development in this first phase, including an early reliance on state-chartered trading companies.[32] The East India Company in Britain, the Dutch East India Company, and the French Compagnie des Indes Occidentales, along with other similarly privileged corporations, undertook the pioneering work of creating colonies and forging international networks of trade. The Compagnie de Saint Domingue, for example, played an important role in colonizing the southern department of Saint Domingue.

The economic theory of mercantilism and colonial policies based on mercantilism formed a major feature of colonialism in its first phase. Adam Smith coined the term in 1776, but mercantilism had long been the cornerstone on which the colonial policies of the European powers rested.[33] The fundamental premise based the wealth of a nation on its reserves of hard currency. Colonies, therefore, were to produce goods that could be taxed, sold, or reexported for the benefit of the mother country or, more particularly, the home government. Mercantilism further entailed the idea not of free trade, but that colonial commerce should be a national monopoly and the exclusive prerogative of the mother country and its government, which hence controlled colonial trade and limited exchanges within the colonial economy itself. In practice, illicit trade often penetrated barriers erected by protectionist, mercantilist policy, and abiding tensions existed between colonies and home countries that expected their due—nowhere better illustrated than in the thirteen colonies in British North America. Nonetheless, mercantilism was the universal economic theory of the eighteenth century that guided government policies toward colonies. Colonialism in the nineteenth century dispensed with mercantilist theory.

Modern colonialism in its second major phase saw a dramatic change in the nature and role of colonies and in the scope of colonization worldwide coincident with the Industrial Revolution. In the nineteenth century mother countries no longer considered colonies as producers of wealth to be siphoned off; now they were seen, on the one hand, as suppliers of raw materials for domestic industries and, on the other, as markets for consumption of European industrial production. This fundamental change in the nature of colonialism represents a shift from old-style merchant capitalism to an industrial capitalism characteristic of the nineteenth century.[34]

Standing at the apex of colonialism in its first, mercantile phase, Saint Domingue thus represents a colony whose successful development ultimately undermined the economic principles on which it was based. The importance to Saint Domingue of trade and merchant capital should already be clear. By the same token, the mercantilist economic policies pursued by the French crown proved profitable for the state but were not in the interests of traders and colonists, for whom free trade was the desideratum. (It should be noted again that production in Saint Domingue has been characterized as "agro–industrial, capitalist and servile."[35]) A fundamental conflict thus emerged between the longer-term capitalist interests of merchants, planters, and colonists in general and the strictly mercantilist policies of royal government. Saint Domingue's extensive contraband trade (with Curaçao, other points in the Caribbean, and with New England) evidences this conflict.[36] So, too, does the famous "partial exclusive" of 1784. This seemingly liberal reform of government trade policy opened the three major ports of Cap François, Port-au-Prince, and Les Cayes as free ports to international trade. But, in fact, as Jean Tarrade has shown, this move maintained, even tightened, trade restrictions with the colony.[37] Royal government was not about to loosen the profitable grip it held on the colonial economy, even as the colony outgrew the initial economic and ideological principles on which it was established. State-supported scientific development of the colony hastened this trend, and, one way or another, revolution loomed for Saint Domingue.

The "Lost" Colony of Saint Domingue

One can understand why historians of science have not studied Saint Domingue before, but if the colony was as important at its height at the end of the Old Regime as has been suggested here, one can well wonder why it has so receded from view. Saint Domingue was not an ephemeral outpost that disappeared mysteriously, like Roanoke. But having been destroyed in the throes of the French Revolution beginning in 1789 and then replaced by Haiti at the start of the nineteenth century, old Saint Domingue has more or less disappeared from history. Several factors combine to explain how such a once great and terrible society as Saint Domingue could largely vanish from view.

A primary reason for the invisibility of Saint Domingue is that, as an outcome of the French Revolution, the French lost Saint Domingue: the colony collapsed in the turmoil of the French and Haitian revolutions. Violence erupted in 1789 in Saint Domingue as word arrived of the outbreak of the revolution in France. The slaves of Saint Domingue began their revolt in 1791, and thereafter peace did not return to the island for nearly thirty years. The English and Spanish became active belligerents in Saint Domingue in 1793, and in 1794 effective control of a reduced and embattled French territory passed to the black insurrectionist leader, Toussaint Louverture. The French attempted to reassert their authority on several occasions, with Napoleon finally sending a large expeditionary force under General Leclerc in 1802. That force was defeated by its black revolutionary adversary, and Haitian independence was declared in 1804.

The loss of the colony meant that France could not tap Saint Domingue and its resources after 1815. The former American colony played no part in the French colonial empire in the nineteenth century. The loss of Saint Domingue extends further, however, for Saint Domingue itself became lost in history; it ceased to exist. The French and Haitian revolutions in Saint Domingue smashed the extraordinarily productive and profitable, albeit inhumane, system the French had installed; as a result, at its birth Haiti ceased to be a factor in the world economy.[38] Of critical importance for this study, the nascent scientific communities and infrastructures in place in Saint Domingue in 1789 likewise vanished. No further maturation ensued; no continuity with later developments followed. The brilliant episode of colonial science in Saint Domingue comes to an abrupt and absolute close with the fall of the Old Regime in France.

It is instructive relative to the "invisibility" of Saint Domingue to consider how Haitians and the French today look back on the colony. While illiteracy remains astoundingly high in Haiti—reduced from 90 percent to closer to 75 percent—Haitians know their national history well and are fairly familiar with its colonial past.[39] Still, if asked, most Haitians will identify Saint Domingue as the town of Santo Domingo in the neighboring Dominican Republic. Further, Haiti today is an impoverished and deeply third-world country, remote from the mainstream and not considered important. Because of that, a tendency exists on the part of more cosmopolitan outsiders to conclude that Haiti (or its predecessors) can never have been important.[40]

To a degree, a recollection of Saint Domingue lingers in the collective memory of the French today. The former colony maintains something of the Wild West, frontier image that it possessed in the eighteenth century, and the character of the rich uncle from the sugar islands has not completely disappeared as a popular stereotype. By the same token, most people in France are also likely to think of the town of Santo Domingo in the Dominican Republic or the Dominican Republic itself when asked about Saint Domingue, and a confusion between Haiti and Tahiti (pronounced similarly in French) is not uncommon.[41]

The colony's name and its precise geography constitute significant obstacles in the way of knowing about the colony. Confusion reigns in this matter of names. Columbus christened the island he discovered in 1492 Hispaniola (or Little Spain), and Hispaniola remains the correct official name for the Caribbean island as a whole today. The Arawak Indians originally called the same island Hayti, or "mountainous land." Bartolomo Columbus founded the *town* of Santo Domingo on the southeast coast of Hispaniola in 1496, and after a while the whole island and region came to be called by that name. Saint Domingue is the French version of Santo Domingo. (Pronounced, Saint Domingue sounds Portuguese, with a drawn-out, nasal *i* in the second syllable of *Domingue*.) Thus French usage in the eighteenth century termed the whole of Hispaniola Saint Domingue, or even, harking back to its swashbuckling days, "the isle and coast of Saint Domingue." Contemporaries likewise referred to the island as having a French part (*la partie française*) and a Spanish part (*la partie espagnole*), with Saint Domingue town as the capital of the Spanish part.[42] In the nineteenth century the names of the western and eastern parts of the island changed to Haiti (1804) and the Dominican Republic (1844), with Santo Domingo (or Saint Domingue in French) still the capital of the latter. The U.S. Marines added to the confusion by calling the whole island Santo Domingo during their occupation of Haiti from 1915 to 1931.[43] Given this overlapping jumble of terms, to refer to Saint Domingue or French Saint Domingue, as we do here, and mean the French colony as it existed on Hispaniola in the eighteenth century will not spark much recognition in most audiences.

However that may be, a crucial factor contributing to a general ignorance of Saint Domingue is that the colony has been overlooked historiographically, at least by scholars writing in English. More will be said about this unexpected obstacle in a moment. A significant literature concerning Saint Domingue exists in the French language, however, and this literature provides a *sine qua non* for the present study.

Small communities of French and Haitian historians have studied colonial Saint Domingue diligently since the early part of this century, and they have produced a large and high-quality body of work in the process.[44] But the French historian Gabriel Debien stands foremost among historians of Saint Domingue. M. Debien's scholarly writings on a variety of topics relating to Saint Domingue span the half century from the 1930s into the 1980s, and all who study Saint Domingue owe Debien a large scholarly debt, not the least for his long-running series of bibliographic reviews that have kept track of a diverse set of sources and that for many years have focused questions for research.[45] A number of French scholars, including Jacques Cauna, Charles Frostin, François Girod, Blanche Maurel, Pierre Pluchon, and Jean Tarrade, directly or indirectly students of Debien, have built on his work and have produced many valuable studies treating eighteenth-century Saint Domingue. Over the years, too, a number of well-qualified Haitian historians have added significantly to this literature, notably Adolphe Cabon, Dantès Bellegarde, Jean Fouchard, and, more recently, Georges Corvington.

Saint Domingue is also a topic for historians working in Francophone communities in Canada and Africa.[46]

A smaller number of researchers writing in French have explored facets of the history of science and medicine in Saint Domingue. The history of medicine in the colony has attracted the most scholarly attention, and one needs to mention works of high importance to the present study by Rulx Léon, Louis Joubert, Zvi Loker, and, most recently and most notably, Pierre Pluchon. Regarding the history of science, despite its flaws, Blanche Maurel's article on the Cercle des Philadelphes, originally published in 1938, remains an important study of that key institution; Pierre Pluchon has recently contributed another useful overview of the Cercle des Philadelphes.[47] The investigation by Georges Anglade into the agricultural project of the Cercle des Philadelphes is also a notable addition to knowledge of that institution. P. Fournier and Alfred Lacroix have produced valuable studies of French missionaries and scientists working overseas in the eighteenth century. And several investigations, notably those by Henri Stehlé, have concerned themselves with the history of economic botany and importing useful plants into Saint Domingue and the Caribbean.

Of course, Saint Domingue is not unknown to Anglo-American historians. Students of American history learn about Saint Domingue and the rum trade with the colonies and states of New England, about the role of the colony in the American Revolution, and about émigrés who fled to America from Saint Domingue after 1793. L. L. Montague, C. L. R. James, Thomas Ott, and David Patrick Geggus are major English-language contributors who have studied the revolutionary period in Saint Domingue and the birth of Haiti, and their works present solid introductions to the colonial background of Saint Domingue. Recent works by Robert Louis Stein on the French slave trade and sugar business in the Old Regime and the monumental recent study of the entire West Indies by David Watts evidence a new level of scholarly attention in English to eighteenth-century Saint Domingue itself. Historians of slavery and race relations, notably Eric Williams and Eugene Genovese, also write knowledgeably about Saint Domingue and Haiti. Still, the most cursory glance at the accompanying bibliography reveals a dearth of titles in English that deal with eighteenth-century Saint Domingue, and none at all in English that concerns the history of science or medicine there.[48]

Unfortunately, the problem is not simply that comparatively few English-language scholars have focused on Saint Domingue in the past. More, the lack of consideration of Saint Domingue in the scholarly literature has led to some highly unbalanced presentations when related topics have been studied in better detail. Sidney Mintz's *Sweetness and Power: The Place of Sugar in Modern History* (1985) provides a notable example. Professor Mintz's study of sugar and its role in transforming the diets and economies of the West is a major work, and it has been widely and deservedly praised. While aware of Saint Domingue and giving full credit to its significance as a sugar-producing colony in the eighteenth century, Professor Mintz limits his study to the British case. Failing to consider the comparable French experience, in effect, results in a one-sided presentation that could all too likely, given the authority of the author, be taken for the full story.[49]

The 1982 volume by Peggy K. Liss, *Atlantic Empires: The Network of Trade and Revolution, 1713–1826,* furnishes another, more serious example of the same problem. Liss's treatment of eighteenth- and early nineteenth-century Spanish, Portuguese, and British empires in the Americas is workmanlike and valuable in its way. She admits having "skimped" on the French, but Saint Domingue and French colonization are simply not part of the account she provides.[50] If Saint Domingue were really as important a European stronghold in America as has been claimed here, this omission is staggering. The historical invisibility of the colony, not to mention the story of its science, could not be more apparent.

Since the present study swims against such a strong historiographical tide, the extent and seriousness of this problem need to be driven home. The 1987 volume edited by Nicholas Canny and Anthony Pagden, *Colonial Identity in the Atlantic World, 1500–1800,* presents a further opportunity to do that. Scholarly in every respect, the authors and editors of this volume are well intentioned in their efforts to think and write about the development of colonial identities in various European outposts in and around the Atlantic. The fact that they fashioned their study explicitly as a cross-cultural investigation makes it even more interesting and appealing—and the complete omission of the French West Indian experience even more startling.[51] The British Caribbean receives treatment, as do the cases of Brazil, Spanish America, British North America, and, notably, Ireland. One would think that the forging of a distinct colonial identity in eighteenth-century Saint Domingue and its fiery transmogrification into Haiti would be considered a major landmark in the very subject Canny, Pagden, and contributors have written a book about. The article by Gilles Paquet and Jean-Pierre Wallot touching on the French in Canada ("A World of Limited Identities"), as good as it is, by its mere presence in this collection downplays and misleads readers regarding the significance of French colonialism in the formation of independent identities in the Americas.

But there is more. In their afterword Pagden and Canny write about the history of independence movements and the development of independent states around the Atlantic. Because Saint Domingue and French colonization in the West Indies do not appear on these scholars' intellectual maps, Pagden and Canny see eighteenth-century Americans in colonial revolt as all sharing a common European identity and culture and united in a "language of contractualism . . . based upon [the] Magna Carta." Further, according to this view, nationalist revolts at the time were not motivated by race.[52] Alas, not to speak of mulatto interests and "identities" in colonial Saint Domingue and in the period of the French Revolution, this interpretation completely ignores that black Haiti emerged from colonial Saint Domingue as the result of violent revolution, to become the second independent republic in the Western Hemisphere. Something might be added about the language of the Magna Carta in forging the Haitian identity, but enough has now been said of the pernicious consequences attendant to the historiographical invisibility of old Saint Domingue.

Ironically, the historian who approaches the study of Saint Domingue and science in Saint Domingue faces an extraordinary wealth of sources on which to base his or her work. The secondary sources mentioned previously and listed in the bibliography more or less speak for themselves. What needs to be emphasized here is the incredible richness of available primary source material—both printed and manuscript.

Numerous works by contemporary commentators concerning Saint Domingue and the colonial question of the day provide an important entrée into the subject.[53] But one eighteenth-century figure, M. L. E. Moreau de Saint-Méry (1750–1819), dominates the contemporary historiography, and readers should know of the extraordinary role he played in preserving Saint Domingue from historical oblivion.[54] A white European born in the Indies, Moreau studied law in Paris, and in 1775 returned to practice at Cap François in Saint Domingue. He became involved in projects to codify French jurisprudence in the Antilles, and from 1784 to 1790 he produced a major six-volume compilation, *Lois et Constitutions des Colonies Françoises de l'Amérique Sous-le-Vent*, a formidable historiographical resource. Moreau then conceived the even larger plan of a complete, encyclopedic history of Saint Domingue. With government support, he traveled and collected information for the project in France and in Saint Domingue. The French Revolution and his own involvement therein prevented the full execution of the project, but as an émigré in Philadelphia, Moreau was still able to produce a monumental two-volume work, *Description Physique, Civile, Politique et Historique de la*

Partie Française de l'Isle de Saint-Domingue (1797–98), and he likewise published an extensive and irreplaceable atlas of the colony (the *Recueil de Vues*) in 1791.[55] Even more, Moreau assembled a vast personal collection of manuscripts, official documents, and ephemera relating to Saint Domingue, and several hundred volumes of these materials, many of which relate to science in the colony and to the Cercle des Philadelphes, are preserved in the national archives of France.[56]

The surprisingly numerous publications of the Cercle des Philadelphes and of its permanent secretary, Charles Arthaud, are all available, if rare. An extraordinary amount of manuscript material concerning the Cercle des Philadelphes exists, given the fact that not a trace of the papers of the Cercle des Philadelphes remains in Haiti.[57] Moreau de Saint-Méry's collections preserve some official correspondence and a long and detailed, if at times suspect, manuscript history of the Cercle des Philadelphes by Arthaud. Many of Arthaud's speeches and public remarks have similarly been preserved. The happy preservation in Paris and in Philadelphia of the papers of Baudry des Lozières, another key member of the Cercle des Philadelphes, only adds to the richness of these resources.

In addition, virtually complete official government records—the French being the best and the brightest bureaucrats of the day—are preserved, and thus another window opens through which to scout the unfolding of events, scientific and otherwise.[58] Further and finally in this connection, a colonial newspaper, the *Affiches Américaines,* appeared twice weekly in Cap François and in Port-au-Prince from 1764 to 1791. With upward of a thousand pages published annually, the *Affiches Américaines* provides a rich and unique insight into the colony at its height and into the world of science, such as it existed there and then.[59] Astonishingly, as much information is available about old Saint Domingue as about any other comparable region of the world in the eighteenth century.[60] That is just another surprise about this very surprising colony.

Part I

❧❦

Eighteenth-Century Saint Domingue: The Old Regime in the Tropics

MAP 2. "Carte de la Partie Française de l'Isle de St. Domingue." Hand-colored manuscript map, unsigned. [1785.] APS, Map collections.

◆§ ONE §◆

Material Factors

THE CREATION OF European overseas colonies inevitably, perhaps primarily, involved grappling with the constraints of the material world. Contact with the mother country had to be maintained over oceanic distances, local geographies had to be mastered, new towns and settlements created, agriculture established, new climates adjusted to, new flora and fauna encountered and tamed, new disease patterns surmounted, and so on. The French in Saint Domingue faced all of these obstacles. In addition, unlike the case with almost every other European colony, French colonization proceeded against an especially wild natural frontier, there being no aboriginal peoples left on Hispaniola and no high civilization to confront there. In a word, the material environment imposed a string of conditions that shaped the historical development of Saint Domingue and the contours of the history of science that unfolded there. The French effort to colonize Saint Domingue cannot be understood without first taking into account the material conditions facing colonists.

To begin with, Saint Domingue was far from France, over 4,800 miles sailing or, in the measure of the day, 2,000 *lieues* one way.[1] Even for the large and dependable three-masted transoceanic vessels sailing regularly, that was a long way, and it took a long time.[2] The voyage usually took five to eight weeks, but sometimes passages lasted three months or longer.[3] The normal route required ships to swing down from ports in France to the latitudes of north Africa, cross the Atlantic there, make land at Cape Samana on northeastern Hispaniola, and follow the northern coast to French territory. The return route exited north through a group of islets to more northerly latitudes and a westward wind back to France. This prevailing pattern of sailing gave the advantage to colonization on the north coast of Saint Domingue and helps explain why Cap François became the colony's most developed center.[4]

Sailing was not entirely free from hazards, for several shipwrecks occurred each year among the thousand or so crossings to and from France and Saint Domingue. Costs, too, were high for paying passengers: fares ranged from 160 livres for one trunk and a place among the crew to 800 livres for three trunks and a seat at the captain's table.[5] Sailing constituted the only way to cover the distance between France and Saint Domingue, and the regular operations of transatlantic sailing obviously required accurate navigation. Utter dependence on wind and tide and the slower pace of transportation generally gave to the contemporary mentality a sense of space and time now familiar only to true sailors and the Amish.

The territory of French Saint Domingue covered the western third of Hispaniola and included the three larger islands of la Tortue (Tortuga), la Gonave, and la Vache on the northern, west-central and southern coasts. (See Maps 2 and 3.) The eighteenth-century colony encompassed somewhat less area than Haiti today (10,850 square miles), about equaling the size of Maryland in the United States or Belgium in Europe.[6]

The land's natural division into mountains and plains represents a fundamental fact of the topography and history of Saint Domingue. The colony was about 60 percent mountain and 40 percent plain. Three mountain chains effectively broke the colony up into its primary administrative districts, the northern, western, and southern departments.[7] One set of mountains, running along the northern arm, cut off the north coast. Another set, at the "back" of the colony on the east, separated French and Spanish territories. In combination with the other chains a third set of mountains, along the southern arm, defined a west-central portion shaped like a backward *c* that swept the north side of the southern arm, the central coasts off of la Gonave (where Port-au-Prince grew up), and the south side of the northern arm. The southern mountains also isolated the southern coast on the Caribbean as a separate area for colonial development. The mountains of Saint Domingue and Haiti today range upward of 8,000 feet; many in the eighteenth century had not yet been climbed by Europeans. Mountains imposed formidable barriers to intracolony communication and transportation, barriers never completely overcome and not effectively overcome until the last years of the Old Regime.

Saint Domingue possessed eight major plains, formed in much the same way that mountains divide plains and valleys in Greece.[8] (See Map 3.) Four of these plains assumed a major importance in the development of the colony. The largest, called the northern plain, the Plaine du Nord or the Plaine du Cap, amounted to a thin strip of land along the north coast;

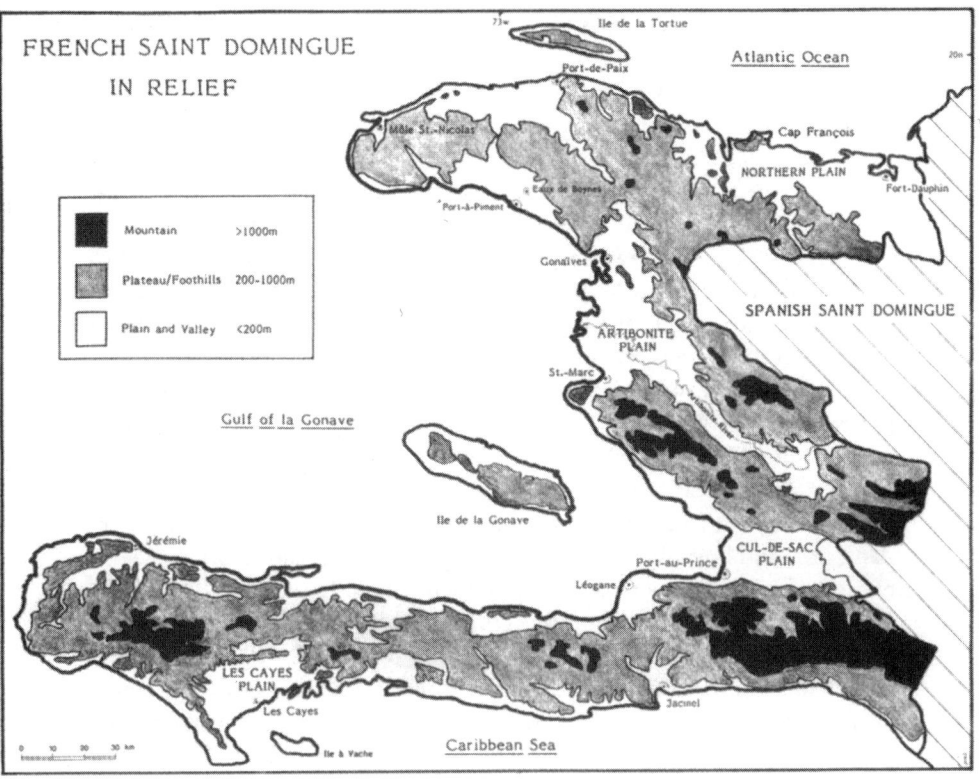

MAP 3. French Saint Domingue in Relief.

it covered an area of 180 square *lieues* or about 1,000 square miles, of which maybe 500 square miles were level and arable. The northern plain became highly developed agriculturally, and Cap François became its urban center and commercial outlet. In the western department the Artibonite Plain of 250 square miles (45 square *lieues*) flanked the Artibonite River that flowed from the interior of Spanish Saint Domingue. Saint Marc served as the main port center for this plain. The third largest plain, the Cul-de-Sac around Port-au-Prince, blanketed 132 square miles (24 square *lieues*) and was very flat. The Cul-de-Sac ended up surpassing all other areas in agricultural development, and after 1787, when political power shifted entirely to Port-au-Prince, it became the decided center of the colony. Several other smaller plains, including that of Léogane, dotted the northern side of the south arm. The fourth largest plain and the major one on the south coast was the plain of Les Cayes, the Plaine des Cayes, 110 square miles (20 square *lieues*). While not as developed as the other

regions, the area around Les Cayes grew rapidly toward 1789 and represented the most promising part of the colony.

French Saint Domingue possessed an extraordinarily long coastline
of over 800 miles, counting bays and coves. That distance exceeded by
over 100 miles the length of the coast of Spanish Saint Domingue, a
colony twice as large in area.[9] Such a long, extended coast gave Saint
Domingue even more of an orientation toward the sea. All was littoral.
Colonization proceeded inward from the coast, and all activity flowed to
and from the coast and its ports. The interior became less developed as
one proceeded inland, and some mountain fastnesses stood as far from
contemporary French civilization as one can imagine.

Western Hispaniola experiences a tropical climate. Maximum tem-
peratures range between 70 and 90 degrees Fahrenheit, although at times
it can be debilitatingly hot. Wet summers and dry winters of varying du-
rations, with upward of 100 inches of rain a year, constitute the two seasons. Western Hispaniola in the eighteenth century was generally verdant, hot, and humid, but the various regions of old Saint Domingue did
exhibit considerable variability in microclimate.[10] The northern coast enjoyed a wetter, more Atlantic-like climate. In the mountain interiors, especially in the north, a truly jungle ecology prevailed. Higher up in the
mountains the weather became more temperate, and pine trees and, once,
even ice could be found. The Artibonite Plain, on the other hand, suffered a desertlike climate and ecology. Only 30 inches of rain fell there a
year, with hot and dry prevailing winds and with giant cacti serving the
function of trees. One colonist called the Artibonite a "little Egypt" with
the Artibonite River as its "Nile."[11] The Cul-de-Sac region, by contrast,
manifested a more moderate climate, ideal for sugar. The southern district on the Caribbean Sea stood exposed to violent weather from the
south.

The combination of mountains and considerable rain meant that Saint
Domingue possessed many, mostly short rivers carrying runoff from the
mountains. At least thirty-two named rivers flowed in Saint Domingue.[12]
The little Dalmarie extended just over 1 mile long; the Artibonite was the
longest river in Saint Domingue, stretching nearly 60 miles in French territory. Flooding often became a dangerous problem, when swollen and
powerful courses would seasonally burst their banks. For example, two
hundred people drowned when the Grande Rivière of Gorge Ste. Rose
flooded in 1722; the Limbé River took out the town of Limbé in 1744.[13]
Engineers undertook river control and hydraulic projects not only for irrigation, as we will see, but also as a matter of public safety.

Other natural disasters struck Saint Domingue. Hurricanes and violent storms regularly hit the Caribbean colony (Figure 1).[14] The storm of

FIG. I. Storm at Léogane: "Vue de la Rade de Léogane." Drawing by Ozanne, engraved by N. Ponce in Moreau de Saint-Méry (1791).

August 16, 1788, for example, damaged or destroyed over a tenth of the houses in Port-au-Prince; four large ships sank in the harbor, and sharks ate six unfortunates. Other storms did comparable damage elsewhere in Saint Domingue on a regular basis.[15]

Conversely, although basically wet, Saint Domingue suffered periodic droughts and attendant disasters for agriculture and humanity. For example, drought struck hard at the Cap-François area in 1726, 1743, 1754, 1757, and 1786, the latter a full year without rain. The Port-à-Piment region went eighteen months without rain in 1779–80, and Moreau de Saint-Méry pityingly describes livestock driven to eating prickly cactus.[16]

Earthquakes, too, posed a serious natural hazard. One hundred recorded earthquakes struck in the period 1700–1793, five of them (1701, 1713, 1734, 1751, 1770) causing significant damage.[17] Earthquakes especially affected the development of Port-au-Prince, which began as a new town only in 1749. The earthquake of 1751 destroyed three-quarters of what had been built by then, and the quake of 1770 destroyed the still nascent town completely. The latter quake supposedly lasted four minutes, killing and burying many in the rubble, and fire ensued. Many people thought of the great Lisbon quake of 1755. An epidemic followed, and because of the continuing fear of earthquakes authorities ordered that Port-au-Prince be rebuilt only in wood.[18]

While more a danger created by man, fire constituted another threat and reality that shaped the development of Saint Domingue. For example, fire completely destroyed the town of Saint Marc in 1724. The fire in Cap François in 1734 burned down half of the town and resulted in the order to rebuild in stone! (Thus the different "looks" of stone and wood in eighteenth-century Cap François and Port-au-Prince.) Lesser fires, like the 1777 fire at the Môle outpost, could still have major local impact.[19] Given the extent of port activity in Saint Domingue, many shipboard fires occurred. In 1781 a spectacular fire began aboard the seventy-four-cannon royal naval vessel, *Intrépide,* in port at Cap François. The ship exploded when fire reached its gunpowder stores, and despite precautions, flaming debris spread fire to the town and to other ships. Forty-two sailors were killed or badly burned.[20] Everyone fought fires when they broke out, especially military and police units. Cap François, Port-au-Prince and other towns had special fire-fighting equipment purchased and maintained at crown expense; the authorities created a special position of director of pumps, equivalent to fire marshal, for Cap François in 1777 with appointments of 1,200 colonial livres.[21] Engineers installed and maintained elaborate fountains and fountain systems in Cap François, Port-au-Prince, and elsewhere in part to fight fires, as well as to supply water for drinking and domestic use.

Starvation regularly resulted as a consequence of natural disasters in Saint Domingue. More remote areas especially suffered from shortages and crop failures, but everywhere people starved when disaster struck, especially slaves. Thirty thousand slaves supposedly perished in the drought of 1775–76, for example. As Charles Arthaud put it in 1788, "Famine is always to be feared."[22]

Disease constitutes a major "material" factor affecting the development of Saint Domingue. The colony suffered many different endemic and epidemic diseases. Chronic and acute illnesses were common and widespread among the population, and unwholesome and unsanitary conditions prevailed generally. Colonists found acclimatization difficult and dangerous, and the colony became known as a diseased and unhealthy place. Death was common and could come suddenly and unexpectedly, a dour note that sounded in all human relations. Slaves and the poor suffered especially, and in certain quarters, patterns of excess exacerbated mortality and drove many to early graves.

To give an illustration of these uncertain and highly insalubrious conditions, note that of one hundred acclimated soldiers sent to Petit Goave from Port-au-Prince in 1777, sixty-six became sick after eight days; after forty-five days, twenty-five had died and only nine remained healthy. Only seventeen of the original one hundred were still alive two years later. In another example, in 1780 disease wiped out half of one pathetic little hamlet, where illness reigned anyway.[23] Suffice it to say that death and disease were commonplace for people living in Saint Domingue in the eighteenth century in ways we usually appreciate only vaguely.

Patterns of disease in Saint Domingue must be seen in the context of the worldwide spread of disease and disease agents that developed after Columbus.[24] Historians are now familiar with the international migration and global exchange of infectious and parasitic diseases between and among Europe, Africa, America, and the rest of the world from the sixteenth century. As a result, port towns (in France, Saint Domingue, and elsewhere) came to share the same horrible epidemiologies.

The bacterial infection, syphilis, existed everywhere, of course. Moreau de Saint-Méry named it the disease one does not name; Charles Mozard, the editor of the *Affiches Américaines,* called it "the most common malady in the colony."[25] Mercury seems to have been an effective treatment, but by no means did it limit the spread of the disease. Related to syphilis, the skin disease yaws "reigned" in the colony. Eczema, scruff, mange, scabies, and skin sores of various sorts were common ailments. Two-thirds of the population was so afflicted, and the medical trade offered a range of palliatives.[26]

Among other bacterial infections, typhoid fever, known as the "mal de Siam," was endemic in Saint Domingue. Despite quarantine procedures, intermittent outbreaks occurred with thousands of deaths. Another major disease in Saint Domingue, bacterial tetanus, produced fevers and lockjaw and killed many newborns, although slaves sometimes blamed tetanus as a cover for infanticide.[27]

Yellow fever, caused by a virus and transmitted by mosquito vectors, likewise took its toll. Smallpox, another virus, was endemic and effected regular ravages. The outbreak in Cap François in 1772, for example, killed twelve hundred when an infected slave ship sank in port and escaping slaves spread the disease through the town. As we will see, the early introduction of inoculation into Saint Domingue, while effective, did little to eliminate smallpox as a danger. A virulent form of viral measles struck adults, often with fatal effects.[28] And, of course, one should not forget ordinary colds and influenzas that regularly swept the colony.

Parasitic infections likewise ran rampant in Saint Domingue. Malaria was the worst, although quinine obtained from "Peruvian bark" and at great expense from Spanish sources provided an effective therapy. Slaves especially suffered horrible dysenteries from various intestinal parasites, notably hookworm. Worms also caused elephantiasis, and several vermifuges and purgatives existed for man and animal. Gangrene produced fevers and death, and amoebic dysenteries only added to the colony's woes.[29]

Animal diseases, including glanders, anthrax, and various eye diseases, threatened and killed animals and livestock constantly. Two thousand horses, for example, died in one especially vicious epizootic outbreak in the Cul-de-Sac in 1751.[30]

New arrivals to the colony, both Europeans and Africans, needed to acclimate themselves to their new environment and were medically so much at risk that an expression, "to pay the clime's tribute," arose to explain what they had to go through, the medical and ecological equivalent of resetting one's biological clock.[31]

Miscellaneous other diseases known on Saint Domingue include tuberculosis, leprosy, and rabies, although the last-named disease seems not as much a threat as in Europe. The practice of smoking tobacco no doubt aggravated respiratory illnesses. As a rule, the elite drank excessively (French wines and brandies and domestic rum), and alcoholism was common. The record mentions madness, depression, and other forms of mental illness. Slaves committed suicide frequently. Finally, for a few whose style of living engendered such diseases, heart attacks and gout round out the list.[32]

Affecting everyone so universally, illness and disease undermined the effort to colonize Saint Domingue. Not surprisingly, physicians, medical institutions, and the scientific study of tropical diseases became deeply ingrained in the structure and functioning of colonial Saint Domingue, as a later chapter will show.

Understanding the role of the material world in the development of Saint Domingue requires recognizing that the colony was established on a frontier, an inhospitable frontier with nature. The Spanish exterminated the aboriginal Arawak Indians soon after arriving in 1492. The Spanish then withdrew from the northern and western portions of Hispaniola, and the land reverted to wilderness.[33] As a result, the ecology of western Hispaniola, while not pristine, was actually wilder in the seventeenth and eighteenth centuries than it had been before the discovery of America. French efforts to colonize Saint Domingue thus proceeded against an imposing wall of raw nature. French Saint Domingue bordered on the bush, and the itchy natural world intruded unavoidably. Beyond the plantations and enclaves of civilization so laboriously carved out by the French and their slaves, Saint Domingue offered a colonial heart of darkness.

Nature offered some riches but mostly presented obstacles to overcome. Pests of various sorts, for example, tormented the everyday lives of the people of Saint Domingue. Depending on the locale, giant mosquitoes swarmed, and flies produced severe bites, subject to infection; Moreau recommended mosquito netting to all who could afford it. Insect attacks on crops posed an especial danger, and ants in particular threatened sugar cane fields. Colonists completely abandoned agriculture in the Côtes-de-Fer region, for example, on account of drought and the blistering "sticky worm." A plague of ants, caterpillars, aphids, and rats combined to assault agriculture around Les Cayes in 1750. Rats and mice infested Saint Domingue, and on occasion populations swelled.[34]

Crabs were also major pests. Large land crabs made holes everywhere, and ate the roots of sugar cane. For its inhabitants, the annual crab migration through the town of Jérémie proved a great and trying confrontation with nature. Every year literally millions of breeding crabs formed a rolling wall and invaded Jérémie on their way to the sea and back. They got in everywhere—even one's bed, says Moreau. They could not be stopped. Their creeping, chitinous noise was infernal. The stench of dead crab was horrendous. Conditions became so bad in 1774 and for three years running that authorities formally ordered all Jérémie residents to dispose of their dead crabs twice a day during the invasion.[35]

On the positive side, the sea provided an important source of food, and local fishing was active. Shellfish (including oysters, conch, sea urchins, and crabs) was plentiful, but the palates of Europeans spurned these foods (excepting oysters), fearing poison. Shellfish did form a notable part of the diets of slaves. But sea life in the form of the teredo posed a particular threat, for these wood-boring sea worms voraciously attacked ships' hulls and made all pilings and piers transitory.[36]

On land, crocodiles were common but seemingly not a danger to man. A large gray lizard upward of 3.5 feet was also common, and apparently quite tasty. Sea turtles and tortoises abounded, and were hunted and eaten by men, crocodiles, and wild pigs. Saint Domingue did not have poisonous snakes, but grass snakes sometimes proved to be a nuisance.[37]

Saint Domingue harbored plentiful game. Sea birds and forest birds provided a variety of fowl. Colonists trapped parakeets and other exotic birds as pets, and they sometimes took rodent piloris and agoutis for food. Numerous feral cattle and pigs roamed the bush, wild from the early days of the Spanish. Hunting feral cattle proved important in the early development of the colony; men still hunted wild cattle and pigs in the eighteenth century, sometimes using dogs, and packs of wild dogs hunted on their own. Slaves in the Tiburon region hunted feral pet monkeys; wild (and domestic) cats acted as a check on rat and bird populations.[38] As we will see, tribes of "feral men" lived beyond the pale in the mountains.

The natural setting and wilderness of Saint Domingue provided defining features of the colony. Nature likewise exerted an effect on European science, for the wild in Saint Domingue represented a call for scientific investigation. The flora and fauna of Saint Domingue were unknown at the outset, and it took a distinguished series of naturalists to catalog their wonders.

If the colonization of Saint Domingue proceeded against a wild, natural frontier, the ecological impact of colonialism was also vast in Saint Domingue, as elsewhere.[39] In the long run, nature could not withstand the power of the human onslaught.

Heavily forested wherever climatic conditions permitted, the land was colonized in the first instance by human conquest and exploitation of the forest. Unlike Haiti today, which has become wholly deforested, deep woods persisted across huge areas of Saint Domingue late into the eighteenth century. Humans interacted with the forest in various ways, first by clearing the forest for settlement. Around towns especially, colonists cut down entire forests. Sugar plantations and field agriculture seem to have been established in more open areas, but the coming of coffee growing marked another major onslaught against native forests. A number of hardwood forests in Saint Domingue supplied good wood for construction, for naval timber, and for the mahogany furniture industry back in France. Mangroves were cut back for firewood. Goats were also a factor in diminishing woodlands. The availability of wood became a problem as the eighteenth century unfolded, and in all areas around human habitation forests soon became scarce. In 1745 authorities ordered the end of uncontrolled cutting of trees for free firewood. As forests receded, the costs of transporting felled trees from the interior to the coasts rose prohibitively, to the point where it actually became cheaper to import lumber from the United States than to cut it in Saint Domingue.[40]

On the whole, the soil of Saint Domingue remained rich and fertile in the eighteenth century, but soil exhaustion did affect older, more settled agricultural areas, and fertilization became necessary by the end of the eighteenth century. Soil erosion turned out to be a serious problem. Wherever colonists cleared sloping land, erosion followed.[41]

Colonists brought with them a small ark of plants and animals that also transformed local ecologies. Many small-scale, private, and accidental introductions occurred as part of the colonizing process. Colonists seemingly cultivated the entire range of European fruits and vegetables in their gardens and orchards: potatoes, cabbages, onions, leeks, carrots, artichokes, strawberries, grapes, and various kinds of trees (fig, apple, mulberry, and cherry). Another range of plants, now thought typical of the Antilles, were likewise brought to Saint Domingue from Africa and elsewhere, including bananas, cassavas, the citrus fruits, mangos, pineapples, and avocados.[42]

Colonists introduced many new animals into the colony. Moreau de Saint-Méry counts 40,000 horses, 50,000 mules, and 250,000 cattle, sheep, goats, and pigs in the colony in 1789. Elsewhere Moreau mentions chickens and both the European and African duck, and a strange buffalo-cow cross existed in the colony, at least in limited numbers. In addition to the previously mentioned monkeys, camels from north Africa added to the colony's fauna in the 1750s, but they scared the horses and would not reproduce. Animals had to be fed, even if ultimately to be eaten, and pens and barns and stock areas had to be set aside. Among the more interesting and useful animals introduced into Saint Domingue were bees. Apiculture began with bees from the Spanish side of Saint Domingue originally brought from Cuba; another introduction brought bees from Martinique. Beeswax and honey were obviously useful and both reported to be of good quality in Saint Domingue.[43]

The ecological impact of bees was no doubt benevolent, but the same cannot be said of a last example: the introduction of French snails into the colony. Presumably for delectation, the Séguineau brothers imported escargot from their native La Rochelle and released them on their coffee plantation in the Arcahaye region. The European snail succeeded all too well in the local struggle for existence, and populations skyrocketed. Snails quickly infested an area of several square miles, and they proved a threat to coffee trees. Concerted action was called for, and the Séguineau brothers went so far as to *pay* their slaves to collect snails, and they gave them salt with which to season and eat snails.[44] The ultimate outcome of the effort to eradicate snails in Arcahaye remains unknown, and the escargot example is wholly minor, but it illustrates well the principles involved in considering the ecological impact of colonialism or, more particularly in this case, French colonialism.

❧ TWO ❧

Historical Development

THE EFFORTS OF THE French to colonize Saint Domingue evolved in a historical context no less than a material one. The contemporary world in Europe, Africa, and the Caribbean provided the fundamental historical parameters affecting developments in Saint Domingue, and to a degree, Saint Domingue came to assume a certain reciprocal importance and historical influence of its own.

French efforts to establish colonies in the Americas date from the early sixteenth century. Following several unsuccessful efforts to establish permanent outposts in Canada, Samuel de Champlain founded the Franco-American settlement at Quebec in 1608, a year after the start of the English colony in Jamestown.[1] French settlement in the West Indies began in 1624 when the party led by Belin d'Esnambuc, appointed colonial governor by Richelieu under the authority of the Saint Christopher Company, landed on Saint Christopher (Saint Kitts) in the Lesser Antilles.[2] A new Compagnie des Isles de l'Amérique took over the French colonial enterprise in the Antilles in 1635 and founded new colonies in Guadeloupe and Martinique. Other, lesser French outposts in the eastern Caribbean grew as extensions of the new colonies in Guadeloupe and Martinique.[3]

French colonization of Saint Domingue occurred comparatively late in this process. Technically, the French may be said to have exerted a weak colonial claim to Saint Domingue only from 1642 when one Le Vasseur received appointment as governor of Tortue Island. But in other respects, the colonization of Saint Domingue did not get under way formally until 1665 when Bertrand d'Ogeron was appointed governor.[4]

The French certainly came late to Hispaniola. Arawak Indians originally inhabited the island, and as many as three million Arawaks lived on Hispaniola when Columbus first landed in 1492.[5] Within two decades the Spanish had completely eradicated the Arawaks and their culture from Hispaniola, the first (but hardly the largest) instance of European genocide in the Americas.[6] Later French incursions in and around Saint Domingue thus proceeded without facing a native population, an unusual situation compared with colonial experience elsewhere, certainly elsewhere in the Caribbean with warlike Carib Indians endemic.[7]

Spanish control of Hispaniola went unchallenged for well over one hundred years. The town of Santo Domingo on the south and east of the island arose early in the sixteenth century as the seat of Spanish power on Hispaniola, but the center of the Spanish empire in the Americas in the sixteenth century quickly shifted to the mainland in Mexico and South America. In the Caribbean Cuba became the primary Spanish outpost and transit point to Spain, and the Spanish presence on Hispaniola diminished considerably as a result. In 1605 the Spanish "foolishly" withdrew from western Hispaniola in an effort to cut off contraband, thereby inadvertently creating a no-man's-land and allowing the land to return to the wild.[8] Into this lush and uninhabited niche slipped the next people to develop Saint Domingue: the pirates.

The rise of pirates and pirate culture in Saint Domingue and elsewhere around the Caribbean in the seventeenth century constitutes one of the great episodes in the history of anarchism.[9] Unfortunately, an extended treatment is out of place here. Suffice it to say that the Île de la Tortue (or Tortuga) off the northwest coast of Saint Domingue formed the center of French piracy in the Caribbean. A diverse group of brigands, many with Protestant backgrounds, established the first pirate enclave there in the early 1630s. Thereafter, the island became the locus of intense, if primitive colonial rivalry among French, English, and Spanish interests. In a complicated series of developments stretching over several decades, various English and French pirates and adventurers, together and separately and with varying degrees of legal backing, established fortified settlements on Tortue. At different times the Spanish—sometimes in conjunction with English or French forces—attempted to quash these settlements, and the island remained deserted at various periods in the 1630s, 1640s, and 1650s.[10]

Tortue, like Port Royal in Jamaica, was a place of contact with the outside world where pirates could revel and be fleeced in peace. By 1650 the population of Tortue grew to 1,150, and thereafter, because of military struggle and because pirates seem not to have enjoyed large groups, offshoots from Tortue soon formed on Hispaniola and Saint Domingue proper. Settlers founded Petit Goave on the southern arm in 1659; Léogane followed in 1663. In 1670 twelve colonists, led by a former pirate with the evocative name Pierre le Long, moved over to form an agricultural and fishing settlement on the northern coast of Hispaniola, thus founding Cap François.[11]

Tales of pirate life became extremely popular among European audiences, but in the long run civilization destroyed pirate culture. French government authority, soon after its arrival in Saint Domingue in the 1660s, adopted policies decidedly opposed to piracy. Safety of trade on the high seas required that marauding be suppressed, and the development of a superior royal navy in the eighteenth century greatly diminished any military need for pirates.[12] Formal colonization likewise proceeded at the expense of the pirates. Everywhere, colonial society and the forces of law and order pressed in on the pirates. Some pirates became small farmers, and melted into the lower orders of white society. Others turned into boatmen (*caboteurs*) who moved goods and people along the coast; still others became salt workers mining salt along tidal shores and river banks in several places in Saint Domingue.[13] Late in the eighteenth century hunters in the interior of Saint Domingue still displayed traces of the buccaneers, and one settlement of old "brothers of the coast" could still be found around Jérémie as late as 1778.[14]

As in so much else, the coming of age of Louis XIV and the advent of Colbert to the finance ministry in France in 1662 marked an important turning point for French colonialism and the beginning of formal colonization in Saint Domingue. The French colonies on Saint Christopher, Guadeloupe, Martinique, and their dependencies (founded in the 1620s and 1630s) were already established and relatively secure by that date, but Colbert reorganized their administration under yet another new company, the Compagnie des Indes Occidentales. With regard to the still unsettled situation in the vicinity of Tortue, Colbert actually paid to transfer the rights to colonize Saint Domingue back to the crown and in turn to the new Compagnie des Indes Occidentales. In 1665 the Compagnie appointed the military officer and adventurer, Bertrand d'Ogeron, governor in Saint Domingue. In his eleven years of rule to 1676 d'Ogeron successfully reasserted French claims in the area and unequivocally established French authority in western Hispaniola. D'Ogeron was the founding father of the French colony in Saint Domingue.[15]

D'Ogeron and other early governors did what they could to colonize Saint Domingue formally. They encouraged immigration, conceded lands, and generally promoted agricultural settlement.[16] For passage to Saint Domingue numbers of poor whites—the *engagés*—voluntarily and involuntarily indentured themselves for terms of three years of essentially slave labor. In the decade 1664–74 they quadrupled the colony's white population, to five thousand.[17] Tobacco, widely cultivated by settlers, formed the main cash crop in Saint Domingue in the seventeenth century. At this stage in the colony's development also officials made the first pathetic efforts to bring women into the colony.[18]

The power of the French state came first in the form of a military governor. Military government in Saint Domingue represented an extension of that previously established in Martinique and Guadeloupe. Military governors regularly received appointment to Saint Domingue from the 1640s, although the separate post of governor-general of Saint Domingue and the Isles Sous le Vent dates only from 1714.[19] From 1688 the French government stationed royal naval and army units in Saint Domingue. Direct military administration increasingly made itself felt in Saint Domingue: policies against pirates began to be enforced, Tortue was abandoned, the seat of the colony moved to the main island, and disillusioned colonists revolted for the first time.[20] Direct French administration of Saint Domingue became only more bureaucratic and effective as time progressed.

While development of the colony in its northern and western departments proceeded under military governorship, in 1698 the government turned over the isolated southern region to a trading company, the Compagnie de Saint-Domingue.[21] Run from France by a board of twelve directors (mostly insiders from the navy and finance ministries), the Saint Domingue Company possessed powerful privileges, notably including all rights of sovereignty and a fifty-year monopoly on trade with the southern department. The company, based in the Caribbean port town of Saint-Louis, extracted its own taxes from colonists, maintained its own police forces, and ran its own courts. It had its own fleet and attempted to meet quotas for importing slaves and new colonists. But the Saint Domingue Company failed miserably. Officials failed to meet immigration quotas, supplies failed to arrive, and colonists actually starved. In 1720, coincident with the failure of Law's bank in France, the company was suppressed. Some of its privileges reverted to the Compagnie des Indes Occidentales, a move that sparked another round of sedition in the nascent colony.[22]

The state of colonial development in Saint Domingue in the second half of the seventeenth century remained weak and should not be overestimated. Reflecting the situation in contemporary Europe, skirmishes between and among the colonial powers of Spain, England, France, and Holland broke out at regular intervals in the Caribbean in the fifty years preceding the Treaty of Utrecht in 1713. In Saint Domingue the Spanish sacked Cap François in 1691 with a force of one thousand men; they massacred the men of the town and, in an unusual move, carried off what women they could find. In 1695, Spanish and English forces again beseiged Cap François and Port-de-Paix.[23] The Treaty of Ryswick of 1697 formally acknowledged French sovereignty over western Hispaniola, but the War of Spanish Succession that followed did little to bring peace to the Caribbean in the first years of the eighteenth century.

With the Treaty of Utrecht, however, and the death of Louis XIV in 1715 a new, more pacific set of circumstances made itself felt in Europe and in the West Indies, and a new era began for the Saint Domingue colony. Moreau de Saint-Méry called this the colony's "second age," and Charles Frostin, Thomas Ott, and other modern historians have rightly labeled the period after 1720 as the beginning of the colony's "Golden Age."[24] In this characteristically Old Regime phase of development that defined the colony until its collapse after 1793, Saint Domingue became fully established as a normally operating overseas possession of France and the Bourbon monarchy. The institution of colonial government and the implantation of a bureaucracy represent important steps in the colonization of Saint Domingue. The rest of this chapter is given over to the official institutions that came to rule in eighteenth-century Saint Domingue.

It would be a mistake to omit mention of the French Catholic church as providing key social and institutional bases for the unfolding of colonization in eighteenth-century Saint Domingue. Saint Domingue grew by parishes, and parishes formed the basic unit dividing the colony administratively. Fifty-two Catholic parishes spread out over Saint Domingue at the end, and each possessed its priest to serve the village and hamlets a parish ordinarily encompassed.[25] Churches were always among the first buildings built in any new town or village, and in Cap François and Port-au-Prince especially, churches were substantial structures, served by enlarged staffs of priests (Figure 2).[26] Several orders of priests and nuns established communities in Saint Domingue. They operated a number of hospitals and at least one convent and convent school; religious orders and organizations owned income-generating property, and they owned slaves.[27]

The church provided the calendar that structured the year.[28] The church, be it noted, constituted an official body in society, and it exer-

FIG. 2. The parish church at Cap François. Fountains at Fort–Dauphin and Cap François. Engraving [by N. Ponce] in Moreau de Saint-Méry (1791).

cised a visible public role when, for example, it participated in civic processions, blessed regimental flags, or led public prayers, as in 1732 for the rain to stop in Cap François. The connection between the French church and ruling power in Saint Domingue reveals itself most clearly in the churches of the larger cities, which had honorary benches formally set aside for the several branches of colonial government.[29]

Not all of the priests of Saint Domingue were virtuous or even religious, and the colony was not known for its attendance to religious duty.[30] Nonetheless, the French Catholic church, through its organization, social cohesion, and political connections, was an important, somewhat overlooked instrumentality active in the colonization of Saint Domingue.

Equally important for the development of the colony in the eighteenth century, a civilian administration came to complement the military government established in the seventeenth century. Civilian government came in the form of a royally appointed *intendant* and the associated offices and bureaucracies of royal civil government. A subdelegation (meaning a bureau or an office) of the intendancy in Martinique and Guadeloupe moved over to Saint Domingue in 1692; the crown appointed a proper intendant in 1705, but created the formal office of intendant in Saint Domingue only in 1719. Later, subdelegations and lesser branches of the Saint Domingue intendancy spread to each of the colony's three departments.[31] The intendant and his office (and the administration of Saint Domingue generally) reported back to Versailles and the (very high) office and person of the royal secretary of state for the navy and the colonies, and thence to the council chambers of his majesty.

The intendant saw to civilian affairs for the crown and coadministered the colony along with the military governor-general. The intendant had particular charge of finances: collecting taxes, managing public accounts, and spending public money.[32] In general the intendant took responsibility for the social and economic welfare of the colony and the smooth operation of its government. But overall the intendant and the governor-general shared administrative responsibility. This circumstance left Saint Domingue with a government "with two heads," and constant conflict rankled governors and intendants.[33] One would think such a dual government ill-advised, but it reflected the deliberate policy of the French crown applied in all administrative regions in France to divide power and have one agent serve as a check on the other. The system of intendant and governor thus made French government and administration in colonial Saint Domingue exactly analogous to what one finds in the thirty or so traditional provinces and affiliated territories in France itself. In effect, Saint Domingue became another French province, the "isles françaises de l'Amérique sous le vent."

After 1695 the official capital of this province and the formal seat of government was always in the western department, and from 1749 that capital was Port-au-Prince. By the same token, Cap François in the north emerged as the largest town and the colony's unofficial capital, the New York of Saint Domingue. During the American War of Independence, the administrators of Saint Domingue moved to Cap François entirely, and they were required to sojourn there at other times.[34] In effect, the colony had two capitals. The long-term strategy of colonial development shifted power away from Cap François, but Port-au-Prince, as discussed earlier, suffered setbacks in its development, and by late in the eighteenth century Cap François remained the leading center in Saint Domingue.

The military itself, overseen by the governor-general, formed a large presence in eighteenth-century Saint Domingue and constituted a significant element in the sociology of the colony.[35] Contingents of the regular French army rotated in and out of Saint Domingue from the later seventeenth century. In 1766, after the Seven Years War, a French colonial army was permanently installed in Saint Domingue. The first force, the Légion de Saint-Domingue, totaled 5,400 regular troops in fifty-four companies, three companies of cavalry, an artillery company, two company-sized construction crews, and officers. In 1773 this army was scaled back to approximately 4,000 men reorganized into two regiments (one each for Cap François and Port-au-Prince) of nearly 1,250, and four companies of artillery totaling some 1,500 additional men.[36] In addition to these all-white troops, standing companies of free mulatto (or mixed-race) soldiers existed from 1762. Given royal designation in 1779 as the *chasseurs royaux* (royal hunters), these mulatto units waged a constant war against runaway slaves, and formed the most fearsome fighting force in Saint Domingue.[37]

Saint Domingue became an armed camp, bristling with military installations and fortifications. The barracks in Cap François (Figure 3) and Port-au-Prince, for example, were among the major structures in the colony, the former sheltering 1,300. Throughout the colony, the military erected and manned coastal batteries, linked by lookouts and signal stations, particularly along firing lines around all ports. The colonial government spent 200,000 livres tournois annually on fortifications, and a director-general of fortifications was appointed for the colony.[38]

The French royal navy also enjoyed a significant official presence in Saint Domingue. The naval commissioner or *ordonnateur,* stationed at Cap François, stood third in rank in the colony after the governor-general and the intendant. He supervised naval affairs generally, and took charge of the fairly extensive installations and operations of the French royal navy in Saint Domingue. Royal soldiers and sailors formed a sig-

FIG. 3. The barracks in Cap François: "Vue des Casernes du Cap François." Engraving [by N. Ponce] in Moreau de Saint-Méry (1791).

nificant part of the sociological milieu of eighteenth-century Saint Domingue: the plight of the common soldier or sailor was generally miserable, and, as Moreau put it, such men defended the colony but not its morals. Much the same point can be made of the 12,000 to 15,000 merchant sailors in the colony at any one time.[39]

An extensive home guard, the *milices,* complemented the regular army and strengthened military rule in the colony. After 1769 the *milices* formed a standing militia, on duty in peacetime. Service was compulsory for able-bodied free males, and in 1789 the colony mustered perhaps 20,000 of these civil guards for its defense. In Cap François alone, 1,600 men made up thirteen uniformed companies of foot soldiers, riflemen, cavalry, artillery, and scouts. Only "free" colonists—that is, nonslaves—composed the militia, for only they could legally carry firearms. But militia companies were nevertheless not all white, and nonwhite components sometimes predominated. For example, 200 whites, 800 free people of color, and 140 free blacks formed militia companies in the remote Mirebalais in 1789.[40]

Saint Domingue in all regards was a highly policed colony and society. Free assembly by slaves especially seemed to threaten "public tranquillity," and the forces of law and order in the colony exercised considerable police power and constant vigilance to suppress any possible disruption. In addition to the military per se, the medieval institution of a mounted constabulary or Maréchaussée operated everywhere in the civil sector. Subservient to higher authority, constabularies differed in number of armed and mounted men and officers, but their role always involved chasing runaway slaves and generally seeing to public order. The Maréchaussée likewise supervised the many public executions.[41] True armed and uniformed police units enforced (or more often, taking bribes, did not enforce) a spectrum of laws governing assembly, markets, sanitation, the bread supply, weights and measures, inns and accommodations, gambling, public entertainments, vagabondage, stray animals, and so on.[42]

The legal and court systems established in Saint Domingue represent another defining element in the constitution of the colony and another important arm of government. Standing atop a fairly elaborate structure of courts, the semi-independent Conseil Supérieur de Saint Domingue was a provincial French parlement in everything but name.[43] Actually, throughout most of the eighteenth century, two Conseils Supérieurs, two parlements, operated in Saint Domingue. The first, established in 1685, remained in the western department of the colony; a second sat in Cap François from 1701. In 1787 and as a serious blow to the Cap, Louis XVI folded these two courts into one, with the seat in Port-au-Prince.

Like French parlements, the Conseils Supérieurs of Saint Domingue performed various judicial functions. They heard and judged civil and criminal cases, both as courts of first jurisdiction and as appeal courts.[44] The conseils likewise registered and recorded—and thereby gave the force of law to—public ordinances and legal edicts issued locally and arriving from France. The Conseils embodied the power structure in Saint Domingue: the governor-general and intendant, de jure members, presided when taking part in court sessions.[45] Other "outsiders" for plenary sessions included senior military and naval officers. The main, robe membership of the Conseils normally consisted of twelve judges (conseillers), with the senior judge ordinarily presiding. Unlike the case in France, these and related parlementary positions were not venal, but they apparently did carry with them personal nobility of the second degree.[46] The Conseils surrounded themselves with the high and serious air of the law. Judges met in mahogany paneled chambers and sat on gold-studded black leather chairs. Their bonnets were scarlet, and their black robes trimmed with ermine. The Conseils of Saint Domingue wore swords beneath their robes, which surprised observers familiar with the customs of the judicial nobility in France.

Other courts operated in Saint Domingue below the level of the Conseils, including ten Sénéchaussée courts. The seneschal, another feudal officer, was technically the personal representative of the king and of the royal court on the scene in Saint Domingue. Each local seneschal and three other judges formed an inferior court hearing upward of twenty thousand cases a year—police cases on Mondays and Thursdays, civil complaints on Tuesdays and Fridays. Scenes of considerable racial conflict regularly broke out in the seneschal courts.[47] Ten admiralty courts sat in so many of Saint Domingue's ports. In these courts, officers of the royal navy with other judges decided maritime cases and handled the formalities of naval business and maritime administration.[48]

Around these courts unfolded the full panoply of the legal profession in Saint Domingue. Royal attorneys and attorneys general prosecuted the king's and people's business before the courts. Official substitutes existed for many positions, and court affairs relied on a small army of clerks, registrars, secretaries, ushers, bailiffs, marshals, sheriff's officers, and other personnel. Among these, translators served in admiralty courts, and court criers wore onyx and ivory rings when leading their respective bodies in public processions. And, of course, cadres of lawyers and notaries made themselves available to serve clients' public and private needs, in court and out.

In addition to the agencies already named, the government established an extensive system of financial receiverships, and general farmers from France made their way to Saint Domingue. Sizable staffs of government physicians, surgeons, and other medical personnel formed yet another element of French officialdom in Saint Domingue. (More will be said about medical agents in a later chapter.) Land tribunals, tax boards, and local assemblies of notables convened regularly, as required. An official civil engineering staff came into being, as did an extensive system of post offices and post-office holders, and any number of inspectorships of this or that. There was even an office of vacant offices![49] Bureaucratically, little distinguished Saint Domingue from any other province in France.

Even though not very powerful or important, the semiofficial Chambers of Commerce and Agriculture represent notable additions to the institutional scene in Saint Domingue. Royal edicts established these chambers, "half for agriculture, half for commerce," in several of the French Caribbean colonies in 1759. This colonial system of agricultural and commercial bureaus arose coincident with a related system of government agricultural societies in France.[50] Two bureaus existed in Saint Domingue: one in Port-au-Prince, the other in the Cap, each composed of four colonists and four merchants. These were exceptional organizations in that they were technically independent of the local administrators and thus a threat to them. Indicative of developing tensions in the colony, the commerce branch of the chambers in Saint Domingue favored free trade and the creation of a mercantile exchange in the colony, and, as a result, the king suppressed the commerce branch in 1766. Reorganized, seven-man Chambers of Agriculture carried on, but they proved largely moribund institutions in the 1770s and 1780s.[51] A revived, private Chamber of Commerce for Cap François, once again permitted in 1784, met weekly as a merchants' and businessmen's association. The local courts called on its expertise. As the revolution approached, the Cap Chamber of Commerce, like its predecessors, began to take radical political and economic positions.[52]

Any survey of government and quasi-government institutions that came to be incorporated in colonial Saint Domingue must also include the colony's official scientific society, the Société Royale des Sciences et des Arts du Cap François—the Cercle des Philadelphes. While hardly the most important institution in colonial Cap François, in many ways the full legal recognition given the Cercle des Philadelphes in 1789 represents the culmination of institutional (as well as scientific) developments in eighteenth-century Saint Domingue.

In the end, French colonization of Saint Domingue involved the establishment of a substantial colonial bureaucracy and institutional infrastructure. The bureaucratic and administrative structure of old Saint Domingue was, of course, entirely typical of Old Regime France. Colonization of Saint Domingue involved the direct transfer of prerevolutionary French bureaucratic models to the West Indies. The apparatus and institutional structures of the Old Regime in France simply extended themselves to the tropics.

An out-of-the-way example vividly illustrates what the coming of French colonization entailed in Saint Domingue. The example concerns the dedication in 1781 of commemorative pedestals for a bridge planned for the Haut du Cap River in Cap François. In a ceremony that Moreau himself labels "as much religious as civic," a public procession set out from the government building in Cap François.[53] The clergy of the Cap church led the parade, followed by a military band. In their turn came Messrs. De Reynaud, the governor, and Le Brasseur, ordonnateur and intendant pro tem, and the Conseil Supérieur. Then came a united corps of military officers, the Sénéchaussee court, and representatives from the parishes of Quartier-Morin, Limonade, and Petite-Anse. Other dignitaries from the town and the surrounding plain appeared next in line, and a large crowd of unclassified folks brought up the rear. The procession made its way to the riverside site by the ferry, where speeches, prayers, fanfares, cannon salvos, and hurrahs marked the occasion. The bridge was never built, and within a few years the engraved pedestals and the columns that came to stand on them sank toward ruin, but the ceremonial gathering there in Cap François in 1781 epitomizes nonetheless the forces and agencies that undergirded French colonialism in Saint Domingue.

Eighteenth-century Saint Domingue, although the very model of a modern major colony, also came to be characterized by trappings of feudal and aristocratic privilege overleaping the Atlantic. Such touches have already been seen with regard to honorary benches in churches, public prayers, the public procession, and the ennobling power of judicial office. Moreau de Saint-Méry remarks that all the great noble families in France owned property in Saint Domingue, and noble rights and feudal rights of domain seem to have operated with regard to at least some noble property in the colony.[54] Colonial government also enforced royal corvées, which could represent a significant drain on plantation manpower.[55] Similar to rich bourgeois in France, planters who succeeded in Saint Domingue sought patents of nobility and to make their way in noble circles in France.[56] In the most extraordinary instances of aristocratic and feudal privilege affecting Saint Domingue, Louis XVI made gifts of the three main islands in the colony—la Tortue, la Gonave, la Vache—to his ministers, the dukes de Choiseul and de Praslin.[57] But these interesting and revealing examples sound grace notes more than the main themes of contemporary colonial development. French Saint Domingue headed inexorably not backward to the European Middle Ages, but forward to the French and Industrial revolutions.

Population and
Sociology

S AINT DOMINGUE was a European colony, but its population and social constitution were like nothing found in Europe. Every New
World settlement forged new sorts of interracial and multicultural societies, but in Saint Domingue, with no native population or culture to
confront and with an extraordinary concentration of slaves, an especially
intense and prototypical pattern of New World social development arose
(Figure 4). In a word, Saint Domingue stood at the cutting edge of black-
white race relations in the eighteenth century, and the colony led the
way in the making of modern interracial and multicultural towns and
societies.

The institution of slavery provided the fundamental parameters that
structured the social order in Saint Domingue. To fathom the organization of colonial society in Saint Domingue, however, one must distinguish carefully between a *racial* axis of black versus white and a *political*
or civic axis of free versus slave. To a similar extent, the axes of rich versus poor and immigrant versus native-born must be kept in mind. Then,
too, one must think of these categories as composed of men and women.
The realities of eighteenth-century Saint Domingue reflect a complicated
pattern of social organization developing along each of these lines.

Racially, three castes emerged: whites, blacks, and mixed-race mulattoes.[1] The legal and political distinction of free versus slave also broke
down into three castes: free white, free people of color (black and mulatto), and slave. The two sets produced by racial and political division
were not entirely congruent, however. For example, the political caste of
free people of color (the *affranchis*) was composed racially of one-third

FIG. 4. "Vue du Port de Nippes." Drawn by Ozanne, engraved by N. Ponce in Moreau de Saint-Méry (1791).

blacks and two-thirds mulattoes. Similarly, a mulatto slave might think himself or herself socially superior to a free black.[2] Operationally, race proved the strongest consideration, and the melding of racial and political status effectively produced three major caste divisions: free white, free mulattoes, and blacks (free and slave).

Table 1 presents an overview of the demographics of Saint Domingue at the end of 1789. Because contemporaries kept records in this way, Table 1 breaks down Saint Domingue's population first and foremost along the axis of free versus slave—that is, according to civic or political status. The previous points notwithstanding, the table makes clear, again, the statistically overwhelming character of black and slave populations in Saint Domingue. Our relatively conservative figures indicate a slave population of 500,000 in 1790, or 89% of the total population. One source suggests that because the head tax on slaves did not include children or persons over forty-five the real number of African-Americans in Saint Domingue might well have been 700,000. Another reliable source puts the numbers of slaves in the colony at 600,000.[3]

The percentages of whites and free people of color (mulatto and black) in the colony, on the other hand, were very small in comparison, at 6 percent and 5 percent respectively. The contrast with the racial and caste makeup of the United States at the end of the eighteenth century is

TABLE 1. The Demographics of Saint Domingue, 1789–1790

	Free Whites	Free People of Color	Slaves
Population breakdowns[a]			
Total = 560,000	32,000	28,000	500,000
Percent of population	6	5	89
Growth rate[b] (%/yr)	1.35	4.5	6.0
Doubling time (yrs)	52	16	12
Racial makeup (%)			
White	[90][c]	[15][d]	0
Blacks	0	35	[94]
Mulattoes	[10]	50	[5]
Other	0	0	[1]
Sex ratios[e] (%)			
Men	80	45	60
Women	20	55	40
Origins[f] (%)			
Immigrant	75	[5]	67
Native	25	[95]	33

Note: Numbers in brackets are estimates.

[a]The figures in Table 1 and in Graphs 1A and 1B derive from the official census of 1788–89 [AN, C^{9A} 162, no. 245], and Abeille, pp. 3–4; Bellegarde (1950), pp. 6, 15; Bellegarde (1923), p. 104; Butel, p. 167; Castonnet des Fosses (1886), p. 20; Cornevin, p. 31; Darby, p. 268; Devèze, pp. 160–61, 189, 256; *DPF,* pp. 28, 84; Doucet, p. 151; James (1963), p. 56; James (1974), p. 551; Julien (1977), p. 65; Julien (1976), p. 162; Houdaille, p. 860; Lokke, pp. 34–35; Montague, pp. 5–6; Ott, p. 9; Pluchon, ed. (1980), pp. 52, 74; Stein (1988), quoting Frostin, p. 88; Tarrade, pp. 44–47; Thompson, 2:25; Wimpffen, p. 250. See also Klein, pp. 295–96. Table 1, produced without benefit of Pluchon (1987), appendix 6, confirms Pluchon's statistics. Not all sources agree, and Geggus (1989), p. 1291n, cautions as to their reliability; Moreau, for example (*DPF,* p. 28), puts the white population at an elevated 40,000. The figures in Table 1 represent conservative, median estimates.

[b]Based on data presented in Graphs 1A and 1B.

[c]All whites were supposedly 100 percent white, but racial mixtures were inevitable, and ultimately one's status as a white was defined socially. Moreau (*DPF,* p. 93) remarks, for example, that his category of *sang-mêlé* was only one–sixty-fourth black (1.6 percent) and indistinguishable from whites; see also *DPF,* p. 99, and Knight, p. 104.

[d]*DPF,* pp. 102, 110, provides the data; Moreau makes clear that one-sixth of the *affranchis* consisted of individuals essentially white.

[e]*DPF,* pp. 119, 722; Houdaille, pp. 863, 865; Saint-Vil, p. 22; Knight, p. 100. Stein (1988), p. 45, puts the ratio of males to females among slaves at 11:9.

[f]Proportions of immigrants versus native-born based on *DPF,* pp. 32, 44.

worth repeating for contrast: 80 percent free white, 2 percent free people of color (black and mulatto), and 18 percent slave.[4]

When one looks at the growth of this population over time (presented in Graphs 1A and 1B), several other notable points emerge.[5] First, that the population of Saint Domingue did not begin to grow substantially until after 1700. Second, the preponderance of slaves in Saint Do-

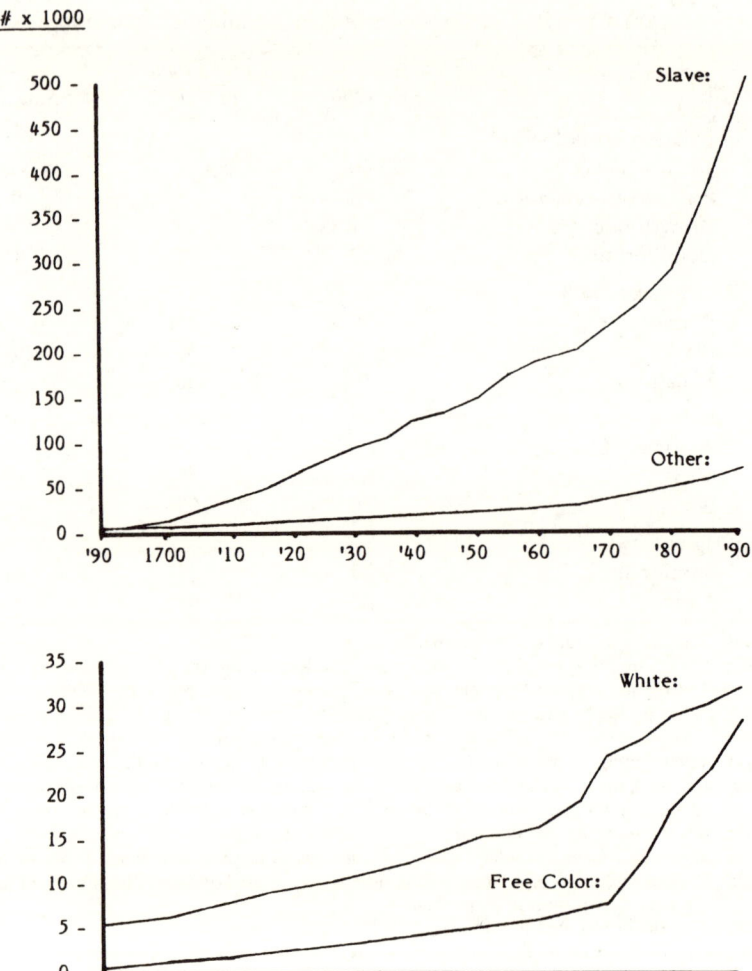

GRAPH I. Population Growth in Saint Domingue

mingue manifests itself from an early date. The number of slaves grew steadily through the 1770s, with a dramatic spurt in the decade 1780–90. By the end of the century slaves were an order of magnitude more numerous than all other groups combined, and the slave population was growing faster than any other group in the colony.

When separated from the demographic trends imposed by slaves, patterns in the relative growth of white and free citizens of color become discernible. Graph 1B evidences continuous growth in the white population across the eighteenth century, with an especially rapid burst immediately after 1763. The number of whites in the colony continued to increase in the 1780s, but less rapidly than earlier and at a much slower rate than any group. The graph likewise reveals that, through the greater part of the century, free people of color (the *affranchis*) constituted the least significant segment of the population. After 1763, however, their numbers increased dramatically and at a rapid rate, essentially the same rate as slaves. By 1789 the number of free people of color rapidly approached that of free whites. This increase in the *affranchie* population resulted in greatly increased racial, caste, and class tensions with free whites, and laid the basis for the tragic unfolding of relations between whites and freedmen in Saint Domingue after 1789.

Societies similar in character to Saint Domingue arose in the many colonies where intensive single-crop agriculture demanded the productive labor of masses of slaves, but Saint Domingue became the largest and most infamous of its type.[6] Contrasted with other French colonies, Guadeloupe followed Saint Domingue, but only at 20 percent its size in terms of population. Martinique came next, at 17 percent the scale of Saint Domingue. French Guiana weighed in at a mere 2 percent of Saint Domingue, and the French presence in the whole of the Indian Ocean amounted to just over 15 percent of that found in Saint Domingue.[7] Spanish Saint Domingue again provides a remarkable contrast in scale and character of population: in 1789 Spanish Saint Domingue had only one-quarter of the population of French Saint Domingue, with a much different racial balance of 40,000 whites, 80,000 free people of color, and only 15,000 slaves.[8]

Slaves formed the largest human segment of Saint Domingue. Two-thirds of the slaves of Saint Domingue were born in Africa; of the one-third born in the colony, most were second-generation Africans.[9] The cultures of Africa, therefore, and their transmutations under the regime of slavery constitute a major factor in the social development of colonial Saint Domingue.

The slave trade was the largest forced migration in history, and it involved the largest flotilla in the eighteenth century. The third largest after the Portuguese and the British, the French slave trade transported approximately one-sixth of the slaves removed from Africa in the eighteenth century. The French completely dominated the market for slaves in French possessions, however, and French demand became so great that slaves originally imported into British possessions apparently found their way into French markets. The French slave trade—a big, if risky business—grew tremendously as the eighteenth century progressed. The 1780s marked the height of the world slave trade, and French slaving activity proved positively frenetic in the decade with 1,100 expeditions bringing a total of 370,000 slaves to the West Indies.[10] Virtually all of the French slavers headed directly for Saint Domingue.

Ninety-nine slave ships arrived in Saint Domingue in 1789 alone, and crammed in their holes breathed 27,000 slaves worth 60 million livres.[11] (The figure of 27,000 imported slaves excludes a rock-bottom 12 percent mortality on "the middle passage.") Most slaves came from western Africa, but the rest of Africa relinquished its share, including Muslim slaves from North Africa. Slave owners overwhelmingly employed slaves, especially new arrivals from Africa (*bosales*), as unskilled agricultural workers. Urban slaves attached to households or other establishments represented a small percentage (perhaps 10 percent) of the total, which is still a substantial number of 50,000 or so. But the great majority of slaves worked on rural plantations, known as *habitations*.

By 1789 Saint Domingue possessed approximately 7,800 *habitations* cultivating sugar, indigo, coffee, and cotton. A sugar plantation required a minimum of 45 slaves, and the average plantation counted 200 slaves. Some plantations (true "plantations" in French) maintained upwards of 1,000 slaves, and Moreau mentions one ranching operation with 1,500 slaves.[12] Studies of plantations reveal a fairly complex organization within slave groups, including skilled workers, unskilled workers, overseers, men, women, the young, and the old.[13] Virtually all whites (excepting only the very poor and outcast) owned slaves, as did free people of color. One could rent slaves, and public and semipublic agencies (including parishes and the king himself) likewise owned slaves. Slaves performed every sort of manual labor, including that of draft animals.[14]

Colonists considered slaves essential to the economic development of Saint Domingue. As one observer put it succinctly in 1792, "Without slaves, no agriculture, no products, no wealth."[15] So closely was slavery tied to the economy and to profit that the "slave-day" arose as a customary unit of measure for an amount of work or for the value of something, such as a piece of property.[16]

Men outnumbered women among slaves, but only slightly. Female slaves were of significant economic value, including as field workers. The population of slaves did not sustain itself, however. Slaves suffered high mortality—one-ninth supposedly died annually, and the birthrate among slaves remained low, in part because of abortion and infanticide. Essentially no formal marriage took place among slaves, and slave owners seem to have made no special efforts to keep families together.[17]

On plantations and in towns, slaves commonly lived in huts or lean-tos. Bedding was typically palm fronds or other vegetation; rags or nothing at all constituted clothing. Slave owners sometimes furnished slaves with their own garden plots and time to tend them, and slaves used these resources to grow staples (potatoes and cassavas) and vegetables for their own consumption and (on occasion) for private sale. One contemporary called the food slaves received worse than that provided for the animals, but modern opinion seems to be that most owners fed their slaves comparatively well.[18]

Slaves were the absolute chattel of their owners. Slaves newly imported into Saint Domingue received brands with the names of their first buyer, their port of entry, and the name given to the slave.[19] Physical punishments and abuse of slaves, including whipping, shackling, and imprisoning, were the *norm* among slave owners. Everywhere private justice prevailed. Even though slaves constituted valuable capital property, private justice sometimes went to sadistic extremes and involved the torture of slaves.[20] (The record notes women sadists.) The small number of sadistically abusive owners coming to the attention of public authorities at most were deprived of their slaves. As it became cheaper to buy an imported slave than to raise one in the colony, conditions of slaves deteriorated as 1789 approached.[21]

The royal edict issued by Louis XIV in 1685, known as the Code Noir, supposedly governed the treatment of slaves.[22] The Code Noir protected slaves to a degree: it granted them Sundays and feast days off. It specified weekly food allowances (in theory including two pounds of beef or three pounds of fish) and two changes of clothes annually. The Code outlawed torture and limited to forty the number of lashes to be administered to a slave in any one session of private justice, and it mandated that slaves could not be abandoned. By the same token, the Code Noir denied slaves civic status, the freedom of movement, and the rights to own property, to engage in commerce, or to carry arms of any kind. The law resolutely denied freedom of assembly to slaves, on pain of death for recidivists. Theft was punished by branding. Penalties for escaping varied: up to one month on the run earned a slave his or her ears cut off and a brand; up to two months got the hamstring cut and another brand; after sixty days it was death. Courts likewise gave the death penalty to slaves who assaulted their masters or any free citizen.

The debased and repressive conditions under which slaves lived seems at variance with their value as capital property. A good piece of "ebony wood" cost a minimum of 2,200 colonial livres in 1789, with native-born Creole slaves worth 25 percent more than slaves brought from Africa.[23] Continued economic growth in the colony depended on an expanding population of slaves, and the inevitable limits on manpower drove up the price of slaves as the eighteenth century progressed. The southern department experienced particularly acute manpower shortages. In 1786, for example, the government awarded a premium of 200 colonial livres for each slave imported into the south. There, decent agricultural workers cost 3,000 livres, and the measliest slave went for 2,800. Everywhere, truly skilled slaves fetched much more, in the range of 4,000 to 6,000 livres, and in one extraordinary instance an owner refused an offer of 15,000 livres for a slave potter.[24]

Despite their highly constrained circumstances, slaves developed and maintained their own cultural existence. Religion and cults brought from Africa provided one important nexus. By all indications, vodun (voodoo) was practiced widely. The rites of vodun increased social cohesion among slaves and helped unite slaves in their great confrontation with whites beginning in 1791.[25] Slave assemblies and dances (*calendas*) occurred regularly with and without official permission. Drinking, stick fighting, gambling, and cockfighting represent other important features of slave social life, tolerated to varying degrees (Figure 5).[26] Slaves buried their own dead in separate cemeteries, sometimes surreptitiously.[27] Funerals took place virtually all the time, and provided further occasions for social intercourse among slaves.

Saint Domingue was a bilingual colony, with French and Creole spoken. French ruled as the language of government. Whites spoke French, as did the rest of the population to varying degrees. A range of French accents betrayed the white speakers' regional origins. But a Creole language arose in Saint Domingue, too, a patois melding of French and African tongues. All Creoles—that is, all persons born in Saint Domingue, black, white, and otherwise—could speak Creole. Slaves spoke Creole exclusively among themselves. Moreau de Saint-Méry, a Creole himself, seems to have become enchanted with the Creole language. "There are a thousand little nothings," he said, "that one wouldn't dare say in French, a thousand voluptuous images impossible to conjure in French that Creole expresses with infinite grace."[28] Only in 1987 did this vigorous and expressive language become an official language in Haiti.

The great number of slaves and their debased conditions created extraordinary social and racial tension in Saint Domingue. The colony embodied, as Pierre Pluchon has put it, a "world of violence."[29] Slaves posed a great and constant danger to the establishment, and the authorities deployed considerable forces of repression to maintain the status quo. Slave revolts occurred in the colony prior to the great uprising in 1791, but the forces of law and order suppressed the last of these (involving 560 slaves) in 1732.[30] The French kept the lid on their slaves for nearly sixty years thereafter, but they could not keep the pot from boiling.

Slaves struck back at the system indirectly through abortion and suicide. Poison and poisoning were direct routes that struck fear in every-

FIG. 5A, B. Slaves dancing and stick fighting: "Danse de Nègres" and "Nègres jouant au bâton." Engravings [by N. Ponce] in Moreau de Saint-Méry (1791).

one.[31] Whites were petrified of being poisoned by their slaves, and in some cases slaves did poison masters. But poison took slave victims, too, probably more. In the famous Macandal case in 1757–58, rebels used poisoning as an instrument to protect and enforce conspiracy in the slave community.[32] The reality of poison was one thing, but the fear it engendered among whites crossed the line into mass hysteria. Poisons became a mad concern that affected legislation, medical experimentation, and scientific research in the colony.

Maroons, or escaped slaves, posed a big problem for authorities. Slaves escaped constantly, and the possibility of escape undermined control on the plantations.[33] Many slaves, in what was known as *petit marronage,* escaped for short periods and returned or were caught. Many escaped slaves formed bands and survived for a while as outlaws on rampages. Plantation and property owners rightly feared incursions by runaway slaves, and they hunted down maroons with a vengeance.[34] Some groups of maroons managed to establish permanent or semipermanent encampments, the most notorious being the maroon colony in the Bahoruco mountains on the Caribbean in the south. (The literature also refers to this famous maroon enclave as Le Maniel.) From the first years of the eighteenth century maroons found a safe haven in these isolated mountains straddling French and Spanish Saint Domingue. An independent enclave emerged composed of runaway slaves, military deserters, and true "forest people" actually born into the colony.[35] Bahoruco maroons represented such a threat to French colonization in principle and in practice that the colonial government mounted at least a dozen military campaigns against them over the course of the eighteenth century.[36] In 1785 French and Spanish authorities concluded a treaty with the Bahoruco enclave, whereby the maroons of the Bahoruco gained freedom and re-settlement in return for their services in hunting other maroons.

Outnumbered sixteen to one by slaves, whites formed a tiny minority of the total population, and only a minority of the white population in turn formed the dominant class. Whites separated themselves along economic and class lines more than slaves or free people of color, and the axis of rich versus poor comes more into play when dealing with whites.

At the top of the pyramid stood the rich and powerful, the so-called *grands blancs*. At its most restrictive the term denotes rich plantation own-ers and, even more, sugar plantation owners. For present purposes, how-ever, *grands blancs* may be extended to include the major merchants in the colony, higher government administrators, and top judicial and military officers. A small number of persons of more middling station (a true middle class) effectively became included among the *grand blancs:* lesser government, judicial, and military officers; attached and independent technical experts of varoius sorts (physicians, engineers); top plantation managers and a rural gentility. *Grands blancs* may have constituted one-fifth to one-third of the white population in 1789.[37]

The so-called *petits blancs* comprised the bulk of the white popula-tion. One can discern several elements within the group of *petits blancs*. Small farmers form one. Slave masters and plantation employees, an-other. Soldiers, royal sailors, and merchant sailors, yet a third. One needs also to include a petite bourgeoisie in towns, white servants, and whites who consorted with nonwhites. In addition Moreau reports that the homeless—people who died as paupers by the roadside or by the sea-side—made up one-sixth of the white population in 1789 (16 percent or some 5,000 people). Moreau considered vagabonds a scourge, and in this he reflected the views of the established order, which regularly dumped vagabonds in prison or the poorhouse, whence they later returned to the streets or the countryside.[38]

Remarkably, perhaps 75 percent of the whites of Saint Domingue were born in France. Immigrants thus composed most of the white population like the black. Jacques Houdaille estimates that 50,000 native French died in Saint Domingue in the period, 1740–91.[39] Several times that number must have passed through Saint Domingue over the years.

White men strongly outnumbered white women in the colony. Some sources put the sex ratio among whites as high as five to one, men over women.[40] Male immigration outpaced female immigration twenty to one. Such a sexual imbalance added to a significant sexual tension in the colony. It fostered prostitution and promoted interracial sexual contact, leading to the rise of the free people of color as an important segment of Saint Domingue society.

Another remarkable characteristic of the population (white and otherwise) was its youth. Few old people survived in Saint Domingue. As Houdaille reports, the age at death for whites in the colony averaged twenty-seven for sailors and thirty-eight for everyone else; the average age at immigration was twenty-three. In some cases people lived into their eighties, but by fifty, apparently, one numbered among the elderly.[41] In a like vein, only a few children lived in the colony. Illness and disease killed many young, and whites frequently shipped their offspring to France.[42] The age profile of the colony thus reveals it to be populated overwhelmingly by people in their twenties and thirties.

White women married early and normally went through a series of marriages. The *grands blancs* doubtless married more, but overall fewer than half the women married. Only three thousand married women lived in the colony, out of a total female population of perhaps 21,000, excluding slaves (6,000 white, 15,000 free women of color). In what may be an extreme example, Moreau describes one hamlet of 200 with only 4 married women in it.[43] Unmarried women commonly entered unofficial liaisons and engaged in prostitution, as we will see. This rarity of marriage underscores Moreau's suggestion of a high degree of sexual tension in the colony.

Whites drank excessively, and alcoholism no doubt ran rampant. In 1789 merchants and colonists imported over 80,000 *barriques* (some 4 million gallons) of wine in addition to a further 12,000 cases of wine, 10,000 *barriques* and *paniers* of beer, and 37,000 cases, anchors, and *paniers* of various brandies and liqueurs. A cuisine emerged for colonists different from that in Europe. Meals for whites centered around hams, stews, salted cod, sausages, fish, game, and (where available) oysters and local truffles. Colonists reportedly ate more than at home in Europe, and cooks prepared food using greater amounts of spices and hot seasonings to encourage the consumption of liquids.[44]

In Europe, Saint Domingue enjoyed a distinct reputation as a place of escape, a place to start over. The record abounds with cases of individuals running away from unpleasant situations in France: the Hecquet son fleeing gambling debts and an unacceptable romantic entanglement; Davezac running away from a father who wanted him to be a priest; Decombaz, the pornographer, escaping the police in Switzerland and Italy. Saint Domingue so commonly served as a route out of trouble that the expression "passer aux isles" came to signify the phenomenon.[45]

Similarly, Saint Domingue came to be known as a place to make a buck, to get rich quickly. Colonists regarded the easy fortune as the "soul of the colony."[46] According to reformers, white colonists sought above all to make their gains as quickly as possible and return to an easier life in France. The ordinary colonist evidenced little interest in putting down roots and developing the colony socially. Saint Domingue thus faced a cultural frontier as much as it did a frontier with nature. Conversely, cultural institutions like the Cercle des Philadelphes had to overcome resistance to the idea that Saint Domingue was anything but a backwater.

Between whites and the predominant slave population stood the free people of color, the third key group in Saint Domingue's population (Figure 6). In terms of political status, the free people of color (the *affranchis*) were legally free, nonwhite citizens. In terms of race, however, the free people of color included blacks, people of mixed race, and, seemingly, whites. Moreau puts the proportion at one-third black, one-half people of mixed race, and one-sixth people of "nuances supérieurs."[47]

African-Americans among the free people of color were usually slaves who had been freed by their masters. In response to complaints about abuse, it became increasingly difficult to free slaves as the eighteenth century progressed, but on some plantations an informal freedom sometimes operated. To believe Moreau, however, the fate of the black freedman (or woman) hardly differed from that of the black slave.[48] Racial prejudice proved stronger than allegiance to political or civic status, and the political grouping of free people of color came to be equated to the racial caste of mulattoes.

People of mixed race arose from the constant concourse of whites with blacks.[49] The mulatto portion of free people of color formed a distinct sociological group, a racial caste socially subordinate to whites and superior to blacks and slaves. To repeat, this group was growing rapidly as 1789 approached and offered a challenge to the dominant white population. Among the free people of color, women seem to have slightly outnumbered men, but, nevertheless, mulattoes maintained the most balanced sex ratio of the colony's three racial castes. Mulattoes, almost by definition born in the colony, formed the only group not to be composed mostly of immigrants.[50]

Many free blacks and mulattoes in Saint Domingue lived in the backwoods, enjoying the recognized right of bushwhacking or homesteading. The rise of coffee production in the colony, especially after 1763, depended on the free people of color and vice versa. The relatively small scale of coffee production, the suitability of Saint Domingue's mountains for growing coffee, and the lack of need for irrigation account for the

FIG. 6A, B. Washerwomen and free people of color: "Blanchisseuses" and "Affranchis des Colonies." Engravings [by N. Ponce] in Moreau de Saint-Méry (1791).

linkage. In some regions, the Mirebalais, for example, free people of color formed the majority and owned almost all of the property. White authority regarded such groups of free people of color with suspicion, however. In towns free people of color lived in segregated neighborhoods, disparagingly (and misleadingly) called "little Guineas." Again, Moreau touts mulatto men as excellent soldiers and hunters of game and men, and he says they also earned livings as tradesmen and craftsmen. Moreau similarly signals a number of free men and women of color as morally among the leading citizens of the colony.[51]

Moreau reserves his highest praise for the mulatto women of Saint Domingue (Figure 7).[52] He lauds these "goddesses," these "priestesses of Venus" far above the colony's white women, not the least because they bathed more. No doubt some mulatto women were respectable wives and mothers, linked to soldier-husbands perhaps or working as housekeepers in white homes. But in cities certainly, most were drawn into prostitution, seemingly at an early age.

Technically illegal, prostitution was tolerated as a necessary evil, especially in a colony with such an imbalance between the sexes. Following the contemporary observer Hilliard d'Auberteuil, Dantès Bellegarde estimated a total of 3,200 prostitutes in Saint Domingue, two-thirds ladies of color and one-third white women.[53] Thus, approximately three out of every twenty women in the nonslave population were prostitutes.

Most prostitutes lived as "housekeepers" with white men; others had

FIG. 7A, B. The dress of slaves and free women of color: "Costumes des Affranchies et des Esclaves." Engravings [by N. Ponce] in Moreau de Saint-Méry (1791).

their own rooms with a "lit de repos dont l'usage n'est pas un problème insoluble."[54] Some owned sumptuous houses, virtual schools for prostitution. Rich and generous clients and lovers provided a life of luxury for the upper echelon of Saint Domingue's prostitutes, and ladies of color engaged in the trade led the colony in fashion. Served by their own loyal slaves, mulatto prostitutes seemingly could not be rivaled in natural beauty, clothing, coiffure, makeup, and perfume. That they kept their looks and enjoyed seducing the husbands of white women provoked bitter jealousies and clashes, which Moreau blames on the white women. By the same token, Moreau does more than intimate lesbian concourse between white and mulatto women. Abortion was commonly practiced, and few children played in these circles. Prostitution and an intense sexuality pulsed everywhere in the colony. As Moreau put it, in Saint Domingue "one is surrounded by a most dangerous seduction."[55] Along these lines, in 1786 the colony's newspaper, the *Affiches Américaines*, published a cautionary letter from a white prostitute newly arrived from France, who complained bitterly of the competition from amateurs and professionals alike and of having been misled in her endeavor.[56]

Whites, blacks, and people of mixed races defined the social and racial character of old Saint Domingue and constituted 99 percent of its population. But, other, smaller groups left their marks on the colony. Jews were prominent among these. At most, no more than a few hundred Jews lived in Saint Domingue at the end of the eighteenth century.[57] Jews were white and free, of course, and lived on par with other whites. Saint Domingue Jews originated from Bordeaux communities established in the sixteenth century by refugees from Portugal, and they came to Saint Domingue via the merchant houses of Bordeaux. European anti-Semitism was transplanted, also. In 1764 Governor d'Estaing severely harassed Saint Domingue's Jews with an arbitrary (local) tax, but a certain official tolerance prevailed, supposedly so long as Jews (and Protestants) did not overtly practice their religions. The southern department formed the center of gravity of the Jewish community in Saint Domingue, with the de Pas family of special importance.[58] The family's founder, Michel Lopez de Pas, a Montpellier physician, arrived as the colony's first royal physician in 1701. The family converted to Christianity in the harassment of 1764, but remained the protector of the Jewish community.

The Spanish exterminated the native Indian population in Saint Domingue, but perhaps five hundred American Indians brought from Louisiana and Canada lived in the colony at the end of the eighteenth century. They were wholly enslaved. A handful of "East Indians" (from India or

FIG. 8. "Vue de Bombardopolis." Drawn by Perignon, engraved by N. Ponce in Moreau de Saint-Méry (1791).

southeast Asia) found their way to Saint Domingue, and they enjoyed full rights like whites, unless they became tainted by African blood, in which case they descended into the mulatto caste.[59]

Acadians from French Canada and an unusual colony of settlers from Germany formed the two most exotic groups to enter into the sociology of Saint Domingue. Both groups arrived at the same time, in 1764, under separate but equally unfortunate circumstances. The dastardly British deported 12,000 Acadians from French Canada after the Seven Years War, and slightly more than 400 came to Saint Domingue by way of New York. The Germans were recruited in Europe as part of a French expedition in 1763 to colonize Kourou in French Guiana in South America. That expedition ended in complete failure, and nearly 2,500 survivors arrived in Saint Domingue the next year.[60]

Both groups initially faced terrible conditions and high mortality, and providing for them severely taxed local resources.[61] The authorities settled the Acadians and the Germans together on the underpopulated northwestern peninsula, but the groups did not get along and were separated. Some Acadians went on to Louisiana; others moved into the Môle area, where they languished as fishermen and herdsmen. The Germans took over the agricultural community of Bombardopolis or Bombarde, as everyone called it (Figure 8). Finally settled, the Germans reestablished the sober, industrious, and prosperous life they had left in Germany. They made Saint Domingue not just a French colony, but truly a European colony.

↝§ FOUR §↜

Industry and Economy

A S NOTED AT THE outset, French Saint Domingue at its height in the 1780s had become the single richest and most productive colony in the world.[1] Eighteenth-century Saint Domingue was decidedly of the highest economic importance to France itself. The colony has been characterized as France's most important "province" and "indispensable" to its overall commercial balance at the end of the eighteenth century.[2] Saint Domingue received two-thirds of French overseas investments, and one-third of all French foreign trade took place with the colony.[3] One person in eight in France—over three million people—earned his livelihood in some way connected with colonial trade, and on the eve of the French Revolution goods worth somewhere between 150 and 170 million livres tournois poured into France from Saint Domingue.[4] At most, only two-thirds of that amount flowed back to Saint Domingue, leaving a positive trade balance of at least 50 million livres. Furthermore, nearly three-quarters of colonial produce initially imported into France was subsequently reexported for solid profits to the rest of Europe, with colonial goods constituting nearly half of total French exports.[5] Taxes and duties on colonial products and their transportation brought additional millions of livres of revenues to the state and various arms of its administration.[6] Saint Domingue embodied a tax base of nearly 1 billion livres, and it more than supported its own administrative costs.[7] The primacy of economic factors driving colonial development in Saint Domingue cannot be denied.[8] To understand the nature and success of the colony, one needs a clear sense of its economic and productive underpinnings.

Saint Domingue's wealth and productivity derived almost entirely from agricultural production and trade of agricultural commodities. Historically, tobacco was the first commodity to come out of the colony. Pirates cultivated tobacco early in the seventeenth century, and tobacco became the key product in the first phase of official colonization after 1665.[9] But the tobacco era and the era of the small farmer generally ended early in the eighteenth century, and an entirely different and much more powerful set of agricultural industries emerged, centered on sugar, coffee, indigo, and cotton. Capping a spurt of growth following the American War of Independence, colonists cultivated as many as 2.5 million acres in French Saint Domingue, with at least 1.5 million acres under "high cultivation."[10] By 1789 over 7,800 plantations in Saint Domingue raised and processed agricultural commodities.[11]

The unit of colonial production was the plantation estate (Table 2). Indeed, the plantation forms the basis of what Franklin Knight terms an "exploitation colony" characteristic of the West Indies, and Knight goes further to label Saint Domingue "the exploitation colony par excellence during the eighteenth century."[12] The activity of production took place according to the famous plantation system, a system of production wherein slave labor, intensified monoculture, and initial processing became combined. As Sidney Mintz has said, the plantation system united "field and factory," and clearly that system of production evolving most especially in Saint Domingue put the colony at the leading edge of world economic development at the end of the eighteenth century.[13]

Raising and processing sugar cane was the foremost agroindustry in eighteenth-century Saint Domingue.[14] According to G. B. Hagelberg, the sugar cane plant "produces larger quantities of utilizable organic matter per land unit in a given time than any other crop."[15] At the peak of their output, Saint Domingue's nearly 800 sugar plantations constituted the world's single largest producer of raw and semirefined sugar, with production from Saint Domingue amounting to about half of the world total.[16] The recent figures by Robert Louis Stein show that Saint Domingue planters exported the equivalent of 192 million pounds to France in 1791. Annual sugar production from Saint Domingue equaled 75 mil-

TABLE 2. Plantations in Saint Domingue, 1789

Sugar	793
Coffee	3,117
Indigo	3,150
Cotton	789

lion livres tournois. Each colonial sugar plantation thus yielded a per annum average of 240,000 pounds of sugar worth 95,000 livres.[17] To be sure, British West Indian sugar production closely rivaled the French, but Saint Domingue sugar dominated the world market, because it cost 40 percent less and because the British consumed domestically most of the sugar they produced abroad.[18] According to Eric Williams, French sugar plantations in Saint Domingue were also cheaper, less expensive to run, and (with more fertile soil) more productive than English sugar plantations on Jamaica and the Barbados.[19] Colonial sugar production drove the slave trade, and, although archaic and highly fragmented, the sugar business was one of the largest in France at the end of the eighteenth century. Sugar cane and the productive labor of slaves fueled the engines of wealth generated in Saint Domingue; a new form of economic activity, "sugar capitalism," developed, and King Sugar played its part in shaping the social character of the colony.[20]

As for the sugar production process itself, slave gangs of unskilled field laborers, accounting for around 60 percent of a plantation's slaves, would continually sow and reap cane (Figure 9).[21] Cane had to be harvested and processed quickly, so that its valuable juice would not deteriorate. Slaves crushed freshly cut cane in rolling machines and extracted the juice. In a series of boiling, cleansing, cooling, and cyrstallizing operations, slave technicians transformed this juice into raw sugar. As muscovado sugar, the resulting product was left to drain of its molasses; as clayed sugar, water percolated through a clay seal set on the pot to purify the product further. Clayed sugar was of a higher quality, but true refining required further crystallization, which almost always took place in France.[22]

The processing operation for sugar required a group of skilled slave laborers. It also required a fairly elaborate and expensive technology. We will have several occasions to point out where colonial development proceeded on the basis of (then) advanced technologies from Europe. This point holds for the processing of sugar. The production process required rolling mills of one sort or another and expensive copper boiling kettles, and this equipment made the sugar business a fairly capital intensive one. Only one-quarter of the sugar plantations of Saint Domingue could afford to install aqueducts and water wheels and power their mills hydraulically. (Water for such wheels came from streams or other irrigation works, about which more later.) Because of finances or circumstance, three-quarters of sugar plantations had to use animal-powered mills, even though animal mills cost more to operate than water-powered mills. Although a few could be found, windmills were not widely employed in Saint Domingue. Animal and hydraulic milling machinery required the

FIG. 9. A sugar mill. From J. B. Dutertre, *Histoire générale des Antilles* (1667).

services of various "engineers" to install, maintain, and improve upon these instruments of the production process.[23]

Coffee constituted the colony's second great commodity. The Dutch took coffee from the Arabs, who had originally domesticated the tree, and in the seventeenth century the Dutch established thriving coffee plantations in Sri Lanka and Java. The Dutch monopoly on coffee collapsed, partly on account of a scientific exchange that backfired. Coffee tree seeds reached the Amsterdam Botanic Garden in 1706, and in seeming innocence the Dutch sent some to the Jardin du Roi in Paris. From these seeds the French turned around and inaugurated coffee production in the Caribbean, introducing the tree into Martinique in 1723 and thence to Saint Domingue in 1726.[24] But coffee did not begin to be grown in the north of Saint Domingue until 1738, and the takeoff in coffee production did not occur in Saint Domingue until after 1765. In that year authorities imported 100,000 trees into the colony; the price and demand for coffee rose considerably thereafter, and production raced to expand.[25] The mountains and climate of Saint Domingue proved ideal for growing coffee; the tree proved especially suitable for small-scale farming; and, the quality of Saint-Domingue coffee rivaled the world's best. By the end of the 1780s, Saint Domingue had become the world's largest producer of coffee.[26]

Production of the vegetable dye, indigo, represents Saint Domingue's third major agroindustry at the end of the eighteenth century. Indigo production very much resembled that of sugar. Slaves cultivated varieties of the indigo plant in fields. The plant required considerable processing to extract the dye, more than sugar, or so experts claimed. The indigo process also required water. Slave workers fermented harvested plants in vats through various stages; the mixture was then beaten and strained, and the residue dried. The beating stage most often took place mechanically using mills and machines of one sort or another. Like sugar, producing indigo dye called for skilled workers, craftsmen, and engineers, in addition to brute slave labor. The height of indigo production in Saint Domingue occurred in the 1740s, declining steadily thereafter, partly on account of disease and insect attacks.[27] The indigo industry nonetheless remained an important element in the economy of Saint Domingue at the end of the century, and preservation of the industry became one of the concerns of the Cercle des Philadelphes.

Cotton was the colony's fourth major agricultural commodity. Cotton may have been indigenous to Saint Domingue, and cotton plantations flourished in the dry regions around Gonaïves and in the Artibonite, especially when irrigated. Production levels reached 1 million pounds in 1768 and doubled to 2 million pounds by 1788; at one point six million cotton plants were cultivated.[28] Growing cotton obviously required a large slave labor force, but there was little processing of cotton in Saint Domingue. Cotton was likewise subject to heavy damage from insects. Saint Domingue furnished one of the major sources of cotton for the French cotton industry, and merchants diverted a sizable portion of Saint Domingue cotton production to Jamaica for final manufacture in England. This illegal commerce was referred to ironically as "the Paris trade."[29]

Cacao might have been a fifth such profitable commodity. Introduced into the colony by Ogeron in 1665, the cacao tree was extensively cultivated through the early decades of the eighteenth century on plantations with upwards of twenty thousand plants. The demand for chocolate in Europe, like that for sugar and coffee, was high, but the cacao tree in Saint Domingue appears to have died out suddenly, seemingly on account of hurricane damage, but possibly also because of disease. By 1736, apparently, cacao production in Saint Domingue had all but ceased, and thereafter Saint Domingue ceased to be an important source of cacao.[30]

The part of Saint Domingue's economy that generated profits and spurred trade, slavery, and development thus centered on the four agricultural commodities of sugar, coffee, indigo, and cotton. One is not surprised, therefore, that experts paid considerable scientific attention to these commodities and to ways to improve their production. Although weak institutions, the Chambers of Agriculture of Cap François and Port-au-Prince took the promotion of agriculture as their charge. In various ways, too, the Cercle des Philadelphes functioned as a learned society for agriculture and promoted agricultural improvements, and colonists responded with a flood of memoirs to various of the Cercle's initiatives. The most notable "scientific" investigator of the agroindustries of Saint Domingue was the physician, Jacques-François Dutrône la Couture, whose volume, *Précis sur la canne et sur les moyens d'en extraire le sel essentiel* (1790), was published at public expense. Dutrône's work provides a detailed chemical analysis as well as a practical manual for sugar production, and it includes remarks on indigo.[31]

The "essential salt" of Dutrône's title opens a small door on the ways in which science and scientific theory became entangled with agriculture and production in Saint Domingue. An even better example reveals itself in a series of memoirs in the papers of Baudry des Lozière's preserved at the Bibliothèque Nationale in Paris.[32] These anonymous and mostly untitled manuscript memoirs concern sugar and indigo production. They may have been written by Baudry himself, or they could have been circulating manuscripts he collected. A common chemical analysis of sugar runs through all the memoirs. Reports one:

> Cane is composed of four elements, to wit:
>   1. An elementary earth that is its base.
>   2. Air juxtaposed between the molecules of salt.
>   3. Water, called water of crystallization.
>   4. A fire given the name phlogiston.[33]

At least in the minds of contemporaries, they thought of stages in the production of sugar, indigo, and the other commodities in terms of theoretical, chemical analysis. The point warrants further exploration, but the primacy of theory over practice assigned by the author of one of the memoirs amply makes the point.

> Theory is queen, and practice is nothing but a slave. The first gives rise to the second. Without the flame of theory, practice is forever lost in a thickening obscurity, and only by using theoretical principles will it find its way free. . . . Two or three great geniuses suffice for rapid strides in theory, while practice proceeds more slowly because it depends on too large a number of mostly incompetent hands. The public certainly profits more from a theory that says something than a practice that says nothing.[34]

The reality of applied science in Saint Domingue was doubtless quite different, but the ideology was strong: theoretical natural science could improve practice.

For the daily bread of the colony—French bread—flour was imported from France, but in other ways the colony tried to provide for itself. Plantation and slave gardens existed everywhere, as did chicken coops and animal pens. Colonists grew rice on a small scale for local consumption, but authorities stifled expanded production after 1746 when royal storehouses rejected rice as a substitute for wheat flour.[35] Ranching was an important but hard-pressed local industry. Fifty thousand horses worked in Saint Domingue. Colonists raised cattle, sheep, goats, and pigs—a quarter million of them—throughout the colony, but ranches and local animal husbandry could not match the colony's need for fresh meat, although imports of cattle from the Spanish part of Saint Domingue partially met a chronic shortage of beef. Poaching by slaves proved a problem for husbandry. Colonists paid dearly for hay and forage, which they raised in all open spaces. In lesser ways, such as beekeeping and processing castor oil for lamps and medicines, agriculture and husbandry in Saint Domingue developed to meet the colony's needs.[36]

Shipping became the most important nonagricultural endeavor pursued in Saint Domingue. Shipping was essential to the colony. Slaves had to be imported from Africa, and agricultural surpluses produced on Saint Domingue's plantations had to be exported to France. French ships also brought needed finished and luxury goods, flour, wine, building materials (as ballast), and miscellaneous items that supported an extensive pack trade. The port of Bordeaux dominated French shipping to the islands, accounting for 50 percent of imports into Saint Domingue.[37] The shipping industry involved tens of thousands of sailors and a lot of expensive equipment and port installations, and shipping gave rise to considerable secondary employment in France.

While not without its problems, French Atlantic shipping boomed in the 1780s. The extent of French colonial trade doubled in the period from 1763 to 1787, and at its peak in 1788 the volume of shipping in and out of Saint Domingue reached roughly 1,200 sailings a year: reportedly 687 ships arrived from France, 527 departed for France.[38] These numbers contrast markedly with the 286 ships in and out of Martinique in 1788 and the 114 in and out of Guadeloupe in the same year.[39] The bulk of shipping in Saint Domingue occurred through Cap François, but ships also sailed directly to France from most major towns in the colony. Shipping along the south coast, however, proved irregular and insufficient to meet demand.[40]

In theory, the mercantilist principle of the "exclusive" governed trade between Saint Domingue and France, but in practice illicit trade penetrated protectionist barriers.[41] Trade between Saint Domingue and the American colonies and states of New England was significant, especially for the Americans.[42] The Yankees brought salted cod, livestock, lumber, beer, and cooking oil into the colony mostly in exchange for molasses, the residue product of sugar manufacture, which they distilled into rum. Proponents of free trade did see some successes later in the colony's history. The Môle outpost in Saint Domingue became a free port in 1767, in the attempt to develop the area for the Acadians and the Germans. As Jean Tarrade has shown, however, these moves and the opening of the three principal ports in the colony to international trade in 1784 actually constituted efforts by the French government to tighten up on mercantilist legislation that governed colonial trade.[43] The policy of mercantilism remained in place through 1789, so considerable illicit trade persisted, especially in the south, where commerce in goods (such as cotton) and slaves continued with Jamaica and Curaçao.[44] As has been suggested, the contentious issue of free trade reveals most clearly the fundamental conflict between the mercantilist policies of government and the economic interests of the plantation economy that emerged on Saint Domingue. The formidable success of colonial capitalism in Saint Domingue clearly came to clash with government policies that guided colonial development through most of the eighteenth century.

Every area of the economy in Saint Domingue endured a chronic shortage of hard currency. The southern department especially suffered from the lack of specie, a situation that encouraged bartering and the widespread circulation of IOUs. The colony depended on and was awash in private credit. In addition, having no cash to purchase slaves, plantation owners became evermore indebted to French and colonial merchants and merchant houses to whom they pledged their future production for credit. Economically, the colony swam in sea of debt, and individual plantations did not always make money.[45]

In addition to shipping, a number of smaller industries and crafts kept the colony going. In his census, Moreau counted 370 lime kilns, 182 distilleries, 36 brick and tile manufactories, and 29 potteries in Saint Domingue. Elsewhere he mentions rope and harness makers, basket weavers, and tanning operations in the colony. Steps seem to have been taken to introduce the silk industry into Saint Domingue, but the unidentified silk machine imported into the colony never became operative.[46] One wonders about the circumstances behind the effort.

Any survey of industrial, agricultural, or technological development in eighteenth-century Saint Domingue must take account of the extensive hydraulic-engineering works that grew up in the colony. We have seen how water proved very desirable in the processing of sugar and essential for processing indigo. We have mentioned that flooding represented a serious problem in the colony and that authorities undertook hydraulic works to control rivers. We will have occasion to see further how the provision of fresh water in towns also became a major focus of concern and activity. All that notwithstanding, the largest use of water by colonists and the largest water works turned out to be for irrigating fields and crops.

The agricultural miracle rendered by irrigation and hydraulic engineering needs to be emphasized. Irrigation utterly transformed large areas of Saint Domingue.[47] Desert areas became verdant, and crops were preserved in times of drought. The combined effect of the sun's tropical heat and water from irrigation created intense and ideal growing conditions for the commodity crops raised in Saint Domingue. At one point in his narrative, speaking of the Cul-de-Sac region, Moreau de Saint-Méry waxes impassioned over water's effect:

> [The Cul-de-Sac] would have disappeared a long time ago as an area for agriculture, if nothing was done to combat the annual six-month drought and the hot winds that destroy all vegetation. Water created all these miracles. Water alone can conserve them. Water enriches the soil. It preserves, increases, embellishes and perfects the products of the earth. It is a powerful agent to turn machines. . . . In a word, without this blessing from nature, which the industry of man has learned to appropriate for himself, these vast manufactories, these worthy and admirable fields, these immense riches would have never existed.[48]

One Joseph Ricord undertook the earliest irrigation works in Saint Domingue in the Cul-de-Sac region in 1731. He irrigated two plantations. A more significant set of works involving twenty-four plantations appeared in Léogane in 1737 and remained operative until destroyed by the earthquake of 1770. The irrigation system for four plantations installed in Acquin in 1739, also by Ricord, provides another early example.[49]

Over the succeeding decades several dozen large-scale hydraulic-engineering works were undertaken for the purposes of irrigation and flood control, and more were on the books in 1789. Projects typically involved dams or dikes for intake from streams or rivers, canals, aqueducts where necessary, bridges, culverts, and sluices to divert water onto fields. An unexceptional example is the system installed along the Rivière

Froide at Anse-à-Veau in 1787, which involved a stone dam 100 feet long and 14 feet deep and a canal 6,000 *toises* or more than 7 miles long.[50]

An irrigation system was costly and no minor matter, especially when it involved more than one plantation. Local syndicates of plantation owners usually built and maintained irrigation works. The colony's administrators approved of each undertaking and sometimes gave active, financial support. Using slave labor, public and private civil and hydraulic engineers built the various works. Irrigation works always proved contentious affairs, however, as touchy plantation owners constantly argued and litigated over costs and water rights.[51]

The Cul-de-Sac region around Port-au-Prince led the colony in the extent of its irrigation. Various irrigation systems existed in the Cul-de-Sac from the 1730s, as mentioned, and a major new installation was completed in 1755. But two later irrigation projects dwarfed these earlier efforts. One, for the Grande Rivière, begun in 1773 and completed in 1785, irrigated 8,000 *carreaux* or 22,000 acres. The project cost 3 million livres, but, according to Moreau, it was worth many millions more. The other project for the Rivière Blanche, completed in 1787 and forty years in the building, watered 5,000 *carreaux* or 14,000 acres. In all, hydraulic systems irrigated fifty-eight plantations and some 36,000 total acres, nearly half of the Cul-de-Sac plain.[52] That flat and well-watered plain, with its lush and regularly laid-out fields and its straight and well-maintained roads, must have been a remarkable sight in 1789.

Irrigation works seem not to have been used on the northern plain of Saint Domingue, perhaps because it received more rain and had a wetter climate. The Arcahaye area to the north and west of Port-au-Prince saw a major system installed in 1742. By 1789 that system had expanded to include eighty irrigation canals of sufficient note to require bridges across them. Other major irrigation projects arose on the plain of Les Cayes in the south.[53] Various projectors installed six large irrigation systems between 1757 and 1786, providing water for 7,000 *carreaux* or 19,500 acres, and planned for additional works to water another 2,600 *carreaux* or 7,200 acres. In 1789 irrigation systems served ninety-four plantations around Les Cayes, with service projected to expand by twenty-six more. Half of the Les Cayes plain received irrigated water.

The number of hydraulic works and the extent of government involvement increased considerably as the decades progressed. Coordination and an overall supervision of irrigation and river control in the colony was badly needed. In 1783 the government established the official position of *ingénieur-hydraulicien du roi* in the colony and appointed J.-J. Verret, a French engineer who had previously proposed a plan for irrigating the Artibonite, to fill the post. Verret supervised the planning and execution of hydraulic works in the colony in an effort to cut down on delays and disputes.[54] Not unexpectedly, Verret and his rival hydraulic engineers in the colony, Jean Trembley and Bertrand Saint-Ouen, became notable associates of the Cercle des Philadelphes.

The development of irrigation in the important Artibonite region proved less successful than elsewhere. Nonetheless, the Artibonite is the most interesting case because it involved the greatest degree of government participation and the most advanced application of technology to the problem of irrigation.[55]

More so than in other areas, the Artibonite went through a whole series of initiatives to implant irrigation works (1744, 1749, 1751, 1753). Nothing resulted from these efforts because of the strength and capriciousness of the Artibonite River. In 1755 a syndicate finally installed an effective system, involving forty plantations and 4,000 *carreaux* or 11,000 acres, but the system collapsed when its dikes failed in 1761. Another construction effort began in 1766, and in 1768 the colonial government rewarded a local planter, M. Bertrand père, for his pioneering efforts to irrigate his own plantation.[56]

In 1773, however, a new plan was launched for irrigating the Artibonite, this time on the initiative and with the backing of the highest levels of government in France. According to Moreau, the minister in France pushed for irrigation of the Artibonite as a means to promote economic development. The minister supposedly welcomed the plan of a colonist, M. Courrejolles, and forwarded it to the Académie des Sciences in Paris, where it received approval, at least in regard to its theoretical component. Trudaine de Montigny, intendant in the Finance Ministry and Academy honorary, released two engineers and their salaries from the Ponts et Chaussées to assist in the work, and Perronet, director of the Ponts et Chaussées and Free Associate of the Paris Academy, consulted on the plans. Capping this significant governmental effort, the king himself promised 1.2 million livres to pay for the project. All went well until the Artibonite River flooded again in 1780 and wiped out the works that had been constructed to date.

This disaster resulted in a redoubling of efforts to control the Artibonite River and irrigate the Artibonite Plain. Three separate initiatives and detailed proposals emerged in 1781. J.-J. Verret launched one from Paris, where he submitted it to the Paris Academy of Sciences. Jean Trembley, a local plantation owner and manager originally from Switzerland, made the second proposal in a lengthy treatise on the subject, and Trembley pressed for funding both from the colonial administrators and privately from Europe.[57] The third proposal—the successful one—came from J.-J. Bertrand de Saint-Ouen, son of the planter rewarded for his irrigation works in 1766. One notable element united all three of these proposals: to varying degrees, they all envisioned using steam engines to pump water for irrigation.

Saint-Ouen succeeded in getting himself appointed royal commissioner for irrigating the Artibonite and adjacent areas in 1781. Based on his detailed plans, work commenced in 1783. The government agreed to provide a steam engine for trials. The firm of the Perrier brothers built a Watt-style engine with a separate condenser in 1784 in Chaumont, France. Duly installed on the Bertrand plantation in Saint Domingue, the machine cost 72,000 livres, and an engineer came from France especially to superintend its operation. On November 11, 1786, Saint-Ouen ceremonially inaugurated the pumping system in the presence of the colony's governor, La Luzerne. The pump produced a lift of 23 feet and could irrigate 2,000 *carreaux* or 5,600 acres. The steam engine burned approximately 30 cubic feet of coal a day at a cost of 175 livres. Saint-Ouen died in 1787, and the irrigation project died with him. Moreau de Saint-Méry saw the Perrier engine not long thereafter. It had become derelict, the victim of unstable soil near the river bank. Into this century visitors continue to stumble across the ruins of this emplacement, mute testimony to unsuccessful efforts to tame the Artibonite in a bygone age.[58]

⋞ FIVE ⋟

The Urban Context

S AINT DOMINGUE was not heavily urbanized. Only 8 percent of the population lived in real towns with populations of over a thousand. This percentage, remarkably low in itself, is also low compared with rates in Europe and in France, which were 15 percent urban at the end of the eighteenth century.[1] Even though cities and towns in Saint Domingue hardly figure demographically, they were of the utmost importance in the life of the colony. As economic and cultural centers, they are crucial to understanding the character of eighteenth-century Saint Domingue and the story of science in the colony (Figure 10).

Eleven towns with populations greater than one thousand arose in Saint Domingue. Three stood in the first rank as veritable cities: Cap François, Port-au-Prince, and Les Cayes. Their populations in 1789 were 18,500, 9,400, and 5,650, respectively.[2] These three urban centers were also the official seats of government in the three departments of the colony, and together they accounted for 70 percent of Saint Domingue's trade. According to the contemporary classification embedded in Map 2, five towns followed in the second rank: Saint-Marc, Gonaïves, the Môle, Port-de-Paix, and Jérémie. Three third-ranked towns followed: Fort Dauphin, Léogane, and Jacmel. Towns of the second and third ranks were secondary, regional centers of commercial or military importance, but still towns. All towns had sizable resident populations, streets, stone buildings, churches, squares, fountains, military and government installations, and other accoutrements of civilization. All the towns of Saint Domingue were also port towns.

Racially, Saint Domingue's cities and towns were more white than

FIG. 10. "Vue du Cap François, prise de derrière la Prison." Drawn by Fernand de la Brunière, engraved by N. Ponce in Moreau de Saint-Méry (1791).

the countryside. On the average white colonists comprised fully 26 percent of the urban population, which contrasts strongly with a complementary rural population only 4 percent white.[3] Free people of color constituted 8 percent of urban populations, a figure likewise greater than in outlying areas, which counted 5 percent mulattoes and free blacks. Conversely, urban populations did not surpass 66 percent slave on average, a high figure, to be sure, but significantly lower than the 91 percent of the rural population that was slave. Proportionally, slightly fewer whites and slightly more free people of color lived in Cap François.

A continuum existed in the level of urban development in Saint Domingue. Cap François, Port-au-Prince, and the rest started out as the meekest encampments, and all developed along the same lines. That is to

say, the smallest villages in Saint Domingue might have turned into full-fledged towns, too, had growth continued in the colony. Still, one needs to distinguish towns per se from villages and lesser hamlets.

In addition to the eleven towns mentioned, approximately fifty villages with populations around 300 grew up around the colony.[4] Thirty-two of these villages were centers of parish life and had churches; another twenty or so villages lacked churches. Villages were not necessarily insubstantial. The parish village of Aquin, for example, had fifty-four houses in it, five with two stories. By the same token, Moreau describes the parish village of Petit Trou in the south as having "forty shingled houses, of which only three of four are passable; the rest is a mass of small huts." On a level below that of villages, another dozen or so tiny hamlets and docking sites dotted the colony. Moreau describes one of these, Le Bourg, east of Jacmel: "Ten mediocre houses make it up. A few miserable whites live there, conducting a pathetic trade in cloth and consumables for slaves. The population is composed of twenty-two individuals of all ages, sexes, and color nuances."[5]

But only 2 percent of the population lived in such villages and hamlets.[6] In other words, nine-tenths of the inhabitants of Saint Domingue lived spread out on rural plantations of the sort described in the previous chapter. And again, black slaves constituted over 90 percent of the plantation population. Among whites in the countryside, some socializing did occur in the form of visits shared among plantations, local soirées, and entertainment offered to guests from towns.[7] The majority of inhabitants in Saint Domingue lived in deep colonial isolation.

Before turning to towns directly, let us pause to consider colonial transportation and how towns, villages, and rural plantations became interconnected. On the whole, intracolonial communications were not very good. Those systems that did emerge did so comparatively late in the colony's history, and shortcomings to transportation and communication continued to block colonial growth and development to the very end. Where and to the extent effective transportation and communication existed in Saint Domingue in the eighteenth century, the situation compares favorably with that found in Europe.

Coastal shipping provided the earliest form of transportation in the colony, one that remained significant throughout the colony's history. Movement along coastal waters fitted Saint Domingue, given the development of the colony as a series of ports and embarkation points along the coastline. Small coastal sailing vessels (and canoes) and their crews, known as *caboteurs,* operated everywhere in the colony, piloting goods and people. In addition, scheduled passenger vessels sailed along coasts and between port towns. A regular network of such sailings developed, especially along the north coast. Still, movement by sea was slow and difficult. The trip fom Anse-à-Veau to Port-au-Prince, for example, 70 miles over water, took five to six days via *cabotage.*[8]

Although lagging behind sea transport, land transportation became fairly well developed in Saint Domingue by 1789. Various means existed to move people and goods on land. Where possible and if they could afford it, travelers and merchants preferred horse and carriage on the high road or shipment by wagon. Travel along trails on horseback and by mule served in other instances. Where roads or trails deteriorated, travelers had to proceed on foot, and in some wilds only hunters and guides could get around.[9] In more civilized areas people also moved about in sedan chairs with slave porters.

Apart from paving sometimes found in towns, all the roads in Saint Domingue were dirt roads. Especially good roads and local transportation systems characterized the three principal plains of Saint Domingue surrounding Cap François, Port-au-Prince, and Les Cayes, not coincidentally the three areas of greatest economic vitality. On these level plains civil engineers easily laid out and built a rectangular grid of serviceable straight roads. Similarly, interregional transportation by road was sometimes effective, especially along the sweep of coast around Port-au-Prince.[10]

Still, many areas of the colony were isolated and cut off. As Moreau put it, in Cap François one spoke of the Caribbean coast of Saint Domingue as one might speak of the Andes. The Môle peninsula, for example, remained inaccessible by land until very late in the century and then only by horse trail. Everything had to be brought in and out of the Mirebalais by mule and mule train. For want of roads, many people got lost and died in the wilderness of Saint Domingue in the eighteenth century, sometimes as a result of being deliberately stranded. The numerous and good local roads in the Boynes area, strangely enough, were due to one M. Brabant, who built them because he was once stranded and lost in the woods.[11]

Overland travel faced many obstacles. Bumpy, stony, and more than liable to potholes, roads turned dusty when dry, muddy when wet. Often two carriages could not pass on carriage roads, and road quality— uneven at best—could change quickly.[12] People and goods faced real dangers in fording rivers and streams, particularly in times of flood, and every year victims drowned and circumstances disrupted travel and communications. Highway robbery and death along the highway were not unknown. There was even the danger of quicksand.[13]

The mountains of Saint Domingue provided by far the greatest obstacle to transportation and communication. The northern mountain chain in particular proved a major barrier delaying development of the principal north-south road linking Cap François and Port-au-Prince, the keystone of a system of royal roads under development. Pressure for the development of this road came from the desire to effect a north-south overland link that did not pass through Spanish territory. The first such route, joining Cap François and Gonaïves, was not achieved until 1751, when the "coupe de Gonaïves" carved a 100-foot stairway through living rock to cross the northern mountains. From 1781 the "coup de Plaisance" provided a second and more convenient route, but one serviceable only by horse or on foot. Spurred by the removal of the judicial Conseil Supérieur from Cap François to Port-au-Prince in 1787 and the resulting necessity to speed couriers and maintain improved communications, travel by carriage between these two cities became possible only in 1787. Elsewhere, in the southern mountains, passage across Cavaillon to Les Cayes remained very difficult until late in the century; the project to upgrade into a full-fledged road the mountain trail between Port-au-Prince and Jacmel on the Caribbean did not begin until 1785, and work was continuing when the colony began to fall apart.[14]

Be it noted: roads were expensive. The royal road between Cap François and Port-au-Prince cost 1.2 million colonial livres. Slave and corvée labor built colonial roads, and buliding and maintaining royal roads was so vital that in 1789 the administrators proposed a permanent staff of eleven highway engineers and superintendents.[15]

The system of royal roads coalesced in the latter 1780s. In 1789 it was possible to go by carriage from Ouanaminthe in the far north and east of the colony to Torbec in the southwest past Les Cayes, a distance of 132 *lieues* or 320 miles. Similarly, one could go from Ouanaminthe to Petit Trou in the south and east, a distance of 112 *lieues* or 270 miles.[16] Passenger stage coach service began along parts of this system. Service along the north coast between Cap François and Fort-Dauphin commenced in 1785; that between the Cap and Port-au-Prince started only in 1789. Informal relay stations existed every 5 leagues, or about every 12 miles, but plans for formal posts never got going. Accommodations in public inns along the road were uncertain and of very poor quality.[17] Regulations reserved places on coaches for whites and for free people of color—slaves were not permitted this sort of travel. For whites, the fare for a one-way carriage trip from the Cap to Port-au-Prince amounted to 396 colonial livres.[18]

River ferries were quite common in Saint Domingue, given the number of its rivers. At many fordings, individuals operated small private ferries, and several larger official ferries ran as well. Three ferries on the Artibonite River, for example, charged fixed tolls. The ferry in Cap François was unusual as an urban ferry, providing the main crossing of the Rivière du Haut du Cap and the key link between Cap François and its outlying agricultural areas. The government leased ferry rights to a series of holders, a lease very profitable to the colony and to lease owners. The toll was 15 *sous* for whites and 7.5 *sous* for slaves, but tripled for blacks on horseback! On a busy Sunday, the Cap ferry collected 2,400 livres, well above the 300 livres it cost daily to lease and operate the ferry. Slaves pulled the rickety Cap ferry across the river on a rope.[19]

River ferries proved inconvenient, slow, and themselves obstacles to transportation. Wherever possible, and if resources were available, colonists and the government built bridges. Seventeen good bridges linked Port-au-Prince and Grand Goave along the southern road, and a strategic bridge of 150 feet at Miragoane linked the south with the western and northern departments of the colony. Two bridges spanned the Artibonite, one 130 feet long. Nothing came of the several plans to construct a bridge at the ferry site in Cap François in the colonial period.[20]

Finally in this connection, the postal system tied the colony together and to France. Regular transatlantic mail began in 1763 with two sailings a week. Royal policy in France vacillated between mounting an official state postal service for the islands versus leasing postal operations to private entrepreneurs for quick money. At one point, twenty special ships plied the Atlantic carrying the mail back and forth between Saint Domingue and France. Within the colony itself, various private farms took leases on the mail for various periods. In 1789 the postal lease cost 160,000 colonial livres a year and was said to be worth 650,000 livres. A system of colonial post offices grew from twenty-one stations in 1773 to fifty-six in 1791. Intracolonial mail deliveries occurred usually once or twice a week, although Léogane and Port-au-Prince had daily mail service.[21] There were post boxes, and stamps were required.

While settlements and some towns arose in the seventeenth century, urbanization in Saint Domingue was decidedly a phenomenon of the eighteenth century, indeed the latter part of that century. Only three of the colony's eleven towns received official foundation before 1700: Léogane in 1674 when d'Ogeron established residence there; Port-de-Paix in 1685 with the transfer of the capital from Tortue; and Jacmel in 1698 under the aegis of the Saint Domingue Company.[22] Even though settled in 1670, Cap François did not become an official town until 1711. The towns of Les Cayes and Saint-Marc became established in 1720 as regional trading centers, and Fort-Dauphin began in 1724 as a military outpost on the northeast coast.[23] The ultimate capital of the colony and its second-ranking town, Port-au-Prince, technically was founded in 1743, but the town did not begin to be built until 1749. The other towns of Jérémie, Gonaïves, and the Môle followed in 1750, 1751, and 1760.[24]

Especially early in the eighteenth century, towns were rudimentary affairs, and the dates of foundations of towns do not make evident just how much urban expansion occurred in the last decades of the colonial period.[25] The population of Cap François, for example, grew at 7 percent a year, tripling in the seventeen years between 1771 and 1788; the cost of living in towns rose proportionally.[26] Graph 2 reveals this late expansion of the colony's main urban centers.[27] The Cap grew substantially and

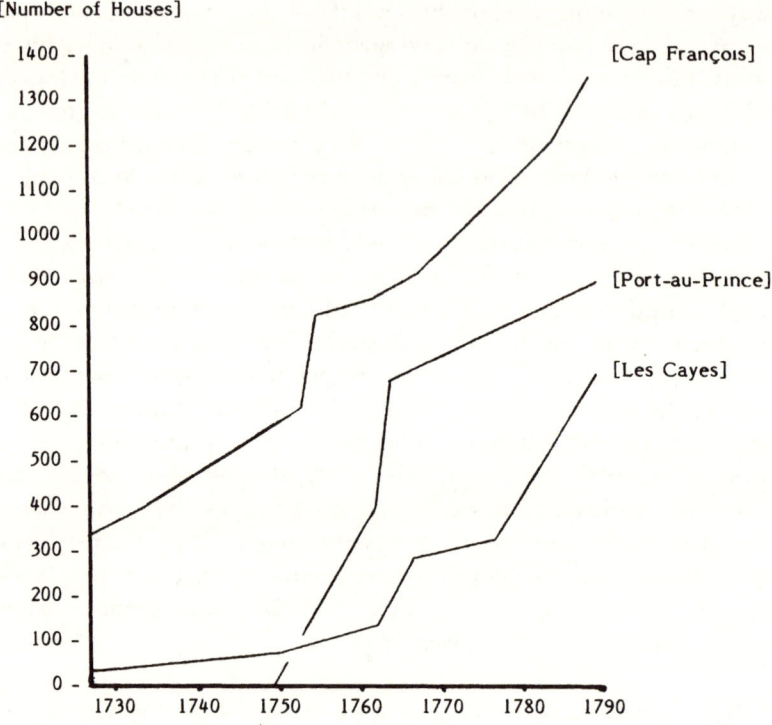

GRAPH 2. Urban Growth in Saint Domingue

continuously, especially after 1763. Les Cayes shows itself to have been the most rapidly growing city at the end. Port-au-Prince grew rapidly in its early phases, but the effects of the earthquake of 1770 clearly retarded its development.

All the towns in Saint Domingue were planned towns. Colonial surveyors laid out settlements, and formal town plans guided urban growth and development. All towns grew up along the same lines, differing only in the extent of their development.[28] A number of elements of European-style urban planning went into the making of any town. In addition to dock and port areas, present in every colonial town in Saint Domingue, the parish church and an adjacent place d'Armes in every case formed the initial center of town growth. (Such a juxtaposition reveals the two pillars on which colonization first proceeded.) The French in Saint Domingue went so far as to plant formal ranks of elm trees around such squares, even in the midst of the jungle.[29] Town blocks and streets were laid out, residential housing erected, markets created, fountains and water systems installed, and garbage dumps, cemeteries, and other areas set aside. In the process, a complete urban infrastructure developed, one that found its model in contemporary France. The French in Saint Domingue replicated the towns of provincial France, and colonial towns formed so many additional urban provincial centers for France. Within the urban infrastructure of Saint Domingue came an increasing governmental, military, and middle-class presence and all the trappings of contemporary French civilization. Everywhere in Saint Domingue's towns, urban space and social space conjoined in mutual self-definition.

Cap François was the most highly developed town in colonial Saint Domingue, and all the elements of eighteenth-century French urban planning are visible in it. As one colonial, Baudry des Lozières, remarked, "The Cap sets the tone; it is the Paris of our island."[30] Reproduced as Map 4, a delightful manuscript map of Cap François drawn around 1787 provides a very accurate view of the town at that date.[31] In addition, Moreau de Saint-Méry devotes over three hundred pages of his *Description de la Partie Française de Saint-Domingue* to an incomparable walking tour of Cap François in 1789. Moreau's atlas of the colony, his *Recueil de Vues,* also provides additional scenes of Cap François and contemporary urban life. These sources allow for an extraordinarily precise sense of how Cap François developed.

The first section of Cap François grew up around the dock area, just above the word "Cap" in the label, "Baye du Cap," on the map. The dock area of Cap François bustled. Ordinarily, one hundred or so ocean-going ships rested in port in Cap François, although during the American War as many as six hundred ships could be found there.[32] Red flags marked the entrance to the port, and an official pilot always boarded to guide ships in, if only as a requirement for maritime insurance.[33] An armed patrol boat, as well as pilot boats, moved among the ships. Regulations required all arriving captains to report under guard to the port captain, in addition to other offices. The port captain supervised all port activity, including the payment of port fees and deciding where ballast should be unloaded. Artillery batteries formed a notable feature of the dock area, and the royal navy reserved the Carénage neighborhood to the north along the quai d'Argout for its docks and hangars. Many warehouses, several wharves, and at least one lighthouse dotted the docks. Rowboats, cranes, and carts littered dockside. Hundreds of soldiers, sailors, and stevedores made an explosive mix in the context of so many cafés, billiard halls, and inns, and fights erupted regularly. The police manned guardhouses.

The church and the fronting square of the place d'Armes were situated in a second section of town above the dock area at the points marked C and Q on the map. The church in Cap François was the largest in the

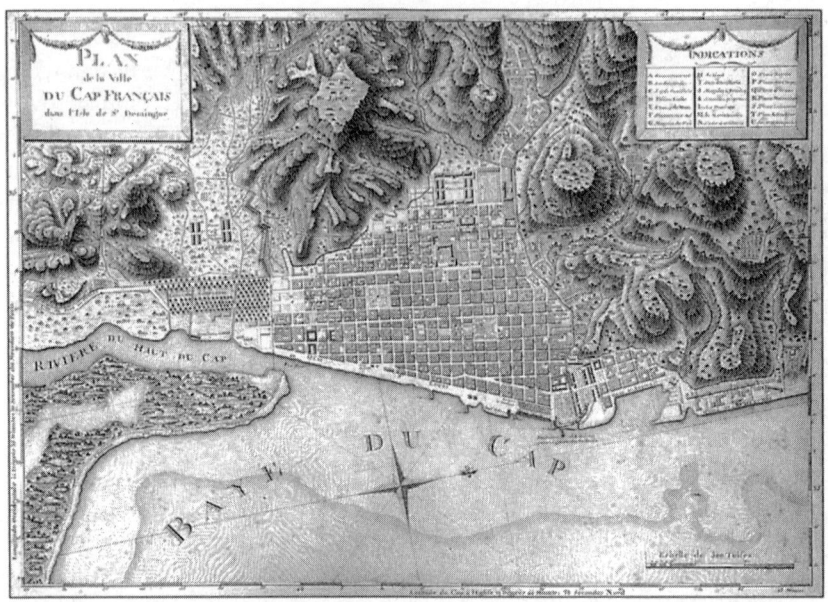

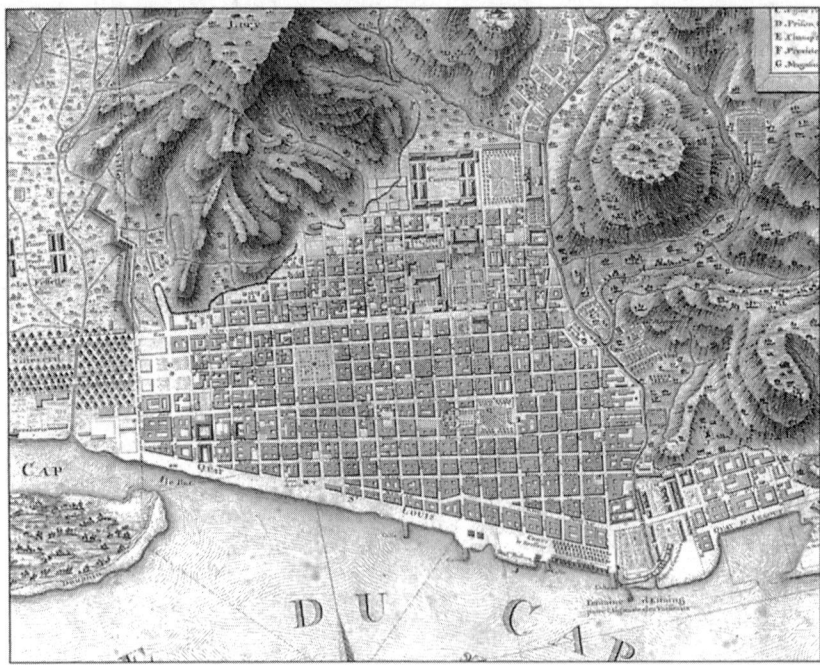

MAP 4. "Plan de la Ville du Cap Français dans l'Isle de St. Domingue," and detail. Hand-colored manuscript map. [By Dupuis?, 1787.] APS, Map collections.

colony, a modern and, according to Moreau, quite unspectacular structure.[34] Built in 1771 at a cost upward of 1.8 million colonial livres to replace the previous church that collapsed, the Cap church had a substantial facade: two stories, 84 feet long, of Nantes stone, and decorated with two statues of the apostles and the arms of France. The church itself was over 200 feet deep. With a floor of black and white stone, the Cap church possessed an organ, two chapels with several bad paintings, a clock that kept poor time, several crypts, a set of bells, and a secret passage in the back that led to the original town cemetery behind the church.[35] As noted, regulations formally reserved honorific benches in the Cap church and the other churches in Saint Domingue for the officers of the colonial administration in its various departments. As elsewhere in the colony, religious sentiment appeared lukewarm in the Cap, with reportedly only seven or eight masses said a year, but parishioners flocked to the church during Holy Week. Priests said a special mass for free blacks, the so-called *messe des nègres,* after regular parish mass.[36]

Across the street from the church stood the place d'Armes. With the obligatory elm trees, a painted fence, turnstiles, and gravel walkways, the place d'Armes was a tiny Luxembourg gardens. One of several in the town, about which more will be said momentarily, the square was dominated by a 20-foot high stone fountain (see Figure 2). The place d'Armes fountain, with antecedents dating back to 1735, carried the arms of the king and of the Cap, and Moreau remarked that it would not be out of place in any town in France. The blocks immediately around the church and the place d'Armes became the most fashionable residential area for whites in Cap François. A market for whites grew up around the place d'Armes, and the rue du Gouvernement became a street for fancy boutiques and luxury shopping.[37]

In all, fifty-six streets in Cap François crossed on a grid, thirty-seven by nineteen, creating 260 town blocks.[38] The first paved streets in town dated from 1776, and there were street signs and street numbers. Sewage ran in the streets, which otherwise became congested with carts and horses, goats, pigs, barking dogs, and various specimens of humanity. To get around in the streets, dozens of coaches, carriages, and sedan chairs were available for hire.[39]

Of the nearly 1,400 houses in Cap François, only 150 were built out of wood. The rest were of stone and mortar, as dictated after the fire of 1734. Seventy-nine public buildings existed in Cap François. The colony counted about 300 two-story structures and 3 or 4 three-storied buildings.[40] A number of small rooms of 15 to 18 feet square with high ceilings and windows fitted with shutters composed the typical house. Rooms normally surrounded a courtyard that possessed its own usually stale well or cistern and huts for city slaves. Homeowners whitewashed their houses on the outside.

As Cap François expanded outward from the docks, the church, and the place d'Armes, new neighborhoods came into existence. Free blacks and mulattoes tended to live in the southwestern part of the town, known as "Petite Guinée."[41] The authorities gave over the place de Clugny, marked *P* on Map 4, as a market for nonwhites. There, on a pretty Sunday, as many as 15,000 slaves might congregate to buy and sell fruits, sweet potatoes, cassavas, seafood, fowl, meat, preserves, dry goods, and flowers.[42] The place de Clugny was a site of public executions, and it, too, came equipped with a stone fountain to serve the neighborhood. To the south, another square, the place Royale, marked *O* on the map, arose later in the town's history. In 1789 officials erected a fountain in this square also, but the area was still under development at the end of the colonial period.

A key area of Cap François arose in the northwest section toward the back of town. With the largest single building in town, the government compound dominated the area just east of the barracks and enclosed a full city block of park.[43] The building, known as *le Gouvernement*, was originally the home of the Jesuits in Saint Domingue, but government agencies took it over in 1763, after the expulsion of the Jesuit order from France and French territories.[44] Various offices of colonial government and administration, previously located around the dock area, moved into the building. Rooms were set aside for the governor, the intendant, the Conseil Supérieur, the Sénéchaussée, the Admiralty, the public prosecutor, and miscellaneous bureaus. A chapel, a library, and several larger reception areas, meeting chambers, and courtrooms all could be found in the government building. The main courtroom of the Conseil Supérieur came completely fitted out with, among other items, a large circular bench for judges with padded seats covered with Russian leather and tacked down with small gilded nails. The decoration in the Admiralty's hearing rooms included fleur-de-lis wallpaper. The cupola atop the building encased a large clock.[45] A grilled iron fence and its portal surrounded the park in front of le Gouvernement, a park otherwise planted with orange trees. In his *Receuil de Vues* Moreau has left us with a wonderful view of the government complex from across the street in the place Montarcher.

The place Montarcher, although small, was the most fashionable square in Cap François, and we will have occasion to refer to it again, as the town theater was found there. In the middle of the square stood an overlarge, pretentious fountain with four ionic columns adorned with

FIG. 11. "Place et Fontaine Montarcher, Devant le Gouvernement, au Cap-François, Isle St. Domingue." Engraving [by N. Ponce] in Moreau de Saint-Méry (1791).

the arms of France and those of the marquis de Vallière and the seigneur de Montarcher, the colony's administrators in 1772 (Figure 11).[46]

Encountering yet another fountain raises the question of water supply to towns. The provision of fresh water was essential to the growth of cities and towns in Saint Domingue and to the general success of the colonizing effort. That the town of Jérémie, for example, had no water of its own and had to haul water into the town, put obvious limits on urban growth there.[47] The inherent difficulty of hauling water and the fact that vendors sold water in towns further evidence the point. Colonists sunk wells wherever possible, but adequately maintaining an urban water supply required large-scale waterworks.[48] Colonial administrators and government hydraulic engineers went to considerable effort and expense to develop and secure an adequate supply of fresh water for urban communities.

The most elaborate water system in the colony arose in Cap François. In the northern part of the town, above the place Clugny, three sets of pipes linked together to form an integrated conduit system. Two of these took their source from the stream in the ravine to the north of town; the other from the ravine behind the barracks on the west. Using aqueducts and underground pipes imported from France, this system brought fresh water to eight of the nine public fountains and water taps in Cap François. Another set of pipes, originating in the ravine on the south of town, brought water to the fountain in the place Royale. The cost of just one part of these works ran to 900,000 colonial livres. Especially in its early years and in droughts, the water system in Cap François went dry all too easily, and it required constant, expensive maintenance. Early on, a full-time superintendent of fountains and waterworks was appointed under the chief engineer of the northern department.[49]

Providing fresh water for ships was a special dimension of water supply in Cap François and other towns in Saint Domingue. Until 1781 ships' hands had to land at the dock at Cap François, traipse up the rue de la Fontaine, get their water at the place d'Armes, and return to their ships—an arduous endeavor. In 1781 authorities extended the conduit system and built the d'Estaing fountain into the bay (Figure 12). An added set of pumps guaranteed the flow. The d'Estaing fountain now allowed sailors to get their water directly by pulling up alongside in small boats. In 1789 another fountain was brought to the dock area, but despite these facilities, the problem of watering ships remained acute, and yet another dockside fountain was planned before the colony collapsed.[50]

Other, similar fountains and water systems arose in the other towns in the colony (see Figure 2). Port-au-Prince developed a fairly elaborate one, costing 800,000 colonial livres; it would have cost more, except that the king's slaves built the system.[51] Port-au-Prince began as the most poorly supplied town in the colony and ended up the best by dint of its excellent water works.

Every town needed a cemetery, and one was located in Cap François at the south end of town along the central rue Espagnole and west along the entrenchments that marked the end of Cap François proper.[52] The La Fossette cemetery busied itself with six hundred Catholic burials a year, almost two a day. Slave gravediggers completely turned over the ground in the cemetery every three years. The slave diggers lived in sheds on the premises. A chaplain manned a chapel. Wooden gates opened into the cemetery. Prior to 1780, painted black, they featured a morbid picture of a corpse and a water clock against a background dotted with white tears and the inscription, "Huc Tendimus Omnes"—"Here Tend All Our Footsteps." The colony's governor in 1780 found the scene too "philosophical" and had the gates painted over in gray. Located where it was, the La Fossette cemetery was the first thing one encountered on arriving at the main entrance to Cap François on the south. With or without the *memento mori,* the cemetery presented a lugubrious vista.[53]

Dealing with garbage and sewage was another necessity faced in the

FIG. 12. "Fontaine d'Estaing au Cap-François." Engraving [by N. Ponce] in Moreau de Saint-Méry (1791).

colonial urban environment. Next to the La Fossette cemetery, just outside of town, stood La Fossette proper, the official town dump. The government leased street cleaning and garbage collecting, operations worth 49,000 livres. Ten mule wagons, each with two slaves, a basket, shovel, and broom, dumped their collections in La Fossette.[54] But such formal efforts at public sanitation hardly kept pace with the realities of keeping the city from choking on its own wastes. The ravine on the north side of town became the major spot where citizens dumped garbage and sewage. That ravine essentially functioned as an open sewer to the sea. The south end of the dock area between the ferry and the wharves likewise constituted a public latrine and another place of filth and stench. Townspeople dumped sewage into the street, and in less frequented streets stinking garbage piled up. In Cap François running water flushed sewage from the latrines in the thirteen-hundred-man barracks, but criminals on the public chain gang emptied the military latrines in Port-au-Prince; citizens kept their distance when the procession passed daily.[55] In a word, sanitary and olfactory conditions were terrible. Civilization and the press of people in colonial towns struggled to deal adequately with the inevitable accompaniments of their activities.

Several public baths and laundry facilities existed in colonial Cap François and other towns in the colony. Scattered around the docks and in the back of town, the several baths at the Cap also did laundry. Made of wood or marble, the baths ran hot and cold water. One could sub-scribe at the baths, and Sundays and holidays were especially busy. That public baths were not sex-segregated increased their appeal and gave rise to some scandal. The laundries in Port-au-Prince were notorious as a place for sexual concourse with willing washerwomen.[56]

Towns had to be fed, which meant that animals had to be slaughtered and slaughterhouses set aside. The authorities located the abattoir in Cap François (a shed 150 by 30 feet and associated pens) just south of the ferry along the Rivière du Cap. Every day at the Cap abattoir butchers slaugh-tered twenty-two beef cattle together weighing upward of 6,000 pounds, ten sheep, twenty-five goats, and thirty pigs. Other towns in the colony slaughtered animals in like proportion to their populations.[57] The land around the abattoir turned into a quagmire; flies buzzed everywhere; the stench passed on the breeze into town, and animal remains simply added to the detritus of urban life.

The colonial authorities tightly controlled baking and the provision of bread. An official royal storehouse dispensed flour to royal and private bakers. Twenty-five bakeries in Cap François went through seventy barrels of flour a day, or nearly 12,000 pounds; thirty-five bakeries in Port-au-Prince baked the same amount of bread each day. Cap François burned 3,000 cords of wood a year (or 8 a day) at a cost of 108,000 colo-nial livres.[58]

Several other noteworthy urban institutions arose in the neighborhood around the government building and its enclosure. The imposing mili-tary barracks was erected behind le Gouvernement in the 1750s.[59] Eighty rooms in the barracks regularly quartered thirteen hundred men; another seventy-five rooms housed seventy-five officers. The barracks contained twelve kitchens, two prisons, and a chapel. Stables were set up next door to the barracks on the south, and on the north the champs de Mars served for drill and parade.[60]

The convent faced one side of the park in front of the government building. Letters patent issued to the Congregation of Notre Dame in 1731 established the convent in Cap François.[61] By the 1780s, eighteen nuns and their novices lived communally in the convent. They wore black habits. Forty to fifty other women, most separated or separating from their husbands, lived as pensioners in the convent, since the convent was required to take in any divorcing or compromised woman who wished. The nuns ran a small convent school, teaching the elements of reading, writing, and arithmetic to school girls from town. The convent supported itself in part with a plantation on the outskirts of town.

Two blocks from the government building in Cap François, the town prison abutted the ravine on the north.[62] The law separated black and white prisoners. Captured maroons were held with other black "criminals," including those condemned to the public chain gang. The chain gang—a form of punishment that began in 1741 as a liberal reform over a sentence to the galleys—went around by itself. White prisoners included common criminals, debtors, vagabonds, and the insane. Debtors received easier treatment, and Moreau claims that, except for the foul and smelly latrines, life at the prison was comparatively comfortable and prison supervision fairly lax. The public executioner lived in the prison, which also had one room set aside as a judicial chamber to avoid transferring prisoners for arraignment. The same room served earlier in the century as a torture chamber.

Next to the prison up the rue de la Providence and facing the barracks across the champs de Mars stood one of the three poorhouses and infirmaries for the poor in Cap François. The hospice in question was the one for white men, and was known as the Providence des Hommes.[63] Begun in 1740, the men's hospice received formal letters patent in 1769. The institution provided an asylum for the homeless, penniless, old, insane, infirm, and dying of Cap François. In its early years the men's Providence served as an outlet for the Bordeaux orphanage, and itself took in abandoned children. After 1775 the administrators assumed direct control over admissions to the poorhouse, and it was no longer enough simply to be homeless or penniless to get in.

The men's poorhouse could accommodate nine hundred, as a result of expansion and service as a military hospital during the American War, but ordinarily only ninety to one hundred unfortunates found shelter within its gates. Brothers of the Hospitaller order attended to the general operations of the hospice and supervised its rooms and infirmary wards in conjunction with the royal medical staff. The Providence had an administrative board consisting of the governor-general, the intendant, two judges from the Conseil Supérieur, the attorney general, two members of the Chambre d'Agriculture, and four notables elected in the parish. Substantial private charity supported most of the expenses of the poorhouse.[64] The Providence possessed certain privileges in the town that brought in additional revenue. It built and sold all coffins, transported the dead, and handled burials in the non-Catholic cemetery. Men from the poorhouse cleaned the streets and prepared sites for public punishments and executions. The church maintained an offering box for the Providence, and proceeds from one performance a year at the town's theater went to benefit the Providence.

A hospice for white women, facing the gardens of the government building across from the convent, was established in 1739, and in the

1750s its administration merged with that of the men's hospice.[65] Public hospices for the sick and destitute also developed in other of Saint Domingue's towns.[66] Destitute free blacks and mulattoes were not allowed in the public hospices for whites, so a freed slave named Jean Jasmin began a hospice for free people of color in the 1750s, and Jasmin's establishment was well along its way toward formal letters patent in 1789.[67] Private hospices for the treatment of slaves also arose in Cap François. Makeshift hospitals existed for slaves on plantations, but prior to the 1780s, no special facility dealt with city slaves or slaves in transit. Then, in 1782 M. Durant, a naval surgeon major, established his Maison de Santé in Cap François with the official approval of the administrators. In Port-au-Prince a surgeon, M. Robert, ran the Maison de Santé, a similar operation with beds for fifty slaves.[68]

The official Royal Hospital, known as the Hôpital de la Charité (Figure 13), was a major landmark in the urban and cultural milieux of Cap François, but its complex of buildings was not situated in the town proper. The hospital, run by the Brothers of Charity, did open in town in 1698, but relocated the next year to a large plantation property outside of town on the road south.[69] Given letters patent in 1719, the hospital expanded considerably during the war years, 1777–82. Indeed, while the provision of medical care in Saint Domingue remained technically independent of the military, without the military the Cap hospital and the substantial system of colonial hospitals would not have grown up for the civilian sector.

Twelve medical wards arose on either side of the broad tree-lined avenue leading into the Cap hospital. Eight hundred to one thousand patients received treatment at any one time. One pavilion (on the right) was for military officers and private patients. It had two stories, fourteen rooms for patients, and its own baths and kitchens. The other wards were mostly for ordinary soldiers and the common ill from Cap François. The brothers of Saint Jean-de-Dieu resided in the main building of the hospital complex at the head of the avenue. Other buildings on the grounds contained kitchens, bakeries, storerooms, baths, stables, latrines, guardrooms, a pharmacy, a chapel, a records office, and lodging for medical staff and personnel. One room was set aside as the *salle de discipline*. The Royal Hospital premises held two cemeteries, one for Christians and one for Jews. Huts for the hospital's slaves were set near the pigsty. The hospital owned and worked a sugar plantation on the property that produced a small income. The Cap hospital possessed perhaps the major private botanical garden in the colony, one far more important than that of the Cercle des Philadelphes. A fish pond on the hospital's grounds produced large carp sold in town. The brothers kept a beehive,

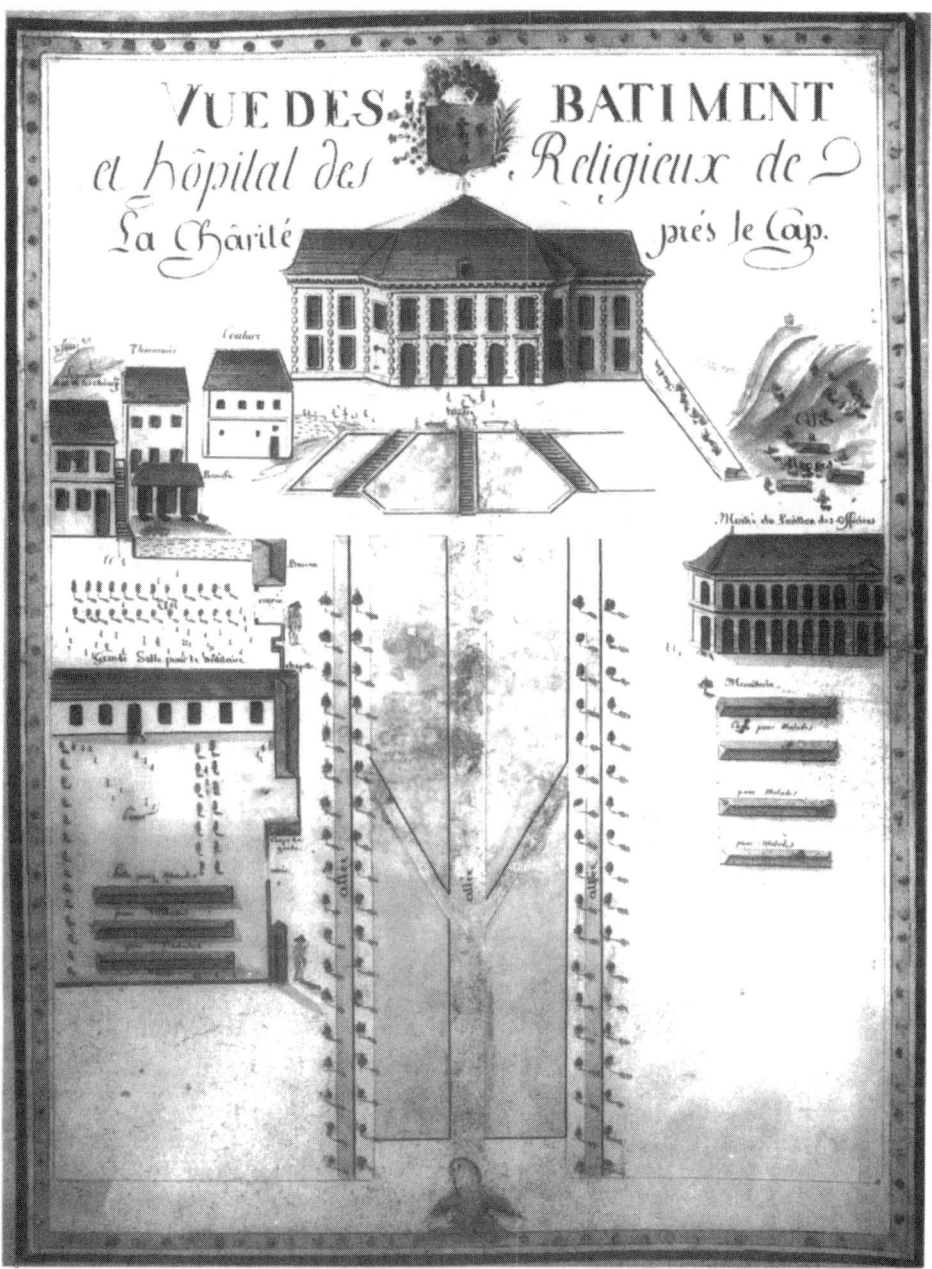

FIG. 13. "Vue des Batiment a Hôpital des Religieux de La Charité près le Cap," unsigned, hand-colored drawing, [1780s].

and the hospital seems the only place in the colony where wine was made.

A corps of royal physicians and surgeons superintended the medical care offered at the hospital. More will be said about the colony's medical personnel in a later chapter, but it suffices to note here that regulations required royal physicians and surgeons to visit regularly from town in carriages provided by the hospital. Medical staff residing at the hospital included physicians, surgeons, and pharmacists in training and their counterparts from France who supposedly spent a year in the hospital before being allowed to practice in the colony. Several of the Brothers of Charity had also been trained as medical personnel.[70]

As the studies of Pierre Pluchon reveal, nearly twenty-three hundred patients received treatment in the Cap hospital in 1786, suffering a hospital mortality rate of just over 8 percent, a rate down considerably from previous years.[71] In theory, patient treatment was quite good: segregation by ailment, one person to a bed; dressings changed twice a day; substantial rations of meat, bread, and wine; and personnel on duty twenty-four hours a day. One wonders about the realities of contemporary hospital care. Moreau does complain about a laxity of supervision and desertions from the hospital.

The Royal Charity hospital in Cap François formed part of a whole system of such hospitals in Saint Domingue. A Royal Charité hospital began in Léogane at the same time as the Cap hospital, 1698. The Léogane hospital also received letters patent in 1719, but remained a small hospital of only fifty beds. It treated 383 patients in 1786, and had a mortality rate of nearly 16 percent. The large hospital in Port-au-Prince originated in 1751, under a completely secular, military administration. Destroyed in the earthquake of 1770, authorities rebuilt the Port-au-Prince hospital as the largest medical establishment in the colony, with over 3,000 patients treated in 1786. In 1766 yet another hospital arose in Les Cayes with fifty beds. Staffed by an independent medical corps, hospitals in these towns accommodated both civilian and military patients. In addition, the colonial administration created at least half a dozen smaller military hospitals and spas to serve the needs of scattered army outposts in the colony. In this way, an entire system of official colonial hospitals emerged in Saint Domingue.[72]

The theater was the crowning glory of urban environments and of cultural life in eighteenth-century Saint Domingue.[73] Eight towns in the colony had theaters. The largest, in Cap François, held 1,500 people. An audience of 750 fitted into the theater in Port-au-Prince; theaters in Léogane and Saint-Marc held 400 each. With respect to theaters, the colonial towns in Saint Domingue compared well with their provincial counterparts in France.

Resident orchestras and companies of professional actors were supported uninterruptedly in Cap François and Port-au-Prince and sporadically for various periods in Saint-Marc, Léogane, and elsewhere. Twelve men and eight women made up the troop of the Comédie du Cap. The orchestra in Port-au-Prince had eleven musicians. Other personnel in theaters included painters, machinists, prompters, wig handlers, tailors, porters, and clerks. Top actors and actresses in the colony performed under contracts for upward of 12,000 colonial livres annually.[74] Budgets for a season could run as high as 300,000 livres; the colonial administration and interested local backers subsidized the theaters. In towns with and without professional companies, local amateurs gave performances, and in some instances syndicates of interested citizens mounted regular seasons. Troops on tour from elsewhere in the colony and from abroad also performed in smaller town theaters.

Approximately two thousand formal theatrical performances, operas, concerts, and recitals took place in the colony until such activity stopped in 1791.[75] Two or three performances usually occurred each week during the season. Offerings spanned the range of contemporary and classical (i.e., seventeenth-century) comedies and tragedies, light opera, song and dance. Orchestra concerts proved less successful, with townsfolk already familiar with orchestras and soloists from operatic performances. Molière's *Le Misanthrope* opened the public theater in Cap François in 1764, and Rousseau's pastorale, *Le Devin du Village,* played there the same year. Beaumarchais's *Le Marriage de Figaro* opened in the Cap theater just weeks after its premiere in Paris in 1784. Figaro was played by the grand talent of the colony, Chevalier, who had no need to study the part, says Moreau snidely, born to it as he was.[76] A local playwright, one M. Clément, achieved a modest success with his play "Monday at the Cap, or Payday," which apparently contained much patois and local color.

Urban space and social space came together everywhere in Saint Domingue's towns, but nowhere more so than in its theaters. The Comédie du Cap provides the prototypical example. The original theater on the rue Vaudreuil began in 1740 as a private venture. In 1764 the theater became a public institution, and in 1766 a new town theater was built, appropriately, on the place Montarcher directly across from the government building, at the very heart of town.[77] The exterior resembled an ordinary house on the square, except for the two balconies over the entrance. The interior of the theater, painted in yellow and blue, was comparatively large, 120 by 40 feet. The proscenium featured two giant sculpted satyrs on the sides and the arms of France at the top. Painted on a blue curtain were the masks of tragedy and comedy and two sylphs holding a banner with the motto, "Castigat Ridendo Mores"—"Reprove Custom through Laughter." Below ranged the orchestra pit and the parterre for the standing mass of theatergoers. Three tiers of boxes and seats, twenty-one boxes to a tier, circled the hall on three sides. The governor-general and the intendant officially commanded two double boxes, one on either side of the hall toward the front. Banners with military and seafaring emblems marked these honorific boxes, and sentries stood guard during performances. Three grilled boxes ran along either side of the parterre. Military officers sat on a special bench at the back of the parterre. The Conseil Supérieur, before it left Cap François in 1787, rented and claimed exclusive rights over the boxes in the first balcony on the right.

Custom reserved ten boxes at the top of the theater for free people of color. Accessed by a separate stairway, seven of these boxes were for mulattoes, and three were for free blacks. Thus, black mothers and their mulatto daughters had to separate during performances. The arrangement provides a most vivid reminder of the extent of racial segregation in the colony.[78] In this and every way, the colonial theater embodied a social, institutional, political, and cultural nexus of the highest order.

The theater was popular because it provided cheap entertainment and a needed escape for all but the largest class of society, slaves. Moreau praises the theater for promoting civic sociability, as well as for the simple recreation it provided, especially for sailors and travelers. White creoles especially loved the theater, according to Moreau. Sexuality must have been palpable at the theater: the stage often came alive with it, and assignations of all sorts began at the theater, including those between the prostitutes in the back boxes and young men in the balcony below.

Dances were also held at the theater. Civic and military groups rented the theater for private dances and soirees. Public dances with paid admission took place twice a week. Formal, late-night masked balls created too much scandal, and they gave way to the redoubt, a paying dance that ran from five to nine o'clock in the evening, more convenient for men after work.[79] Free people of color could *watch* dances from their boxes in the theater.

Like the theater, the Vauxhall also served as a place for dancing and socializing. The Vauxhall or Wauxhall, an unusual institution in the French colonial towns of Saint Domingue, derived from the Vauxhall Tea Gardens in England that opened in 1732.[80] Some Vauxhalls in Saint Domingue did resemble tea houses or, more properly, coffeehouses, with light refreshment, social games, and reading. The Vauxhall in Cap François was built especially as such in the Carénage neighborhood in the north of town in 1776. It had a coffee room and a lounge, like other Vauxhalls, but it also had a dance hall where dances were given like those at the theater. The Cap militia once held a dinner for five hundred at the Vauxhall, but business fell off, the added attraction of fireworks notwithstanding. Free people of color took to dancing at the Cap Vauxhall on Sundays, and when they wouldn't let whites in, officials closed the place down.

Bars and gambling houses represent other notable, unofficial institutions in the social and cultural makeup of Saint Domingue's towns. Only thirty *cabarets* received licenses in Cap François, but hundreds more operated everywhere illegally. Billiard parlors likewise flourished, and regulated inns put up strangers in town.[81]

Other amusements intertwined themselves in the cultural fabric of Saint Domingue's cities and towns. Soldiers marched, military bands played, and the cavalry paraded. Magicians, tumblers, and high-wire artists performed in towns across the colony, and the Punch and Judy show no doubt played familiar scenes along the docks. An unusual attraction was the traveling wax museum. One show in 1787 featured likenesses of Voltaire and Rousseau in addition to Louis XVI and Marie Antoinette; another in 1789 displayed General Washington in military garb.[82] Horse shows and dressage events took place, but seemingly colonists did not race their horses.[83] Cockfights were held surreptitiously, and the more aggressive citizenry indulged themselves in fighting and dueling.[84]

The *cabinet littéraire,* or subscription reading club, constituted a more highbrow feature of the social and cultural topography of Old Regime Saint Domingue. Five roughly similar *cabinets littéraires* existed in the colony, and they functioned something like men's clubs.[85] They had reading rooms where members could peruse French journals and newspapers subscribed to by the *cabinet.* They offered small library collections, and served as bookstores for people who wanted to order books through their offices. They sometimes had game rooms for cards, backgammon, and billiards. The club at Les Cayes served meals. Subscription ranged from 100 to 600 colonial livres a year. Cap François supported three *cabinets littéraires,* one with eighty subscribers.

The press formed another significant institution shaping the cultural landscape of Saint Domingue. Despite the great need for one, the printing press came late to Saint Domingue.[86] A printer from Dijon, Joseph Payen, did secure a privilege and operated presses briefly in Cap François and Léogane in 1724 and 1725. His *Code Noir* was the first work printed in the colony. But Payen ran into difficulties with a recalcitrant governor, and he got caught pushing dirty books and pictures. A worker struggled to continue the press after Payen's departure, but he and it died together shortly thereafter. Another attempt to establish a press in the colony occurred in 1742, this time with support from the colonial administrators, but nothing resulted. Only in late 1763, after entreaties from the Chamber of Agriculture and Commerce and with administration approval, did a printing operation finally begin in Cap François, run by the printer Antoine Marie from Nantes. Pursuant to his privilege, Marie opened a second press in Port-au-Prince in 1765. The presses in Cap François and Port-au-Prince operated under different privilege holders until the end of the colonial period. The earthquake of 1770 destroyed the press in Port-au-Prince, and it took a while to get it going again. The colony's third press began operations in Port-au-Prince in 1788.

The press came to Saint Domingue first and foremost as an aid to colonial government and administration. Prior to the advent of the press (and afterward), copyists struggled to keep up with the demand for forms, documents, pamphlets, announcements, and the like from the administrators, the two sovereign courts, lesser tribunals, tax and revenue offices, and individuals who had dealings with these bodies. In printing the colony's newspaper, the *Affiches Américaines,* the presses of Saint Domingue performed a quasi-official service in publicizing official news. The same holds true for the official *Almanach de Saint-Domingue,* published annually from 1765.[87] Only after the colony's business got handled, did anything else get published in Saint Domingue. The administrators seemed quite touchy with regard to the press and censored it strictly.[88]

Printing was one of the most advanced craft industries to be found in the colony. The presses themselves, the type, and other equipment necessary to the industry were imported from France, not the least on account of the trade "exclusive" with metropolitan France. Seventeen printshop workers toiled in Cap François in 1791, and the printing trade in the colonies seems to have preserved a feature from its counterpart in France, that of passing through the hands of widows.[89] With the cost of labor and materials high, it remained cheaper overall to have items printed in France rather than in the colony. With certain exceptions, the Saint Domingue press printed only works of immediate necessity for local consumption.

Nonetheless, the colonial press in Saint Domingue was very active, fully comparable to provincial presses in France. The colonial press produced a substantial body of work over the years, including all but one of the numerous publications of the Cercle des Philadelphes. Roughly ten different editors and publishers were active in the colony up to 1789, and catalogs of newspapers and serials produced in the colony list fifty different titles appearing between 1765 and 1793, although most appeared after the lifting of press restrictions in 1789.[90] In the early years, however, the colony did not support local journals or reviews. One literary and poetry review was published briefly in 1769, but it failed. Another review, the *Journal de Saint Domingue,* devoted partly to belles lettres and partly to agriculture, science, and economy, ran for fifteen months up to January 1767 but failed for want of subscribers.[91] A medical gazette appeared for a while in the 1770s.

The publication of the colony's official newspaper, the *Affiches Américaines,* represents the greatest achievement of the press in Saint Domingue (Figure 14). The series began at Cap François in 1764.[92] The newspaper was published weekly in quarto, and at the outset formed an annual volume of around four hundred pages.[93] To keep it under their watchful eyes, in 1768 the administrators transferred the paper to the capital in Port-au-Prince, but a weekly supplement to the regular *Affiches*—in effect a separate edition of the newspaper—continued to come out at Cap François under such titles as the *Avis du Cap* or *Feuille du Cap.* In 1783 Charles Mozard was appointed editor and publisher of the *Affiches Américaines,* and he ran the *Affiches* through 1791. Although censored, Mozard's lively colonial voice speaks to us still.[94]

In 1785 Mozard adopted a folio format for the *Affiches,* then averaging about eight hundred pages a year. In late 1786 he dropped the folio format but added a second weekly edition in quarto for Port-au-Prince. In 1789 the *Affiches* issued twice weekly from Cap François, too. The *Affiches Américaines* ended up being published in Port-au-Prince on Thursdays and Saturdays in editions of eight and four pages respectively and in Cap François on Saturdays and Wednesdays in editions of the same size. An annual volume ran to over one thousand pages.[95]

The *Affiches Américaines* reported official news and practical information first and foremost, especially government ordinances and announcements. Port activity and commerce also received considerable attention. Every number of the *Affiches* detailed arrivals, departures, and loading of ships in port. Names of prospective passengers were also listed, as required by law. The paper gave current prices for imported goods such as wine, flour, salted meat, and oil and posted current shipping charges for the major export commodities. Another prominent feature of the *Affiches Américaines* was the section devoted to listings of captured maroons (whose masters were sought) and escaped slaves (whose masters were seeking them). These notices appeared alongside similar ones for lost and found horses, mules, and other animals, which gives one pause to think that contemporaries saw only a difference of degree and not kind between lost slaves and lost animals. Under the rubric of "political news," Mozard and the *Affiches Américaines* presented a digest of European news derived from the foreign press and from informed parties arriving in Saint Do-

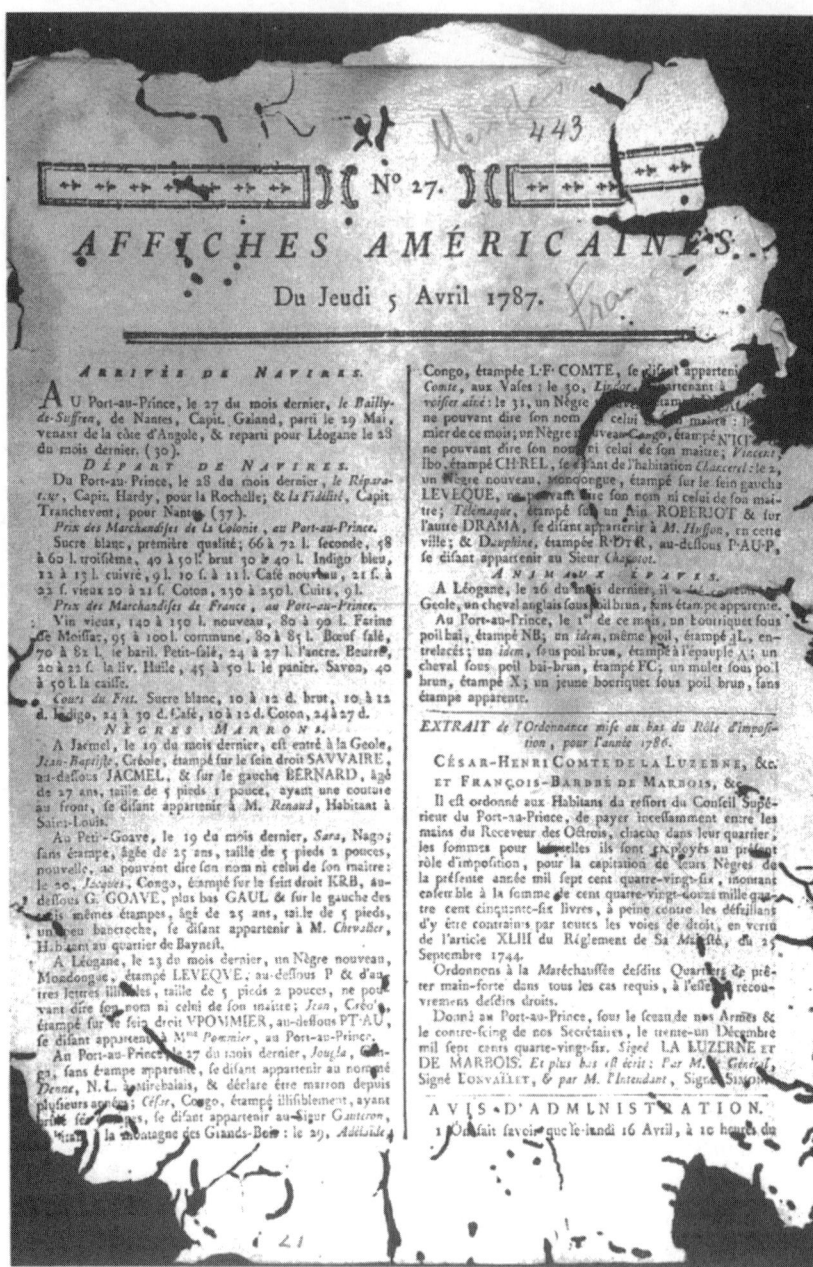

FIG. 14. The *Affiches Américaines*, no. 27, April 5, 1787.

mingue that week. Although such news dealt largely with the comings and goings of various monarchs and nobility, reports stemmed from everywhere in Europe and could sometimes be important.[96] Mozard gave news and reviews of the theater, to which he got in free, and for a price he published private commercial announcements, wherein a whole world of people, goods, and services reveals itself. Beyond all that, Mozard published and commented on a stream of material that came his way: a prospectus for this or that, an extract from another journal, a letter from a colonial correspondent, almost always on a safe and sane subject like natural history. When the Cercle des Philadelphes came into existence, it got complete coverage in the *Affiches Américaines*. Mozard published only a little poetry and belles letters.

Fifteen hundred colonists subscribed to the *Affiches Américaines*, each paying 66 colonial livres a year after 1787. The colonial government and bodies in its administration took a number of subscriptions. Advertising charges brought in an additional 40,000 livres, for a total income of 139,000 livres. Expenses amounted to 89,000 livres. The net profit of 50,000 colonial livres was divided between the printers in Port-au-Prince and Cap François.[97]

In a manner similar to the *cabinets littéraires*, presses like Mozard's in Port-au-Prince and the Imprimerie royale on the rue Royale in Cap François also sold imported books and periodicals, as did two or three other straightforward bookstores and paper-supply stores in the colony. The existence of these bookstores suggests that colonists bought a lot of books and did a lot of reading. It used to be thought that books were rare and largely unread in the crass and uncivilized atmosphere of Saint Domingue. However, in 1954 Jean Fouchard, examining the *Affiches Américaines*, demonstrated that reading and the book trade were far from negligible in the colony.[98] Fouchard took note of estate sales and sales by parties returning to France, and he identified several private libraries in Saint Domingue on the order of three and four hundred volumes. The owners of these and lesser libraries included a judge at the Conseil Supérieur, a physician, a surgeon, a merchant, several plantation owners, and a marquis.

Agents sometimes prepared sale catalogs of colonial libraries. From them and from the lists of volumes on sale in various bookstores that regularly advertised in the *Affiches Américaines*, one can specify what colonists read.[99] Two points of note emerge. First, comparatively few novels or light works are to be found on these lists. The overwhelming bulk comprises sober and solid tomes devoted to ecclesiastical history, secular and modern history, classical French theater and literature (e.g., Corneille, Molière, Boileau, La Fontaine), jurisprudence, science, and medicine. Thus overall, and this would be the second point, the book collections in Saint Domingue did not differ in character from those in bourgeois-professional and robe circles in France in the same period.[100]

Science became a subject of particular interest in the colony, and, to judge from their advertisements, certain bookstores seem to have specialized in scientific titles. In addition to several editions of the *Encyclopédie*, including the original, a set of some thirty-three volumes of *Histoire et mémoires* of the Paris Academy of Sciences existed somewhere in the colony. Someone sold Buffon's complete works in forty-three volumes in 1784, and one could have Voltaire's (and Mme de Châtelet) *Éléments de Newton*. The stout-hearted found Lalande's astronomy available (abridged), as well as multivolume standard courses in mathematics by Bézout and Camus; one differential and integral calculus text is mentioned. Other works in chemistry, *physique*, natural history, and medicine were sold and presumably read.[101]

Pornography became a thriving business in Saint Domingue. Some titillating material passed under the guise of medical books, but explicit pornography was also widely available.[102] Colonists particularly favored *Margot la ravaudeuse* and *Le paysan perveti*. Although the authorities closed down Payen's shop in 1724, trade in pornographic books and prints was not altogether subterranean, but received at least tacit approval from officialdom. Without it, the pornographer and owner of a *cabinet littéraire*, Gabriel Decombaz, and his associate would not have been allowed their thinly veiled advertisement: "We have the honor to offer our services for all the commissions one wishes to give us. Our contacts, principally in Paris, allow us to complete any order quickly and exactly, no matter what its nature or type."[103]

The press was a key institution in the development and the character of the colony and, in the end, every sort of contemporary book and journal found its way to Saint Domingue.

Many goods and services were available in Saint Domingue's towns. For example, musical instruments, instrument makers and repairers, sheet music, and musical instruction could all be had. (Elite whites commonly made music privately in urban and country parlors.) Private teachers offered instruction in a variety of subjects. One could take lessons in the fine arts of dancing, singing, instrument playing, and painting, the latter taught by traveling portrait artists. Foreign-language teachers offered instruction in English, Spanish, Italian, and German, and one could learn shorthand.[104]

The state licensed and tightly controlled apothecary shops, but they existed and sold a wide variety of products, including an *Eau anti-*

vénérienne and the powder of Mlle Aubin Genier from Marseille. One to-
bacconist sold antiscorbutic cigars.[105] A number of dentists in town advertised along the lines, "The teeth are the mouth's ornament, and
without them a lady's smile lacks grace."[106] One who advertised himself
as a dentist, M. Moretti, was also a royal tightrope walker and tumbler.
That he sold opiates and "sponges" suggests the presence of drug addic-
tion and contraception in the colony.[107] Architects, engineers, mill-
wrights, royal clockmakers and repairers, handymen and others all presented themselves and their services in the newspaper. One plaster artist
cast busts and models. Three persons offered themselves as dowsers.[108]

One social institution, typical of contemporary urban settings in Europe
and elsewhere in the Americas, was distinctly absent in Saint Domingue:
schools of higher education. Unlike the Spanish and the English, who
established colleges and universities in their colonies, the French had a
deliberate colonial policy against such institutions. The policy fell in line
with other emphases on the mother country, and it was designed to dis-
courage tendencies toward colonial independence, so alarmingly seen in
the thirteen colonies in British North America. Any talk of public education or higher education was seditious and dangerous talk.[109]

The authorities did permit private secondary schools, akin to the
French *collège,* however. A number of these schools competed in the
colony in its last years in what amounted to a small boom in colonial secondary education. These schools—all seemingly boarding schools—ca-
tered to the adolescent children of the elite. One school, perhaps more of
a military academy, existed in the colony in 1780. The Maison d'Éducation of Messrs. Hervé and Legrand opened in Port-au-Prince in July 1786
"under the protection of government." In January 1787, Renault's Académie d'Éducation opened in Cap François. Tuition was 1,800 colonial
livres, and students wore approved school uniforms. In May 1787,
M. Dorseuil opened his Maison d'Éducation, also at the Cap. With four
professors in addition to Dorseuil, the school took in 35 boarding students and 110 day students. Also in the Cap, M. Guynemer ran a boarding school for amanuenses. In 1788 a boarding school for young ladies
began in Cap François. It offered religion, supervision, and instruction fit
for a demoiselle, and a full pension ran to 3,300 colonial livres.[110] That
was apparently still cheaper than what one had to pay in France.

Contemporary material provides a good idea of the curriculum
taught in these schools. For girls, instruction was in reading, writing,
history, geography, dancing, and music. Boys got a lot more, at least at
the hands of Dorseuil and his professors.[111] Reading and writing came
first, along with drawing, fencing, dancing, and music, supposedly nec-

FIG. 15. "Vue du Cap François, Isle St. Domingue, prise du Chemin de la Petite Anse." Drawn by Fernand de la Brunière, engraved by N. Ponce in Moreau de Saint-Méry (1791).

essary arts for which additional payment was required at other schools. Teachers taught mythology and sacred and secular history, as well as mathematics through geometry, supposedly according to Euclid, Clairaut, and Bézout. Geography and cosmography were likewise subjects, and a little of the *trivium*'s grammar, rhetoric, and philosophy showed up, the latter using Nollet's *Physique!* The curriculum manifested a practical, business-oriented slant: lessons in commerce, "which many young colonials wish to pursue," were given twice daily. Foreign-language instruction was available only if fathers paid extra.

Masons and the Masonic movement made themselves strongly felt in Saint Domingue. The Masonic movement, so characteristic of the Old Regime, embodied that noble search after brotherhood and Enlightenment that knew not class or nation. Ritualized fraternities for an elite class of men, Masonic lodges fissioned as social centers everywhere in Europe and America.[112] As the researches of Alain Le Bihan have shown, twenty lodges arose in eighteenth-century Saint Domingue with perhaps another forty colonial chapters dependent on them. The Masonic movement, while decidedly an urban phenomenon, thus also extended to small towns and hamlets in the colony.[113] There were a thousand Masons in Saint Domingue.[114] One out of every twenty-five white males and probably more like one out of every three or four sociologically eligible white males was a Mason.

The first Masonic lodge in the colony started in Les Cayes in 1738, perhaps picked up from the English in Jamaica. The first lodge in Cap François, Saint-Jean de Jérusalem Écossaise, began in 1749 out of the Grand Orient in Bordeaux, almost certainly through the commercial connection. The Masonic movement did not take off in Saint Domingue until after 1763, however, but the number of lodges in the colony grew steadily from that date through 1789, from four to twenty.[115] Two lodges existed in Cap François: Saint-Jean de Jérusalem Écossaise and La Verité, founded in 1767. Moreau de Saint-Méry was "venerable" of the former lodge for four years in the 1770s. The Verité lodge had the largest membership of any lodge in the colony, with perhaps 150 Masons at its peak. Membership in the two Cap lodges fell off in the 1780s, and they merged later in the decade.[116] Two lodges operated in Port-au-Prince in the latter 1780s, La Réunion Désirée and Les Amies Reunis, and the Port-au-Prince lodges gained a reputation for philanthropy. In the American War the Masonic brotherhood in the colony showed its universalist solidarity in the reception afforded certain captured English officers who were also Masons and who gave secret signs.[117]

The Masonic movement shaped the cultural and social landscape of Saint Domingue as elsewhere. The extent of the influence of the Masons is a matter of debate, and the Masons will recur prominently in this narrative in connection with the origins of the Cercle des Philadelphes.[118]

The Cercle des Philadelphes constituted yet another prominent cultural institution in the urban milieu of Cap François. The creation of the Cercle des Philadelphes in 1784 and its elevation to the status of Société Royale des Sciences et des Arts in 1789 capped the effort of the French to colonize Saint Domingue in the eighteenth century. More than the colony's lush plantations, more than its barracks and gun emplacements, more than its churches, law courts, hospitals, prisons, and poorhouses, more than its roads and aqueducts, more even than the theater, which, after all, was also for the hoi polloi, the Cercle des Philadelphes cum Société Royale signified to the world that the colony had arrived, that it was not a backwater, that it was thoroughly modern. Jean Fouchard goes a bit far, perhaps, when he speaks of Cap François in the latter 1780s as "civilization's most advanced beacon in the Americas" (Figure 15), but he cannot have missed the mark by much.[119]

The Cercle des Philadelphes is clearly an institution of the utmost importance to any effort to understand the connections between science and colonialism in the eighteenth century. At this point we have surveyed the complexities of what colonial development entailed in eighteenth-century French Saint Domingue. We can now examine the roles played by science and the Cercle des Philadelphes in the unfolding of the colonial process.

Part II

Science in a New World Setting

French science and medicine formed an intrinsic part of French colonialism in Saint Domingue. The ideas and institutions of eighteenth-century French science and medicine were not somehow grafted onto the colony after the fact but, rather, constituted fundamental forces in building the colony and maintaining its slave-based economy and society (Figure 16). Furthermore, science and organized knowledge were agencies enlisted in the colonialist cause primarily by the French government and not especially by colonists themselves or merchants or others involved in the complex history of Saint Domingue. Contemporary science and medicine seemed to offer the mercantilist state the power of specialized knowledge and the promise of real utility in facilitating colonial development, and by and large only the state possessed the resources required to support and exploit science and medicine in the interests of colonial development. In return the domain of eighteenth-century science and learning became enlarged and enriched through government support and through the colonial experience generally.

With these guiding themes, the second part of this study explores the range of circumstances in Saint Domingue in which "scientific" activities took place. It makes the thematic manifest by showing the concrete ways in which science and medicine served the state and functioned as productive forces in Saint Domingue, and it seeks to detail the reciprocal impact of those activities on contemporary science and learning in France.

FIG. 16. Surveyors at Jérémie: "Vue de la Ville de Jérémie." Drawn by Ozanne, engraved by N. Ponce in Moreau de Saint-Méry (1791).

ᵕᶳ SIX ᶳᵕ

Missionary Naturalists

FRENCH COLONIALISM in the West Indies began in the 1620s, and the history of French colonial science begins then as well, with expeditionary religious missionaries accompanying the first colonists. Such priests either served on voyages or became attached to houses in the colonies. In terms of science missionaries worked primarily in botany and natural history. Some wrote in the encyclopedic and descriptive tradition of earlier Spanish chroniclers, and some were more formally scientific in character. Applied botany was a major aspect of their work. As scientists, seventeenth- and eighteenth-century French missionary priests were autodidacts and served primarily as informers for the scientific community in France. On the whole unsystematic, their scientific work has to be seen in a context of the larger role of such missionaries in the island colonies: to convert and to lead religious communities, to recount voyages and the colonial experience, to popularize colonization back in France, and then to report items of scientific interest.[1]

Religious missionaries functioned as scientific field workers, but it would be a mistake to credit the French church with a significant role promoting colonial science. Rather the French government deserves primary honors for initiating and supporting the activities of these scientific missionaries. Every French monarch and regent from the time of Louis XIII took a personal interest in scientific work in the colonies. Richelieu, in first sanctioning French colonization of the Antilles, bureaucratized procedures by formally assigning a priest to accompany and report on every French voyage. In 1635 Richelieu sent Raymond Breton, S.J. (1609–79), with the initial colonizing party to Guadeloupe.[2] Breton returned to France and in 1665 and

1666 published his *Dictionnaire caraïbe* and *Grammaire caraïbe,* notable ethnographic and linguistic works.

In 1640 Richelieu sent the Dominican priest, J.-B. (Jacques) Dutertre
(also Du Tertre, 1610–87) to the islands.[3] That Dutertre was a former soldier
and adventurer made him that much more fit for the rough, pirate world of
the contemporary Antilles. Dutertre stayed sixteen years in Guadeloupe and
the nascent French colonies. His one-volume *Histoire générale des Isles de . . .
l'Amérique* appeared in 1654; this work he later expanded into a four-volume
Histoire générale des Antilles (Paris, 1667–71). Dutertre concerned himself primarily with providing a general description of the French Antilles and at-
tracting colonists there. He devoted one part of his large history to botany
and cultivated plants. Notably, the work is illustrated. While not a great or
systematic example of pure science, Dutertre's effort was valued and relied
on as a solid piece of applied economic botany.[4]

One of the most interesting and well known of the early French missionary naturalists was the Dominican priest, J.-B. Labatt (1666–1738).[5]
Labatt gave up a position as professor of philosophy and mathematics at
Nancy to volunteer for service in the islands. He spent the years
1694–1706 traveling throughout the colonial and pirate worlds of the
French Antilles. Mostly he stayed in Martinique, where he ran a sugar
plantation for his order. He was a "secularized priest" and an enthusiastic
observer of the world around him. He published his *Nouveau voyage aux
Isles de l'Amérique* in an eight-volume duodecimo edition in 1722 and a
two-volume quarto edition in 1724. In his botanical sections Labatt, like
his naturalist brothers, shows special interest in economically useful
plants, and he presents virtual monographs on sugar, cotton, cacao, ginger, vanilla, and other commodity products. As an applied botanist, the
merry Father Labatt seems to have specialized in making wines and distilling rum.

The first of the missionary naturalists to be permanently stationed in
Saint Domingue was the Jesuit curé, J.-B. Le Pers (1675–1735). Le Pers
came to Saint Domingue in 1704 and remained there until his death in
1735. He served as a parish priest in the north of Saint Domingue, where
he supervised the building of ten churches. He studied the botany and
natural history of the colony using Tournefort as his guide, and he compiled a large manuscript of eighteen years' worth of notes and observations. In 1730 a Jesuit colleague of Le Pers, P.-F. Xavier de Charlevoix
(1682–1761), published this manuscript in France in two quarto volumes
with government support as the *Histoire de l'Isle espagnole ou de Saint
Domingue.*[6]

The use of French missionary naturalists for the scientific investigation of French colonies and other foreign lands received an important

stimulus and extension late in the seventeenth century, again from a scientific arm of central government, the Jardin du Roi in Paris and its superintendent, Guy-Cresent Fagon (1638–1718). The Jardin du Roi originated formally in 1635 and served primarily as a facility for medical and pharmaceutical teaching and research.[7] Colbert assimilated the Jardin into royal administration, and Fagon harmonized its functioning with the Parisian medical establishment, turning the Jardin into a first-class scientific center. As one of his major accomplishments, Fagon launched a series of botanical research expeditions that tapped talent in the religious community.

The most noted of Fagon's agents and protégés was the Minim monk, Charles Plumier (1646–1704). Plumier was a botanist of the first rank and by far the greatest naturalist to study the French Antilles.[8] He made three government-sponsored trips to the West Indies between 1689 and 1697. After the first of these Louis XIV appointed Plumier *botanist du roi* and gave him a pension. A disciple of Joseph Tournefort, professor of botany at the Jardin du Roi, Plumier published his detailed and exact botanical drawings and descriptions according to Tournefort's system of plant taxonomy by genus. The state paid for the handsome engraving and printing of Plumier's several strictly scientific works. Among others, these included his *Description des plantes de l'Amérique* (1693), with 103 folio pages and 108 illustrations; a formidable monograph on American ferns, *Tractatus de filicibus americanis* (1705), with 146 folio pages and 170 plates; and an eight-volume *Botanicon americanum* (1705), with hundreds of folio plates. The latter two works appeared posthumously, as Plumier died in 1704 while on yet another expedition for Fagon, this time to Peru in search of quinine. Plumier left behind some 6,000 drawings of botanical and zoological specimens and twenty-two volumes of unpublished manuscripts at the Jardin du Roi. One noteworthy minor work of Plumier, a paper in the *Mémoires de Trévoux* in 1703, proved that the source of the very valuable red cochineal dye was indeed an insect.[9]

Fagon had another important emissary in Louis Feuillée (1660–1732), a colleague of Plumier in the Minim order. Fagon sent Feuillée on two expeditions, one to the Antilles in 1703–06 and one to South America between 1707 and 1711. More of an astronomer and a mathematician, that aspect of Feuillée's career will be discussed in the next chapter, but Feuillée's botanical work was not negligible.[10] He identified many new species of fish, bird, and reptile on his voyages, and with state subsidies he published an illustrated *Histoire des plantes médicales* to accompany his *Journal des observations . . . faites sur les côtes orientales de l'Amérique* (1714, 1725). Feuillée claimed a greater objectivity than some of his naturalist predecessors, saying he would report only what he saw with his own eyes, but such principles did not prevent him from depicting a fantastic South American monster that was part calf, horse, and human child![11] Feuillée passed through Saint Domingue in February and March 1705, stopping and making observations at Saint-Louis on the Caribbean coast. He traveled on a pirate ship with a pirate crew and was attacked by other pirates and the Spanish.[12]

The tradition of the naturalist priest continued in Saint Domingue with the Dominican, J.-B.-M. Nicolson (1734–73).[13] Nicolson served as Dominican superior in Léogane from 1769 until his death in 1773. On royal command from Louis XV, Nicolson's family published his manuscript posthumously in 1776 as the *Essai sur l'Histoire Naturelle de St. Domingue* (376 pages, 10 plates). The first part of Nicolson's work presents a general description of the colony, but the bulk of the work concerns botany. Nicolson complains about an uneven quality to earlier botanical research in Saint Domingue and laments that colonists do not even have names for most plants.[14] Nicolson adds a limited study of the fauna of Saint Domingue and concludes with a chapter on archaeological investigations of the extinct Arawaks.

To a limited degree the tradition of the naturalist priest cum vocational scientist continued after Nicolson. The abbé de la Haye, for example, curate in the Dondon region, was a skilled botanist and naturalist active in the 1780s, and membership lists of the Cercle des Philadelphes show three other colonial priests as associates.[15] But an era of sorts came to an end with Nicolson. Indicative of changed circumstances, the Cercle des Philadelphes planned to publish de la Haye's botanical manuscript, and secular amateur botanists, such as Lefebvre-Deshayes, came to rival naturalist priests. The pace of botanical work in the colony actually increased in the latter part of the eighteenth century, but, notably, this work was carried out by secular, full-time, more professional scientists who held paying positions as botanistes du roi and formed part of the secular state bureaucracy.[16] The era of the missionary naturalist essentially ended by midcentury. But great credit for pioneering and basic research in botany and natural history in Saint Domingue and the tropical colonies must go to the founding fathers reviewed here, Plumier above all. In addition to disseminating practical knowledge that aided colonial development, French missionary naturalists of the seventeenth and eighteenth centuries made notable scientific contributions, later assimilated into the master syntheses of Linnaeus, Buffon, and Cuvier.[17] While not all saintly and not without an essential level of secular support, men of God wrote the first chapter in the scientific study of Saint Domingue.

Before leaving the subject of botanical investigations in and around Saint Domingue, three other naturalists should be mentioned briefly. The first, Sir Hans Sloane (1660–1753), never traveled to Saint Domingue, but he cast an English shadow over botanical research in the West Indies. Sloane did spend fifteen months in Jamaica in 1688–89, and he later published his *Voyage to . . . Jamaica* in two large folio volumes with 274 plates (1707, 1725). This substantial work and Sloane's later position as president of the Royal Society of London helped establish the reputation of the English as tropical botanists.[18]

French botanists in the Antilles patriotically adhered first to Tournefort and then to the "natural" system of Bernard de Jussieu, but students and disciples of Linnaeus and his sexual taxonomic system came to Saint Domingue in the eighteenth century and had their own impact on the course of botanical research coming out of the colony. Maria Theresa of Austria sent the baron Nicolas Joseph de Jacquin (1727–1817), a follower of Linnaeus, on a botanical expedition to the Antilles. De Jacquin, later professor of botany at the University of Vienna, was one of the most distinguished botanical travelers in the eighteenth century, and he spent 1757 and 1758 in Saint Domingue.[19] A small volume of his researches, *Enumeratio systematica plantarum . . . in Insulis Carribaeis,* appeared in Leyden in 1760, and he published a larger presentation of American plant species according to Linnean principles, *Selectarum stirpium americanarum historia,* in Mannheim in 1763 (284 pages and 183 plates, some in color).

The Swede, Olof Swartz, was an actual student of Linnaeus. The master sent him to the West Indies, and he spent 1784 and 1785 in Saint Domingue. Swartz returned to Sweden and published his *Nova genera et species plantarum* in 1788 and his *Observationes botanicae . . . plantae Indiae occidentalis* in 1791. Swartz later published two other volumes devoted to West Indian flora.[20]

The botany and natural history of Saint Domingue became well studied in the eighteenth century. While early colonial botanists never denied the practical and, indeed, emphasized it to greater or lesser extents, the initial cataloging and classifying of the natural riches of Saint Domingue pertained as much to "pure" science. Not too directly practical, botanical collections and compendiums were passive in a way and served primarily to communicate knowledge. By comparison, a striking new era in the history of colonial botany began when French government authorities initiated conscious efforts at applied botany of immediate economic utility. A later chapter details that important development. But Saint Domingue and the tropics generally proved an initial unknown to European science and colonizing powers in more ways than just the botanical. Saint Domingue was also of considerable astronomical and cartographical interest, and paralleling the botanical expeditions by missionaries and others discussed here, Saint Domingue became the point of focus for yet another series of scientific expeditions undertaken to the New World.

ᴥ§ SEVEN §ᴥ

Expeditions to
Saint Domingue

S CIENCE BECAME institutionalized and developed semi-independently
in Saint Domingue in a variety of ways yet to be explored, but
throughout the eighteenth century Saint Domingue served European science in one way simply as an overseas outpost, a distant place for observations, an exotic venue for research. This aspect of the history of science
in the colony involved scientists coming to Saint Domingue to make observations and returning to France to complete and publish their work.
The thrust of their work was practical, to buttress colonial development
in one fashion or another, but the scientific community and the center of
scientific activity remained in Paris. Saint Domingue existed as a place on
the periphery where one went to explore and take data.

The early history of botany in the colony discussed in the previous
chapter already makes the point. The government sent Plumier and
Feuillée on expeditions explicitly to collect information, and missionary
naturalists stationed in the colonies generally returned to France or had
their work transmitted to France for publication. In a like manner, Saint
Domingue became a landfall in the Antilles where French astronomers
and cartographers came to make observations and do fieldwork, and a
number of noteworthy astronomical and cartographical expeditions
passed through Saint Domingue in the eighteenth century. Astronomers
carried out their scientific work in and around Saint Domingue with
government support and under the auspices of the Académie Royale des
Sciences. Such work concerned many of the most interesting topics in
contemporary astronomical and geophysical research, and it addressed
several practical problems facing colonial development.

The story begins in the seventeenth century. In 1672 the Paris Academy of Sciences, prompted by Christian Huygens and J.-D. Cassini, sent its *élève astronome,* Jean Richer (1630–96), to Cayenne on the equator in French Guiana. There, acting under instructions from the Academy, Richer made a remarkable discovery. He found that a pendulum beats more slowly at the equator than toward the poles.[1] Richer reported his results in the volume of *Traitez de Mathématique* published by the Academy in 1676.

Richer's was an odd, empirical discovery. It certainly had practical implications in terms of regulating pendulum clocks, and theoretically it posed problems, dealt with at first by Huygens and others in terms of ethereal Cartesian vortices. After the publication of Newton's *Principia* in 1687, however, Richer's discovery became a point of contention in evaluating the science and world views of Newton and Descartes. Newton stated that, because of rotational forces, the mass of the earth should extend at the equator and flatten at the poles. Cartesian science, on the other hand, held that, because of inward pressure exerted by the spinning ether, the earth should flatten at the equator and extend out at the poles. The earth resembled either an American football stood on end or a slightly deflated beach ball, depending upon one's camp. Richer's discovery thus helped trigger the famous shape-of-the-earth controversy that for decades divided the national science traditions of England and France. Richer's discovery also prompted the earliest astronomical work in Saint Domingue.

Following Newton, we now explain the slowing down of pendulums at the equator by saying that, because the earth *is* flattened at the poles, the equator is further from the center of the earth than the poles, that the force of gravity is therefore less at the equator, and that a pendulum will thus beat more slowly at the equator than at the poles. Scientists did not all understand or agree upon that interpretation at the turn of the eighteenth century, however. They first needed to confirm that what Richer had observed was indeed the case, and in 1699–1700 the Paris Academy sent out an otherwise obscure geographer, one Des Hayes, to Cayenne and to Saint Domingue. Traveling aboard a royal ship commanded by Bernard Renau, an honorary member of the Academy of Sciences, Des Hayes experimented with pendulums and took data on magnetic declinations, tides, and points navigational.[2] He confirmed Richer's finding and suggested an even greater slowing effect than Richer had found. Des Hayes communicated his results to the Paris Academy through the Jesuit, Thomas Gouye, an astronomer and honorary member of the Academy. Fontenelle published them in his reports in the *Histoire et mémoires* of the Academy.[3] In a later edition of the *Principia* Newton cites Des Hayes in support of his interpretation of the slowing of the pendulum at the equator.[4]

The voyage of the Minim monk Louis Feuillée to Saint Domingue and the Caribbean in 1703–06, discussed previously in connection with its botanical component, likewise concerned astronomy and navigation, including the pendulum phenomenon. Feuillée was a protégé of J.-D. Cassini and a corresponding member of the Paris Academy. Previously in 1700–1702 Feuillée had served with Tournefort on a cartographical and botanical expedition to the Levant, an expedition undertaken on the initiative of Academy president, the abbé Bignon, and the royal minister, Pontchartrain.[5]

On his voyage to Saint Domingue in 1703–06, Feuillée recorded pendulum data to confirm Richer. But the primary astronomical purpose of Feuillée's voyage was to coordinate observations of the satellites of Jupiter with the *Observatoire royale* in Paris.[6] The reason for attempting such observations derived from another outstanding issue in contemporary science and technology: the famous problem of longitude, or how to find one's place at sea, a problem of obvious practical importance for navigation and cartography. In 1612 Galileo first proposed that the problem of longitude might be solved by standard tables and sightings of Jupiter's satellites, and Feuillée's voyage represented another test of that fruitless possibility. On his return Louis XIV made Feuillée a *mathématicien du roi,* and on his next trip to South America from 1707 to 1711, again taken in concert with the Paris Academy and the *Observatoire,* Feuillée again observed pendulums, Jupiter's satellites, magnetic declinations, the sun, the moon, and the planets. Feuillée undertook the most advanced work by any astronomer in the Americas to that date.[7]

The astronomical establishment in Paris attempted coordinated observations in Saint Domingue next in 1706, a year after Feuillée passed through the colony to observe Jupiter's satellites. This time the objectives included simultaneous observations of a scheduled lunar eclipse in Paris and in Saint Domingue. Through differences in local times of the eclipse, the distance between the two locales and hence the exact longitude of Saint Domingue could be determined. The Jesuit missionary, P.-L. Boutin, observed the eclipse of April 28, 1706, from Port-de-Paix. He transmitted his observations to the Academy of Sciences, again through the hands of the Jesuit honorary Gouye; the astronomer and *pensionnaire* of the Academy, Phillippe de la Hire, then calculated that Saint Domingue stood 6 degrees or 400 miles further west than previously thought! French astronomers repeated the same procedure the next year for the eclipse of April 17, 1707; the results this time showed Saint Domingue only 2.5 degrees or 170 miles further west.[8] Knowing Saint

Domingue's exact location was essential for navigation and the development of the colony generally. The nonobvious point is that the eclipse observations of 1706 and 1707 represent the first time astronomical methods, rather than navigators' reports, were employed to settle the matter. The eclipse observations of 1706 and 1707 were merely the first steps in a long march to get scientifically accurate maps and charts of Saint Domingue.

The dispute between Newtonians and Cartesians over the shape of the earth and a satisfactory interpretation of the pendulum phenomenon discovered by Richer remained unsettled into the 1730s. In 1735, goaded by the *pensionnaire* Newtonian, P.-L. Maupertuis, the Paris Academy of Sciences acted to settle the issues by dramatically sending out two famous expeditions to measure the earth's curvature. One, led by Maupertuis, went north to Lapland. The other, led by newly elected Academy *associé*, C.-M. de La Condamine (1701–74), went south to the equator in Peru. These expeditions constituted the largest single scientific project in memory, and the results developed over the ensuing years proved instrumental in converting the Parisian academicians to Newtonianism and Newton's explanation of the pendulum.[9]

Saint Domingue is only indirectly connected to the story of these expeditions—the southern party stayed over in the colony for three months in 1735 on the way to Peru. Still, what little can be said about the visit speaks volumes about the character of contemporary science and the state of affairs in Saint Domingue.

After a crossing of two and a half months from La Rochelle, the expeditionary party from the Academy arrived in Petit Goave in Saint Domingue on July 29, 1735, aboard the royal vessel *Portefaix*. And an impressive party it must have been when it disembarked the next day. La Condamine led the party with two senior colleagues from the Academy, Louis Godin and Pierre Bouguer, both astronomers and *pensionnaires*. A young Joseph de Jussieu, then unaffiliated with the Academy but associated with it through his more famous brothers, came attached to the expedition as physician and naturalist. An engineer, two geographer-surveyors, a draftsman, a clockmaker, a surgeon, and six domestics—for a total of sixteen—likewise stepped off the boat in Petit Goave as part of the expedition.[10] With Saint Domingue still in its formative stages in 1735, the travelers from Paris stepped into a fairly primitive environment. The population of Petit Goave at the time numbered somewhat over 600 whites, 250 free people of color, and 2,000 slaves.

The original plan called for only a brief stay in Saint Domingue before changing ships and continuing to a rendezvous with a Spanish contingent in Cartegena on the Caribbean in South America. But delays ensued, and the group reluctantly ended up spending three months in and around Petit Goave. La Condamine wrote positively of his desire to make use of their enforced stay, and the French scientists duly made measurements of pendulums, meteorological phenomena, and the elevation of local mountains. Less sanguine, Bouguer wrote of the scientific work as a diversion from the oppression of the place and spoke wistfully of ever seeing France again.[11] Godin was less affected, but Joseph de Jussieu apparently got very depressed.[12]

"It really is a good thing sometimes to send observers on the spot," wrote Godin, "because those who are here, doctors and others, think more of earning money in sugar or indigo than in enriching or adding to the Institutes of M. Tournefort."[13] Still, the little scientific party from Paris seems to have enjoyed what society it could find in the neighborhood. The academicians particularly welcomed the company of a local physician, one Duhamel, and in 1737 the Academy duly elected this Duhamel its *correspondant*.[14]

By coincidence, a supposed French count, known to La Condamine, was hiding out in Saint Domingue at the time of the expedition's visit. This count studiously avoided meeting any of the academicians, but that did not prevent him from offering a rather unflattering portrait.

One speaks of nothing else but the important expedition of our academicians. I learned from some trustworthy people, whom the babble of these *savants* could not deceive, that they are taking a lot of money with them to South America. They spent a fair bit in Cap François before leaving. To my face a merchant familiar with their destination predicted that if they do not leave off with their haughty airs and this French petulance that so displeases foreigners, they run the risk of finding themselves in some disagreeable situations, despite the protection given them by the Spanish court. Beyond that, everyone rushes to receive them well.[15]

The pride and ostentation of the Frenchmen sounds right, and the ominous prediction proved accurate, as members of the expedition ran into one difficulty after another, splitting up after their work in Peru and separately finding their ways back to France over the following decade and longer. (Godin, for example, took an extended appointment at the university at Lima.) But all that remained before them while they waited in Saint Domingue. At last the royal ship *Vautour* arrived, and the party finally sailed on October 30, 1735, for Cartegena. The Spaniards, Don

Antonio de Ulloa and Don Jorge Juan y Santaclia (known as Don George), had been in Cartegena since July, and they became very impatient and upset over not having heard from the French. The long-awaited rendezvous occurred on November 15, 1735, and the expedition continued from there.[16] The stop in Saint Domingue provides no more than a glimpse of the expedition, its social character, and the difficult circumstances in which it did its work.

Accurately finding one's longitude at sea remained an unsolved problem into the 1760s, despite financial incentives offered by the French Academy of Sciences and the English Board of Longitude from early in the eighteenth century.[17] This problem directly affected the colony in Saint Domingue, for captains had no way other than sailing experience to say how far east or west they were at any time in a voyage. That uncertainty could make huge differences for getting to port.

As is well known, the chronometer represented the practical solution to the problem of longitude, for by taking an accurate and reliable clock set to, say, Paris time and comparing it to local time, one's longitudinal distance from Paris can be calculated from the difference in times. The technique can be used at sea to determine one's position and on land to establish fixed points for making maps and charts.

The English clockmaker, John Harrison, won the race to develop a clock rugged and accurate enough to serve as a marine chronometer. The English first tested Harrison's clock in 1763. But French clockmakers did not lag far behind in crafting chronometers for France, and, indeed, priority over Harrison was claimed for one of those French clockmakers, Ferdinand Berthoud.[18] However that may be, not long thereafter, from 1767 through 1772, the French navy and the Paris Academy of Sciences conducted their own series of four sea trials to test and begin nautical use of Berthoud's and other chronometers.

The first test occurred in 1767 aboard the *Aurore* sailing in northern European waters with the academicians J.-C. de Borda and C.-J. Messier. The second followed in 1768 aboard the *Enjouée* with J.-D. Cassini IV sailing to the Canary Islands. The third and fourth trials went to Saint Domingue. The astronomer, church canon, and free associate of the Academy, Alexandre-Guy Pingré (1711–96), conducted tests in 1769 aboard the *Isis,* captained by count C.-P.-C. de Fleurieu. In 1771 Father Pingré and Borda repeated the voyage to Saint Domingue, conducting tests aboard the *Flore.*[19] The results of these tests demonstrated the perfection of the chronometer for determining longitude, and henceforth navigators regularly employed clocks to traverse the Atlantic to and from Saint Domingue. On a voyage of 3,600 miles one could be confident of one's position to within 20 miles. The problem of longitude had been effectively solved.

Sailing aboard the *Isis,* The Pingré-Fleurieu mission went to Saint Domingue in 1769 primarily to test chronometers, but it had an important secondary assignment—to observe the transit of Venus of June 3, 1769.

The story of international cooperative efforts to observe the transits of Venus of 1761 and 1769 (and a related Mercury transit in 1753) forms an important and heroic chapter in the history of eighteenth-century astronomy, a chapter told so remarkably by Harry Woolf.[20] The passage of Venus across the face of the sun is a comparatively rare event observable from earth; it happens only twice, eight years apart, every century or so. For the 1761 transit, scientific communities in France, England, Sweden, Russia, and elsewhere mobilized themselves and their supporting governments for 120 observations around the globe. For the transit of 1769 Europeans mounted 150 observations in such far away places as Tahiti, Beijing, Manila, Batavia, India, Hudson Bay, Baja California, and Siberia, in addition to areas throughout Europe and the North American colonies. The expeditions and coordinated observations of the transits of Venus, particularly the transit of 1769, constitute the largest "big science" projects in the eighteenth century and were of a scale not duplicated until well into the nineteenth century.

The reason so much effort went into these observations had to do with a major uncertainty plaguing contemporary astronomy: the value of the astronomical unit or the distance between the earth and the sun. Comparing observations of the transit from observers spread out over the globe, eighteenth-century astronomers could and did determine this fundamental astronomical unit and hence the real dimensions of the solar system for the first time with any accuracy.[21]

Saint Domingue's connection to the transit efforts goes back to 1753 when a naval ensign, one M. de la Cardonie, was apparently sent on the *Illustre* to observe the transit of Mercury of that year. Moreau de Saint-Méry provides the single reference to the episode, and no indication remains of what came of this colonial observation of the Mercury transit.[22] Saint Domingue did not function as an observing site for the Venus transit in 1761, but it did in 1769, chosen especially for its western location.[23] Originally scheduled to go to Saint Domingue with Fleurieu, the astronomer and Paris academician, J.-J. Lalande, begged off because he got seasick. Pingré, who had observed the 1761 transit in the Indian Ocean, substituted. Bad weather delayed the Pingré-Fleurieu expedition in port in Rochefort for two months, the Atlantic crossing took nearly three months, and finally the twenty-gun frigate, *Isis,* arrived in Cap François on May 23, 1769, after a further ten-day voyage from Martinique. The party established an observation post, a wooden platform built on a hillock at the end of the rue Saint Domingue just outside Cap François on the north. They chose the site because it got an hour and a quarter more sun than down in the town. Pingré called the climate brutal and the heat stifling. Heavy afternoon and evening rains raised fears for transit day.

The sky proved clear on June 3 when the transit began at 2:30 P.M. The party took four sets of observations. Pingré observed with a high-quality achromatic telescope of 5 feet. Fleurieu employed a Dolland-style achromatic telescope 2.5 feet long. M. de la Fillière, a naval officer, observed with a lower-quality achromatic instrument of 3 feet in length, and M. Saqui Destourès, head of the expedition's marine detachment, used the spotting scope on Pingré's quadrant instrument. Two marines and another naval officer counted out the time clocked by a pendulum as the observations proceeded. A crush of onlookers proved the only problem until 5:30 P.M., when clouds obscured the view with the transit only half over. Even so, Parisian scientists were anxious for the Cap observations, and a special courier sped them back to France, the Academy, and the king.[24] Pingré, Fleurieu, and the rest of the expedition left Saint Domingue two weeks later on June 16, and arrived in France on October 31, 1769.

Pingré's connection to Saint Domingue did not end in 1769. He returned with Borda to finish the clock tests in 1771, as mentioned, and he again took astronomical observations from the "observatory" at the Cap. From 1773 to 1789, with the exception of only a few years, Pingré annually provided a set of astronomical ephemerides calculated for Saint Domingue published in the colonial *Almanach de Saint-Domingue*.[25] In a way, Alexandre Pingré became the astronomer of Saint Domingue.

Not unexpectedly, given the sea link to the colony and the necessity of sailing, the royal navy and the French government placed a high priority on developing accurate maps of Saint Domingue. Every expedition to Saint Domingue discussed thus far had a cartographical component designed to collect information for improving maps, and the desire for better maps likewise motivated the work undertaken in the colony concerning longitude and the chronometer. The French crown maintained a formal depository of charts, maps, and other nautical information within the ministry of the navy and the colonies, wherein worked a corps of royal geographers. The Académie Royale de Marine, established at Brest in 1752, was another institution charged to coordinate and improve French cartography. Reformed and upgraded in 1763, in 1771 the Brest

Marine Academy became formally linked to the Academy of Sciences in Paris.[26] Many of the men involved in the story of cartography in Saint Domingue—Fleurieu, Borda, Chabert, and others—became members of the Brest Marine Academy; some were also members in Paris.

A number of maps of Saint Domingue drawn from a number of different sources appeared in the first half of the eighteenth century. The observations of one Frézier provided the basis for several, including a 1725 map by the elder Delisle, principle geographer to the king. Another appeared in 1731 based on the work of the Jesuit, Le Pers. Experts regarded these early maps as unsatisfactory, however, and beginning in 1750, the naval ministry made a push for better maps of the colony and its watery environs. The work fell primarily to the geographer, J.-N. Bellin (1702–72), who produced a series of improved maps and charts of Saint Domingue and vicinity through the 1770s.[27] But Bellin's maps themselves came in for heavy criticism—"They are all incorrect," said Fleurieu in 1773, and he demanded a fresh attack on the cartography of Saint Domingue. Of particular concern were the straits and passages between the numerous small islands in international waters just to the north of Saint Domingue. These passages had to be negotiated on the return voyage to France, and they could prove quite dangerous.[28]

As a result, the navy ministry sent A.-H.-A. de Chastenet, count of Puységur (1752–1806), on a voyage to Saint Domingue in 1784–85 to take cartographical measurements and to draw up a new set of maps of the colony and especially of the northern passages.[29] In part because he brought his beautiful wife and in part because he was an ardent mesmerist, Puységur's visit had a great impact in the colony, as we will see. Regarding his cartographical work, Puységur was quite accomplished. Assisted by an astronomer-chaplain detached from the Paris Observatoire and using perfected chronometric methods, Puységur spent fourteen months in and around Saint Domingue taking observations and recording data. He returned to France in 1785, and in 1787 the government published his *Le Pilote de l'Isle Saint-Domingue et les Débouquements de cette Isle* and his *Détail sur la navigation aux côtes de Saint-Domingue et dans ses Débouquements*.[30]

Puységur's maps were accurate, of the highest quality, and immensely serviceable to the navy and to captains and navigators. Sailing to or from Saint Domingue in the eighteenth century remained an arduous and dangerous undertaking, but after Puységur, not for want of very good maps. Adolphe Cabon, who has written on this matter, compares Puységur's work with the extensive cartographical efforts of the U.S. Marines after they took over Haiti in 1915. Cabon says the Marines had better instruments, spent more money, and took longer (five years), but produced no better maps of Haiti's coasts than did Puységur in the eighteenth century.[31]

It is telling that, while the French invested significant resources in naval cartography, they expended no comparable effort to develop topographical maps of the colony and its interior. In 1764 the government did send a group of French engineers and geographers to Saint Domingue to conduct a topographic and hydrographic survey, but most died upon arrival, and nothing resulted.[32] Naval and military officials made maps of local installations, and local surveyors and assessors mapped out towns and plantations. Private parties also undertook intracolonial cartography, as in the case of Moreau de Saint-Méry's *Recueil de Vues,* the best contemporary atlas of the colony. Still lacking in 1789, however, was a general topographical map of the colony, its rivers, and especially its interior regions.[33] The contrast with efforts to map and chart Saint Domingue's coasts, ports, and nautical environment reveals French priorities in developing the colony in Saint Domingue.

Although mathematically trained and proficient astronomers and cartographers came to Saint Domingue for short periods in the eighteenth century, not surprisingly very little trace of the exact mathematical sciences can be found indigenous in the colony. Essentially no scientific activity in the colony involved mathematics, and the story of the mathematical sciences in Saint Domingue remains wholly that of visiting European experts.

The *Affiches Américaines* occasionally published word of notable advances in the mathematical sciences reaching the colony from Europe. There, for example, readers learned of Euler's solution [*sic*] to the famous three-body problem and Herschel's discovery of his new planet.[34] In August 1769, no doubt prompted by the Venus transit of the previous June, the editor of the *Affiches Américaines* called for local astronomers to observe a comet visible from Saint Domingue. The editor's doubts that such astronomers existed proved correct in October, when he published his own naked-eye observations.[35] Similarly, in 1776 the editor issued a call in the paper on behalf of Father Pingré in France for colonial observations of an upcoming eclipse:

> Observations from Port-au-Prince, Cap François, and even other parts of the island are desired. If one observes and wishes to communicate observations to the Academy of Sciences or the Marine Academy, it is essential to state one's method of determining the

true time of the beginning and the end of the eclipse, as well as the quality and power of the telescopes one is using.[36]

These requirements, not the least being the instruments, proved too much for the scientific and technical base in the colony, and a mere report of the eclipse appeared in the paper. Still later, an uncertainty was expressed whether a particular solar eclipse would be visible in the colony, and it took an astronomer of the Puységur party then in Saint Domingue to provide the definitive word that it would not.[37]

Colonial surveyors did possess some degree of mathematical literacy. The several official surveyors laid out roads, towns, and plantation properties. Not all of them would seem entirely competent, for at one point, when an ambitious effort was proposed to study variations in the magnetic needle across Saint Domingue in order to improve surveying, commentators pointed out that the disorder in the colony's surveys stemmed not from some esoteric natural phenomenon but from its surveyors.[38] Adam List, royal surveyor at the Môle, seems to have been the exception, however. Praised as "one [!] of the best mathematicians in the colony," List became a colonial associate of the Cercle des Philadelphes, and, in a notable feat of mathematical competence, at a public meeting of the Cercle in 1789 List announced his independent prediction of a solar eclipse for 1799, not an insignificant mathematical and astronomical accomplishment.[39] Moreau de Saint-Méry mentions that Father Archange, who died as a curé in Saint Domingue, published "a big book on astronomy," but the cases of Adam List and Father Archange (if true) stand out more as counterexamples illustrating the narrow limits of mathematical and astronomical competence in the colony.[40]

The relatively impoverished situation regarding mathematics and the mathematical sciences in the colony is readily understandable and contrasts with the case of botany. Although also dependent on and reporting to the mother science in Europe, botanists and naturalists came to be resident in Saint Domingue. A few colonial naturalists did emerge, and, later in the century, government botanists superintended botanical gardens. Nothing like that degree of institutionalization can be said for colonial astronomers, mathematicians, or cartographers.

EIGHT

Medicine and Medical Administration

T HE HISTORY OF medicine and medical administration in eighteenth-century Saint Domingue is a substantial one, and that history occupies a significant place in the overall story of science and colonialism in Old Regime Saint Domingue. Unlike the cases of astronomy and taxonomic botany, which touched the colony directly to only a small degree and which were centered "outside" the colony, medicine and medical science became heavily institutionalized in colonial Saint Domingue. Various departments of royal government spent significant public resources on the development of medicine and medical infrastructures in the colony, and medicine and medical science were seen as absolutely essential to colonial development. Given that tropical diseases presented evident obstacles to the health and welfare of colonists and their slaves and to the prosperity of the colony as a whole, the reasons for the strong government commitment to colonial medicine and for the growth of medicine in the colony are not hard to fathom.[1]

A previous chapter discussed the system of hospitals that grew up in Saint Domingue, a major pillar for organized medicine in the colony.[2] To recall, colonial hospitals—notably those in Cap François, Port-au-Prince, Léogane, and Les Cayes—derived from and were integrated into the colony's military establishment and that without the military, such a hospital system would not have arisen for the civilian sector. The focus in this chapter is on the cadre of medical personnel who served in hospitals and who otherwise concerned themselves with medicine and health care in the colony. It is not too early to point out, however, that the colony's medical personnel maintained its independence from military control or command per se. Technically beholden to the colonial bureaucracy, for all intents and purposes organized medicine in Saint Domingue achieved complete professional autonomy.

Medical practice in Saint Domingue and the French West Indies remained wholly unregulated until the turn of the eighteenth century. Native American medical practices continued in areas where Indians survived, and the pirates brought with them a special class of medical comrades, known as *frateurs,* to treat their wounds.[3] The Dutch pirate doctor, Alexandre-Olivier Esquemeling (or Oexmelin), was the most famous of the *frateurs.* He served with the pirates on Tortue Island in the 1660s, later earned an M.D. degree from Leyden, and in 1678 published his highly successful book on the pirates, *De Americaenishe Zeerovers.*[4] Later in the seventeenth century, as early official colonization proceeded, the *frateurs* turned into a special class of indentured surgeons or *engagés* who served as medical specialists. They received better treatment than the ordinary indentured, and their numbers grew. But such surgeons were largely unschooled, as well as subject to no regulation or control.[5]

In 1701 the crown appointed the colony's original *médecin du roi,* the Montpellier physician Michel Lopez de Pas, and the struggle began to gain control over medical affairs in the colony. A second médecin du roi, one Fontaine, also from Montpellier, took up his charge in Cap François in 1714, and after that no one disputed the principle that medicine in the colony was technically subject to control by governmental and medical authorities.[6] From that point on, the history is one of organized medicine and efforts to police medicine. And, of course, the fundamental tenet of that police was that only licensed whites could be involved in medicine and the delivery of medical care.

The organization and structure of medicine in Saint Domingue developed in complex ways over the course of the eighteenth century, but by and large medical models in contemporary France provided the pattern for organized medicine in the West Indies. A distinction needs to be drawn between the colony's official public health service and various private medical practices. In other words, a few physicians, surgeons, and others became paid employees of the state and colonial government, while the self-same colonial medical bureaucracy licensed others to practice privately within the confines of organized and policed medicine in Saint Domingue. Overlaying this public-private dichotomy, colonial medicine became organized hierarchically and pyramidally, both in terms of the traditional hierarchy of physicians, surgeons, apothecaries, midwives, veterinarians, barbers, and so on and in terms of gradations of rank and status within each of these "guilds." And then, the official medical establishment in Saint Domingue defined a larger world of un-licensed "plantation" surgery and slave medicine that operated extra-legally in the countryside.

At the top of the colonial medical establishment stood the three offi-cial positions of médecin du roi, one for each department in the colony. Technically equals, the médecin du roi in Cap François informally headed the medical community in Saint Domingue. As Pierre Pluchon points out, the title of médecin du roi is a misnomer, for it signifies the official position of chief royal medical officer in the colony and not simply a phy-sician licensed by the crown, of whom several also existed in the colony.[7] Appointment as first médecin du roi or first royal physician was usually based on previous experience in the colony, on the recommendations of local administrators, and, of course, on prior university training and degrees.

Below the position of first royal physician stood that of first royal surgeon, who served as the chief surgical officer in each of the three de-partments of the colony. The first royal surgeon superintended matters surgical alongside the first royal physician. First royal surgeons and ordi-nary royal surgeons—both positions of high status—were trained ex-perts given royal military commissions as surgeons major on the recom-mendation of the naval ministry and posted to Saint Domingue.

First royal physicians and first royal surgeons were paid employees of the state, and they headed the public health corps in each of their regions. This cadre of perhaps two dozen government medical officers, known as the *entretenus,* constituted an official public health service, and it formed the core of organized medicine in colonial Saint Domingue.[8] Over the period 1704–1803, ninety-nine individuals served in an official govern-ment capacity as medical *entretenus:* forty-three physicians, thirty-two surgeons, and twenty-four apothecaries, midwives, veterinarians, and others.[9] The power of government health officials becomes manifest when one realizes that they licensed the colony's private medical practi-tioners, and, of course, it was illegal to practice any medical art without a license.

At any one time about fifty additional medical personnel practiced with royal appointments as royal physicians, royal surgeons, or royal apothecaries but without remuneration and without being directly part of the colonial medical service.[10] This group of medical professionals with royal commissions formed the next rung on the ladder of organized medicine in the colony. They were likewise high-status, trained profes-sionals, but before they could take up their work in the colony they had to be screened by the first royal physician or an appropriate panel and their royal credentials accepted before the local Conseil Supérieur.

Military surgeons major attached to army and naval units formed a special subset of medical personnel in the colony. The army regiments in Cap François and Port-au-Prince both had surgeons major attached, as did artillery units, the militia, and Admiralty offices in the colony. In addition naval surgeons and surgeons major put in at Saint Domingue's ports aboard vessels of the royal navy. But local authorities in Saint Domingue did not receive army and naval surgeons major on an equal footing. Army surgeons major had their hands full; they were few in number, transient in the colony, and presented no administrative threat to local medical structures. Army surgeons major were therefore welcomed in the colony, allowed the privilege of private practice, and incorporated de jure into colonial medical licensing boards. It is understandable that because there were so many of them and because the level of their expertise was so rudimentary, ordinary naval surgeons had to serve an internship and pass a rigorous examination process before being allowed to practice in the colony. But the same rigorous rules also applied to higher-status, more competent, and commissioned naval surgeons major.[11] The explanation for the difference in treatment of army and naval surgeons major lies in the fact that the navy's powerful medical services posed a significant threat to the independence of the medical community in the colony, and therefore local physicians made every effort to restrict the rights of naval surgeons major. But more of that later.

The rest of the medical establishment in Saint Domingue arose and functioned as a complement to the government medical bureaucracy and the top echelons of private and military medicine in the colony. Although estimates vary of the total number of medical personnel, Pluchon's recent figures are certainly to be preferred. For 1791 he lists 26 licensed physicians, 291 licensed surgeons, 24 licensed apothecaries, and 600 to 800 unlicensed plantation surgeons in the colony.[12] Each of these groups will be discussed in turn, but for now the point can be made that the proportion of unlicensed plantation surgeons testifies not only that the control of medicine was ineffective in the countryside but also that organized and professionalized medicine was effectively a thing of the towns.

Thus in 1789 8 physicians, surgeons and apothecaries in the royal health service, 10 regular physicians, 24 master surgeons, 21 ordinary surgeons, 11 apothecaries, 2 midwives, 1 expert dentist, and 1 trussmaker and bandager worked in Cap François. In Port-au-Prince one could find 8 on the medical service staff, 8 regular physicians, 14 surgeons, 4 apothecaries, and 1 licensed midwife. Other towns possessed public and private medical personnel in like proportion.[13] These and other medical personnel in the colony possessed the notable privilege of exemption from service in the militia.

The hierarchical character of medicine in Saint Domingue goes beyond the traditional ranking of physician, surgeon, apothecary, midwife. Oligarchies as well as hierarchies dominated colonial medicine. That is, through their control of licensing boards, a core of top government medical specialists and senior licensed personnel governed the practice of their respective trades and enforced or tried to enforce a medical and economic monopoly by licensing some practitioners and excluding others. As a result, real differences in power and status separated the ranks of first royal physician, royal physician, and ordinary physician. Parallel distinctions can be drawn between and among first royal surgeons, royal surgeons major, master surgeons, ordinary surgeons in the royal navy, surgeons in the merchant marine and slave trade, not to mention plantation surgeons. Royal apothecaries similarly distinguished themselves from master apothecaries, ordinary pharmacists, and illegal peddlers of drugs. The model of medicine found in Saint Domingue as a ranked order of specialties controlled by an established body of sanctioned experts was taken over entirely from metropolitan France, and the same pattern repeated itself in other French colonies.[14]

First royal physicians functioned as leaders of the medical community, chief health officers, primary medical inspectors, and top medical functionaries. Their particular duties, as specified in various ordinances, were fairly diverse.[15] Their primary obligation was to provide medical supervision of hospitals and medical treatment of military personnel. First royal physicians had to tour hospitals three to four times a week and to submit weekly reports to the administrators. In addition to supervising the medical care of the sick and wounded, first royal physicians approved all certificates for medical leave for soldiers and sailors, attested to medical conditions, and authorized treatments at state expense. They also had to inspect all arriving slave ships and to impose quarantines, if necessary. First royal physicians served on all boards to examine and license the colony's other medical personnel, and they also monitored the accounts of licensees. Regulations charged first royal physicians with others in the medical establishment to inspect and superintend pharmacies and the drugstores of surgeons and apothecaries. First royal physicians also gave approbations for advertised drugs, and a few of these duly appeared. First royal physicians received a fairly low pay for the performance of all these duties; the top salary amounted to only 2,400 colonial livres, a figure not raised after 1740. Still, by the long-standing French tradition of the *cumul,* one physician could combine several medical posts within the administration and draw a total salary of 5,000 to 6,000 livres of public appointments.[16]

Twenty-six private physicians practiced medicine for fees in the colony in 1791. The administrators and the first royal physicians interviewed private practitioners arriving from France, who had to present either their university credentials or, if well connected, royal warrants as (ordinary) médecin du roi. Such warrants (*brevets*), obtained subsequent to receiving an M.D. degree, had to be registered before the local Conseil Supérieur. Nineteen of the twenty-six private physicians in Saint Domingue in 1791, or over 70 percent, held degrees from Montpellier.[17]

Licensed surgeons outnumbered physicians in Saint Domingue on the order of ten to one at the end of the colonial period. Surgeons predominated over doctors even more in the earlier stages of colonial development, and throughout the history of the colony ordinary surgeons proved the more important official agents providing health care. A handful of surgeons major and first royal surgeons represented the elite of surgery, but semiskilled craftsmen composed the overwhelming bulk of the community of licensed surgeons.[18]

A series of royal edicts dictated the strict control of surgeons and their subordination to physicians.[19] Only licensed Catholic surgeons could practice surgery, and regulations forbade them to practice internal medicine. From 1739 the rules required surgeons newly arrived in the colony to take a year's residency in one of the colony's hospitals and to attend medical and surgical seminars offered by the first royal physician, but those requirements, although renewed in 1764, fell into disuse. To get a license, surgeons had to present their credentials and be examined by a licensing board consisting of the first royal physician, the first royal surgeon, and two master surgeons. Surgeons swore allegiance to the Conseil Supérieur and the first royal physician, and they had to submit twice-yearly reports on their activities. Regulations controlled surgeons' fees and stipulated fines for various infractions. Surgeons in the colonies (but not in France) could employ drugs in their practice, and the first royal physician and the first royal apothecary annually inspected surgeons's stores of medicaments. The professional power of royal physicians to control surgeons and medical activity in the colony substantially increased in 1764 when the civilian judge from the Conseil Supérieur was dropped from all medical boards.

In Cap François, Petit Goave, and possibly elsewhere, master surgeons organized themselves into guilds. Although they never secured letters patent as a guild or obtained the sought-after privilege of granting their own certificates in surgery, master surgeons did serve on surgical licensing boards, and from as early as 1723 they established a de facto corporate existence as the Company of Master Surgeons.[20]

Before proceeding, this is perhaps the place to touch on the teaching of medicine and medical arts in the colony. One cannot say that a great deal of medical instruction took place in Saint Domingue. Regulations required the original chief médecins du roi to give monthly medical conferences for surgeons, and the reform of 1739 created four surgeon-in-training positions in each of the colony's hospitals.[21] To a minor degree, then, hospitals constituted teaching centers with students and the resident medical and surgical staffs superintended by the first royal physician. The authorities sanctioned several courses in midwifery at various periods, and in 1789 a physician in Saint-Marc offered a private course in medicine and surgery, to which other members of the profession received free admission.[22] Thus, whereas colonial officials excluded the teaching of the liberal arts, they promoted instruction in medicine and surgery to a degree. When one considers that the rules insisted on the local licensing of physicians and surgeons regardless of previous schooling or experience, the extent to which medicine and surgery developed autonomously in the colony becomes ever more apparent.

Apothecaries occupied a lower rung on the ladder, just above midwives, but apothecaries stood not far removed laterally from spice traders and perfumers.[23] Apothecaries appeared in Saint Domingue after 1764, as a specialization growing out of surgical practice. Three royal apothecaries, appointed by local administrators, served in the three departments of the colony, and a handful of master apothecaries and ordinary pharmacists brought the total in the trade to perhaps twenty individuals in the colony at the end of the Old Regime. No apothecary could open a shop or deal in drugs prior to an examination by a board of physicians, surgeons, and master apothecaries. Drugs had to be kept under lock and key, exact records had to be kept, and periodic inspections took place. Outside of Cap François, Port-au-Prince, and perhaps a few other towns, such rules and regulations could not be enforced, and illicit peddling of drugs did develop. Still, in 1780 the law declared strong drugs—especially poisons, including arsenic, rat poison, and corrosives—controlled substances and made them available only from a royal apothecary.[24] A handful of exceptional apothecaries interested themselves in chemistry and the higher reaches of their science.[25]

The colony's surgeons trained local midwives, and a board consisting of all physicians and licensed midwives examined midwife candidates. In 1764 the government sent an additional group of paid, licensed midwives from France to practice in the colony, but they apparently got uppity and were suppressed in 1773. In that year a physician-obstetrician was appointed and given exclusive rights to examine midwives, and the position was upgraded in 1787. The previously mentioned courses in midwifery included one using a mechanical doll, and midwives advertised with the approbation of physicians. The law did not permit slaves and free people of color to act as midwives or medically in any fashion, but the Cap Conseil Supérieur made a legal exception for the experienced mulatto midwife in Cap François, the widow Cottin.[26]

Finally in this connection, the government brought in on contracts at least two veterinarians (*artistes vétérinaires*) from the state veterinary school in Alfort, France. These official veterinarians augmented the small group of private veterinarians working in Saint Domingue. Public opinion strongly supported the nomination of one of the latter, La Pole, to an appointment as royal veterinarian.[27]

Such was the constitution of official, policed medicine in Saint Domingue. The creation of a formal medical establishment vested with increasing powers of professional control represents a major milestone in the development of the colony, but official medicine still faced a hopeless struggle to control unlicensed practice throughout the colony. Outside the world of official medicine, custom and anarchy reigned; unsanctioned practitioners dispensed most of the colony's health care there in the world of unofficial medicine, and the countryside crawled with unlicensed plantation surgeons. Pluchon, again, puts their numbers at six to eight hundred.[28] Plantation surgeons, just a cut above barbers, were generally inexperienced and knew only the rudiments of their trade. They were also likely to be young, under twenty years old. Such "surgeons" contracted to work on a plantation or to rotate among a group of plantations, and to an overwhelming extent the great mass of the colony's slaves and their rural masters depended on these surgeons for their surgical and health needs. Demand for the services of plantation surgeons must have been great, for by putting together contracts with plantations, they could earn upwards of 8,000 livres, substantially more than the top rank of government physicians in the towns.[29]

Medical manuals aided plantation surgeons in their work. J. F. Lafosse wrote a notable manual, *Avis aux habitans des colonies, particulièrement à ceux de l'Isle de S. Domingue,* which was published in 1787. Lafosse, a Montpellier-trained doctor and *correspondant* of the Société Royale de Médecine in Paris, practiced in the Miragoane area from 1777 to 1784 and had extensive experience with what might be termed plantation medicine. His book concerns itself especially with the care and feeding of slaves, and La Fosse makes clear the connection between medicine and the maintenance of slave society:

> I will be happy if the people for whom I have written this essay can gain some profit from what I have said about various illnesses, and if sensible and benevolent plantation owners, whose goals I support, can find herein some means of ameliorating or alleviating the lot of

the beings they control. Every colonist needs to be concerned with the health of slaves, as much out of self-interest as out of a sense of humanity.[30]

Only whites could practice medicine or any of the allied arts. The ban on slaves having anything to do with any form of medical aid was absolute. On pain of death, ordinances made it illegal for slaves to administer even the most rudimentary treatments, including first aid; the authorities initially allowed an exception for slaves to treat snakebite, but they later withdrew even that privilege. The white colonial establishment was especially sensitive about poisons and strictly proscribed slaves from having anything to do with drugs or poisons. Slaves could not touch, possess, or carry drugs; they could not make or distribute drugs; they could not be employed by pharmacists or apothecaries.[31]

Characteristic of the racism of the colony, the same restrictions regarding medical and pharmaceutical practice also applied to all free people of color. Through emancipations, the number of free black citizens grew, and in order to keep medicines out of the hands of *all* blacks, they became off limits to *all* people of color. Free mulatto citizens, free black citizens, and slaves were all forbidden to act in any medical capacity. Medicine legally and to some extent effectively was for whites only.

To the extent possible, of course, the mass of nonwhite colonists in Saint Domingue provided for their own medical care outside of white, Western medicine. A particular sort of medical practitioner, known as *caperlata* or *kaperlata,* arose among and in the service of the free people of color and poor whites. Nothing is known about these "doctors," except that they had a name and that official medicine vigorously condemned them.[32]

More significantly, the hundreds of thousands of slaves imported into the colony brought with them traditional medicines and healing arts from Africa. These became mixed with residues of native American medicine to produce a separate slave medicine and pharmacopeia, to which white physicians and surgeons turned as a last resort. Vodun (voodoo) likewise had a medical component of black and slave medical practice. Specialized healers emerged in slave groups, and women from Sierra Leone were especially renowned for their knowledge of the healing arts.[33] About the medicine practiced among the mass of Saint Domingue's inhabitants, the informed and seemingly unbiased judgment of Louis Joubert is that unofficial slave medicine in general proved just as efficacious as official Western medicine at the time.[34]

The men who served as first royal physicians and as heads of the medical establishment in Saint Domingue on the whole must be judged an active, progressive, and perhaps even valiant group. Two of them, one from the first half of the eighteenth century, one from the second, deserve to be singled out for their work as medical administrators and as medical scientists: J.-B.-R. Pouppée-Desportes and Charles Arthaud.

The king named Jean-Baptiste-René Pouppée-Desportes (1704–48) first royal physican for all of Saint Domingue in 1732. The position of médecin du roi had existed in the colony for three decades, but Pouppée-Desportes was the first medical personality to make himself strongly felt in the colony. Desportes had an M.D. degree from Rheims and came to the colony as a protégé of Dufay and the Jussieu brothers at the Jardin du Roi. He sent ipecac seeds to the Jardin and models of sugar-processing machinery to the Royal Academy of Sciences in Paris. The Academy elected him a *correspondant* in 1738. Desportes, as first royal physician, was the driving force behind the major hospital and medical reform of 1739.[35]

Historians of medicine have styled Pouppée-Desportes the "veritable founder of tropical pathology."[36] Desportes died in the colony in 1748, and only in 1770 did his family publish his three-volume work, *Histoire des Maladies de S. Domingue,* posthumously. The first two volumes cover chronic and epidemic diseases in Saint Domingue; in his work Desportes was the first to distinguish and describe yaws and yellow fever. He gave the third volume of his *Histoire* over to an "Abrégé des Plantes Usuelles de S. Domingue." This part of the work dealt primarily with medicinal plants and plant remedies, but Desportes also intended it as a botanical supplement to Plumier.[37]

Charles Arthaud was undoubtedly the most notable physician in the colony in the second half of the century. Born in 1748 in Pont-à-Mousson, near Nancy in France, Arthaud received his M.D. degee from the medical faculty at Nancy in 1769, publishing an unexceptional dissertation of experiments on Hallerian sensibility in arteries.[38] Arthaud arrived in Saint Domingue in the early 1770s on family business, possibly having to do with his brother, Jean Artaud [*sic*], later chief civil engineer in the northern department of the colony. Charles Arthaud began a private medical practice in Cap François, and he sought to make himself useful in the colony and to work his way up in the colonial medical establishment. In 1776 he published his *Traité des Pians* (*On Yaws*), and in a public-spirited gesture in the same year he revealed his remedy for gout. In 1777 Arthaud again received notice for reviving a drowned female slave using the dockside apparatus newly installed in Cap François for reviving the drowned.[39] Arthaud was quick to attack and quick to defend in the hothouse of local medical controversies.[40] In 1777 Arthaud returned briefly to France in connection with a malpractice suit, becoming an associate of the Société Royale de Médecine and the Académie Royale de Chirurgie during his stay. Back in Saint Domingue, Arthaud served as chief physician at the hospital at Léogane during the American War. In 1783 he returned to Cap François as interim first royal physician and then as the permanent chief médecin du roi.

Arthaud made it to the top of the pyramid. He became rich, owning a sugar plantation in Limbé and goodly property in Cap François. He married Moreau de Saint-Méry's sister-in-law and had three children, but his wife and children all died within a space of eighteen months, probably before Arthaud returned to the Cap in 1783.[41] Beginning in 1784 Arthaud became the driving force behind the Cercle des Philadelphes, and in the later 1780s Arthaud edited and published a number of medical and scientific works both in his own name and on the part of the Cercle des Philadelphes. Arthaud will recur often in this narrative.

Official medicine in Saint Domingue formed an important pillar on which colonization rested, and colonial medicine proceeded under the aegis of and, indeed, as part of colonial government. Still, the development of an independent medical establishment in Saint Domingue has to be seen as a long-term victory for local colonial medical structures over a powerful and expanding medical bureaucracy attached to the state secretariat and ministry of the navy back in France.

As the French royal navy grew in the seventeenth century, a naval medical corps arose to attend to the health and medical needs of naval officers and staff, royal sailors, and merchant sailors. As Bernard Broussolle and Philippe Masson show in their study, a series of royal ordinances stretching from 1673 through 1789 put into place a substantial and elaborate medical organization on land and at sea under the control of the ministry of the navy and the colonies.[42] In Saint Domingue the colony's civil intendant reported on medical affairs to the same state secretary and minister of the navy, but that channel remained separate from and not subservient to the medical services proper of the naval ministry. This bureaucratic separation notwithstanding, the naval medical corps based in France possessed great expertise in colonial medicine and had strong interests in the state of affairs medical in Saint Domingue.

The naval medical corps in France centered on the three ports of Brest, Rochefort, and Toulon in France. Naval hospitals and associated medical-surgical colleges in each of these ports provided the primary institutional base of the naval medical corps, the hospital in Rochefort becoming a major installation with twelve hundred beds.[43] By the end of the eighteenth century the paid staff of the naval medical corps grew to 285. Ninety percent of that corps served at sea, including 54 surgeons major, 27 second surgeons, 45 surgeon's aides, 45 student surgeons, and 85 apothecaries. They provided the medical personnel for royal naval vessels and for hospital ships. An additional 29 men served on land and ran the naval hospitals, the medical colleges, and the service itself, with 6 landlubbing physicians heading the organization.[44] The naval medical bureaucracy added to its functions and its power when it gained responsibility for approving all surgeons sailing with the merchant marine, as well as with the royal navy. Another 1,000 or so volunteer medical personnel complemented the paid staff of the naval medical corps; these men served in the navy as volunteer surgeons and apothecaries as an alternate way to complete a medical apprenticeship; they later would enter an appropriate medical guild on land. In times of war, naval medical authorities resorted to forced conscription of surgeons and medical workers, mobilizations that swelled numbers under their administration even further.

The emergence of such a large and powerful medical arm within the royal navy and the naval ministry naturally provoked opposition from civilian medical communities in France and in Saint Domingue, jealous that the navy not infringe upon their privileges and practices. First in 1711 and more forcefully in 1738, royal edicts affirmed the independence of organized medicine in Saint Domingue. The key regulation said that members of the naval medical corps could only treat naval personnel and their families. It further stipulated that naval medical personnel be under the civilian control of the intendant while in the colony. The much more rigorous reception afforded naval surgeons major over army surgeons major noted earlier formed part of the same power struggle. Any disputes should have been settled at that point in 1738, but the continuing growth of the naval medical corps precipitated ongoing struggles and turf battles to the end of the century. By and large, the medical establishment in Saint Domingue won these, and victories solidified the independence of organized medicine in the colony.[45]

Trouble began again in the 1760s after Pierre-Isaac Poissonnier combined several royal commissions into a powerful position as head of the naval medical corps and naval health services in France. In 1763 the king appointed Poissonnier inspector of port hospitals and then director general of medicine, pharmacy and botany in the colonies. These appointments brought 14,000 livres tournois annually, and the latter post in principle allowed the elder Poissonnier a large say in colonial medical affairs, particularly in nominations to major royal medical appointments in the colonies as well as in the navy itself. In 1768 Pierre Poissonnier hired his brother, Antoine Poissonnier-Desperrières, as adjunct administrator for medicine. The younger Poissonnier had served from 1748 to 1751 as médecin du roi in Saint Domingue, and he had published a *Traité des fièvres de l'île de Saint-Domingue* in 1763 and a *Traité sur les maladies des gens de mer* in 1767.[46]

The two Poissonnier brothers formed an expansive and imperial "diumvirate" in control of all medicine in the royal navy and with theoretically great control over medicine in the colonies. By the same token, by the 1760s the medical establishment in Saint Domingue was already set in its ways, jealous and defensive. Further conflict was inevitable.

Pierre Poissonnier sent his first emissary to Saint Domingue, one d'Hormepierre, in 1763 as chief medical inspector of colonial hospitals. The local medical community repudiated d'Hormepierre completely, and nothing resulted from this effort at general medical inspection before d'Hormepierre died in the colony in 1765.[47] The Poissonnier brothers next tried to exercise direct appointments to medical posts in Saint Domingue by having the king order the installation of one Duchemin de l'Étang in 1775 as médecin du roi at Les Cayes Saint Louis on the Caribbean. Duchemin's reception in the colony was forced and resisted.[48] In December 1777, Duchemin began publishing the first medical journal in the colony, the *Gazette de Médecine et d'Hippiatrique*. Both Duchemin and his journal were immediately and vigorously attacked in the *Affiches Américaines* in a series of letters by none other than Charles Arthaud.[49] Arthaud assailed Duchemin across a spectrum of points, such as the use of opium and the utility of veterinary students, but the medical subject matter was not really the issue. The aggressive defense seemed to work, for Duchemin's journal closed in February 1779 after only eight numbers, and the man himself disappears from the record.

The local medical establishment in Saint Domingue vigorously and successfully resisted the Poissonnier effort to implement a supposed requirement for all médecins du roi in the colonies to correspond with the medical offices in the naval ministry. Arthaud called the proposal "repugnant," and the Poissonniers let it drop.[50]

The Poissonniers had greater success with their agent, Jean-Barthélemy Dazille (1732–1812).[51] Dazille started out in the 1750s as a simple naval surgeon, spending two decades in the colonies, mostly in Saint Domingue. He then studied medicine in Paris and eventually got an M.D. degree from the faculty at Douai. In 1775 he joined the Poissonniers in the navy ministry, and, to avoid trouble, he was appointed as *honorary* médecin du roi in Saint Domingue. Dazille spent the period 1775–84 in the colony. Not surprisingly, he specialized in the economically key areas of medical conditions on plantations and the medical treatment of slaves. At the time Arthaud became first médecin du roi at the Cap, the ministry recalled Dazille to France, where he served in the ministry as an expert on colonial medicine and inspector of colonial hospitals.

Dazille dedicated his *Observations sur les Maladies des Nègres* to the state minister for the navy, Sartine. First published in 1776, an expanded edition appeared in 1792. Dazille, with some justification, claimed his to be the first formal study of illnesses among slaves, and the work has received praise as sane and reasonable in its approach, especially in matters of hygiene.[52] Dazille's subject was of high importance for colonialism and colonials in Saint Domingue, and state support for Dazille and his research promised to pay off handsomely. In his work Dazille did not mistake the significance of his subject or the reason the Poissonniers sent him to Saint Domingue.

> The introduction of slaves is the major and fundamental means for a colony to prosper, and the conservation of these unfortunate beings is what makes that means effective. [To be concerned with the health of slaves] is to occupy oneself with that which is useful to colonists in particular, to the commerce of the nation in general, and to the prosperity of the State.[53]

With similar practical and amoral ends in mind, the authorities recruited Dazille to work up and publish his *Observation Générales sur les Maladies des climats chauds* (1785) and his *Observations sur le Tétanos* (1788). These were both published at government expense and with the approval of the Société Royale de Médecine, both Poissonnier brothers serving on the committee. As a result of considerable knowledge, experience, and backing, Dazille compiled the most significant record of work in colonial and tropical medicine of anyone in the eighteenth century.[54] Dazille became a *correspondant* of the Société Royale de Médecine in Paris and a national associate of the Cercle des Philadelphes in Cap François.

As an agent of the naval ministry, however, Dazille was not well regarded in medical circles in Saint Domingue, nor did he act benevolently toward the medical establishment in the colony or toward Arthaud in particular. Dazille writes of the "pains, difficulties, jealousies, malice, and enemies" he encountered in dealing with the physicians of Saint Domingue. As inspector of colonial hospitals, Dazille strongly and consistently attacked Charles Arthaud and Arthaud's medical supervision of the Charity Hospital in Cap François, and as part of his proposed reforms for hospital administration in Saint Domingue, Dazille suggested the creation of a second chief médecin du roi to serve alongside Arthaud.[55]

But Arthaud remained insulated from Dazille and the Poissonniers. Arthaud reported to the intendant and through him to the state secretary for the navy, who forwarded Arthaud's reports *down* to the elder Poissonnier for his edification and that of the Société Royale de Médecine.[56] Although involved in Saint Domingue throughout the century, the medical branch of the naval ministry never gained power over medicine and medical administration within the colony. Arthaud embraced hospital reform only when it became convenient in 1791.

One of the more interesting and revealing medical establishments created in Saint Domingue was the health spa known as the Eaux de Boynes. Located in the interior of the northern arm of the colony north of Port-à-Piment, the Eaux de Boynes replicated the model of European spas and hot springs, and its extraordinary development testifies to the advent of medical (and, indeed, all) institutions in the colony as the unfolding of a social process.

The hot sulphur springs of the Eaux de Boynes were absolutely unknown to any colonist before 1725. A slave tracking runaway cattle in the forest outside of Port-à-Piment discovered them that year.[57] The waters gained a reputation for their curative powers, and over the years the springs became a site of pilgrimages and miraculous cures. A veritable shrine arose around the waters, with nearby trees bedecked with discarded crutches and written testimonials to the water's healing effect. A log of cures was kept from 1739.

In 1772 the Eaux de Boynes became a public institution by dint of a gift to the crown from the proprietor of the land, and from that point the guiding hand of the government shaped the development of the facility. The naval commissioner (ordonnateur) in Saint Domingue pressed to turn the springs into a formal spa, and the governor-general and intendant supported the proposal. Officials named the waters after P.-E. de Boynes, then secretary of state and minister of the navy and the colonies, and they undertook further development at public expense.

Naturally, once the waters had officially fallen into the hands of the state, the first thing was to get medical and scientific opinion on the "true" virtues of the resource. The Cap François physician, J.-L. Polony, and P.-F. Chatard, royal apothecary at the Cap, performed the first chemical analyses of the waters immediately in 1772. A later director of the Eaux de Boynes, Joseph Gauché, made subsequent analyses, which the Cercle des Philadelphes published. One may doubt the significance or validity of these forays into chemistry, but contemporary medical authorities indeed pronounced positively on the medical benefits to be derived from the waters. Treatment supposedly proved effective for dozens of different maladies: dropsy, rheumatism, gout, paralysis, skin diseases, apoplexy, nerve conditions, bowel obstructions, diarrheas, tumors, asthma, parasites, and so on and so forth.[58]

Official approval and development by governmental authority brought the paid posts of director of the spa and medical inspector. The state stood to gain from the institutionalization of the Eaux de Boynes because the facility could offer a less expensive treatment for convalescing military personnel, and it would keep in the colony the 40,000 livres a year spent by French colonists who "took the waters" in the Spanish part of the island.[59] The French government spent over 600,000 colonial livres of royal funds in upgrading the Eaux de Boynes and creating a military hospital on the premises. The installation operated after that on contracts with the government to provide medical care to military personnel and on fees paid by private clients.

Mismanaged in its early years, the Boynes establishment solidified after 1786 with Joseph Gauché as its director and the omnipresent Charles Arthaud receiving the stipend as medical inspector.[60] At its height the Eaux de Boynes consisted of seven different springs, one frequented especially by the ladies. There were sixteen public baths, steam baths, and mud baths. Fifteen wooden buildings composed the spa complex that also included a public square and a fresh-water fountain. The military hospital had sixty beds, its own hot springs, and a wall to keep in potential deserters. Another building housed the bath for the poor; it was crowded, pathetic, and the scene of the religiosity that characterized the spa earlier in its development. Excepting linen, everything one needed for a visit to the spa could be found at the Eaux de Boynes itself. Government engineers and slave crews cut a 3-league carriage road (over 7 miles) from the coast at Port-à-Piment, and a passenger boat sailed every fortnight from Cap François.

The Eaux des Boynes was not the only spa and hot springs in French Saint Domingue—it was only the most developed. Several similar colonial springs all progressed along the same road from initial discovery, through chemical analysis and medical inspection, to official sanction. The spring at Dalmarie on the southern peninsula provides one example. In 1760 the military consigned soldiers and sailors at hospital rates to the care of one Martin, a rich surgeon living at the site; everything had to be brought into the Dalmarie spa. Similarly, the nearby springs at Tiburon were discovered only in 1759. The Tiburon site never received official authorization, but by late in the century upward of one hundred people could be found taking the waters there. Finally, also in 1759, a *médecin-*

botaniste, J. Guyon de Chabanne, having proved the medicinal virtues of a sulphur spring in the Cul-de-Sac, received the concession to develop the spring and to provide medical care.[61]

The several hot springs in Saint Domingue and their elaboration, however partial, into the social and medical institution of the health spa testify to the social character of the development of medicine and medical institutions in the colony and to the centrality of European models in that development.

Inoculation against smallpox represents a significant area of progressive medical work in Saint Domingue. Lady Mary Wortley Montagu and others introduced the technique of smallpox inoculation into England and Europe from Turkey in the 1710s and 1720s. Inoculations were tried in Saint Domingue as early as 1745, nearly two decades before the Paris medical faculty approved the practice, and three decades before it became widely accepted in France.[62] In 1767 M. de la Chapelle became the first in the colony to administer inoculations on a large scale to slaves. R.-N. Joubert de la Motte vigorously promoted inoculation from 1768. Joubert held an M.D. degree from Angers and letters of correspondence from the Société Royale de Médecine in Paris and the Académie des Sciences, Arts et Belles-Lettres in Dijon. The king appointed Joubert royal botanist and director of the royal botanical gardens in Port-au-Prince.[63] Likewise an original colonial associate of the Cercle des Philadelphes, Joubert de la Motte was a notable medical and scientific personage in the colony, and his voice carried weight in the matter of inoculation.

After the smallpox epidemic of 1771, the Cap François physician J.-L. Polony successfully inoculated himself, his wife, and his children. In 1772 one Fournier de Varenne inoculated ninety-eight people on his plantation and wrote up a small tract on the technique that circulated in manuscript thereafter.[64] But a further turning point in the already advanced level of inoculation in Saint Domingue came in 1774 with the arrival of Siméon Worlock, brother-in-law of the English inoculator, Daniel Sutton. For a modest per capita fee Worlock inoculated plantation slaves by the thousands, no doubt becoming rich in the process. Worlock became a naturalized citizen of the colony in 1779. He was a *correspondant* of the Société Royale de Médecine in Paris, from which he won a minor prize for his work on an animal pestilence, and he became a resident associate of the Cercle des Philadelphes in 1787.[65] Inoculation was not universal in Saint Domingue by late in the century, for the slightly risky procedure still had its critics, and smallpox epidemics continued to occur through 1789 and presuambly thereafter. Still, armed with an effective and inexpensive technique, Worlock and other inoculators easily improved the medical and economic conditions of the colony. The revealing connection between slavery and the early and widespread use of inoculation in Saint Domingue should be clear. At very little cost inoculation protected and preserved the valuable capital property of slave owners. Inoculation strengthened the slave system in Saint Domingue, and it provided a model for the social utility of science, medical and otherwise.

Insofar as physicians and others in the colonial medical community in the 1770s and 1780s subscribed to any medical theory, they shared the medical philosophy of the Société Royale de Médecine in Paris that held to environmental causes for diseases. The Société Royale de Médecine was an innovative, activist organization chartered in 1776 over the objections of the Paris medical faculty.[66] The Société Royale organized physicians and others all over France, and it carried out an extensive Baconian program of empirical medical research designed to demonstrate the effects of climate and the environment on the etiology of disease.

A number of physicians important in the story of medicine in Saint Domingue were also members and *correspondants* of the Société Royale de Médecine in Paris: Arthaud, Lafosse, the Poissonnier brothers, Dazille, Joubert, and Worlock; and the research program of the Parisian Société Royale influenced medical and scientific research undertaken in the colony by the Cercle des Philadelphes. Arthaud's association with the environmental medical philosophy of the Société Royale shows itself in his 1787 paper, "Observations sur les Constitutions des Saisons & sur les Maladies qui paraissent en dépendre."[67] Lafosse's 1787 manual, *Avis . . . sur les principales causes des maladies,* is likewise full of environmental and climatological determinism.[68] Such opinions made a difference in a legal case in Saint-Marc in 1786, when a judge accepted the professional judgment of the local royal physician and surgeon that polluted water negatively affected the air and one's health. Based on this scientific and medical advice, the judge then issued a complicated ruling that regulated water use and garbage disposal in Saint-Marc.[69] The idea that the environment precipitates disease seems so well accepted that François de Neufchateau, a judge on the Conseil Supérieur and honorary member of the Cercle des Philadelphes, felt no reluctance in going further to link environmentalism with the social constitution of the colony. In his *Études du Magistrat* of 1787 he wrote:

> The mark of the climate imprints itself on men and consequently on their social relations [*leurs rapports entr'eux*], on the bonds they establish, on the contracts they form, and hence on the affairs the magistrate must judge. Here in the tropics *nature* herself under-

goes changes that must produce their effects in society and in legislation.[70]

The commitment of colonial physicians and others in the later 1770s and 1780s to the environmentalism of the Société Royale de Médecine in Paris appears even more clearly when contrasted with the views of medical authorities earlier in the century. Pouppée-Desportes, writing before 1748, did say that climate was the first cause of disease, but he added that human temperaments were also important factors.[71] N. L. Bourgeois, writing before 1776, tried to show that the air in Saint Domingue did not cause disease and that disease developed as a function of temperament or humors. But by 1787 Bourgeois's views had clearly become so antiquated that a nephew, P.-J.-B. Nougaret, saw fit to make substantial changes in editing his uncle's original manuscript for publication. Nougaret says he made the changes because "in the first place it seemed to me that to do so would harmonize with the useful aims of the Société Royale de Médecine, which has asked for memoirs on the diseases of different countries."[72]

Poisons, the close control of drugs and poisons, and the extreme anxieties felt in Saint Domingue regarding poisons are issues that have come up before. The constellation of concerns in the colony over poisons prompted a minor but revealing tradition of medical research regarding the scientific nature and agency of poisons. In this instance the elite's fear of its slaves clearly directed problem choice in research. In 1757, for example, coincident with the hysteria surrounding the Macandal affair, medical personnel at Fort-Dauphin carried out experiments on dogs to prove the lethal effects (by coagulation) of clandestine slave concoctions. In 1763 the Choiseul ministry ordered the colony's physicians and surgeons to investigate and report on poisonous plants in the island territory. Another ministerial order in 1774 gave a pronouncement on poisons from Poissonnier-Desperrières.[73] Our friend Arthaud carried out two pieces of poison research. In 1778 he conducted experiments poisoning dogs and sent the results to the Société Royale de Médecine in Paris.[74] Arthaud also wrote a paper "on the effects of the bite of the crab-spider of the Antilles," which appeared in Rozier's Journal in 1787. The small crab-spider was supposedly as venomous as a viper and reputedly killed horses and cattle. In his paper Arthaud reported his microscopic observations of the animal and his experiments proving their bites fatal to chickens.[75] Regarding the connection between colonial fears over poisons and this tradition of useful medical research, no ambiguity exists about a last instance, for in 1758 at least, colonial physicians experimented with poisons on human subjects, presumably killing the two condemned slaves put at their disposal.[76]

NINE

Economic Botany and
Animal Economy

A SERIES OF WORLDWIDE plant transfers following the discovery of America transformed the economies of Europe and the New World. The introduction and spread of wheat, corn, and the potato as food staples supported an expanding world population. Other crops, notably sugar and coffee, spread around the world and formed the basis of the plantation system and colonial empires in the eighteenth century.[1] The significance of plant transfers to the colonial development of Saint Domingue should already be obvious. Sugar, coffee, and indigo—the colony's three most significant commodity products—were not native to Hispaniola, yet at its height Saint Domingue became the world's leading producer of at least the first two of these products. Without plant transfers and government policies of promoting the cultivation of economically useful products, Saint Domingue and other similar colonies would never have existed.

This chapter explores a more restricted point, however: that, as plantation systems and colonial economies developed in the eighteenth century, the great mercantilist powers invested significant resources in programs of applied botanical research and development. Following on initial botanical explorations of their territories, the French, the British, and the Dutch actively sought to identify new, economically useful plant (and animal) products and to introduce them into large-scale production, as had been done previously with coffee and sugar. To carry out programs of applied botany, colonial powers created intercontinental systems of colonial botanical gardens with staffs of paid scientific experts. Colonial botanical stations systematically exchanged plants and plant products in coordination with scientific centers in Europe. They undertook research of various sorts, including experimental farming. Allied to such programs of botanical research and development came the notion of breaking the botanical monopolies of other nations, given that wherever possible, as a corollary of the mercantilist policy of "exclusive" colonial trade, governments maintained nationalist economic monopolies over valuable plant and plant products.[2] Programs of direct applied botany sought to achieve an immediate or near-term economic benefit for the nation funding the enterprise, and in this area the knowledge of scientific experts promised great returns on the investment.

The applied botanical garden of the later eighteenth century succeeded two earlier types of European botanical gardens: medical botanical gardens and scientific botanical gardens. Formal botanical gardens emerged in the sixteenth century from medieval herb and apothecary (*physick*) gardens.[3] For the most part, university-centered medical faculties controlled early botanical gardens. Created for the purpose of promoting pharmacy and pharmaceutical research, medical botanical gardens were oragnizations ancillary to medicine. To a degree, they served to teach botany and pharmacy.

Gardens devoted more to the scientific study of botany arose next. Europe possessed sixteen hundred botanical gardens at the end of the eighteenth century, most of the scientific type.[4] The most important of these gardens, the Jardin du Roi in Paris (1635) and the Royal Gardens at Kew (1759), were creatures of the state and not the university. They aimed to further new knowledge in the scientific study of the botanical world and more particularly to work out rational classification systems. The major scientific gardens served as collection centers that received specimens from outlying areas, and they offered considerable instruction in the scientific aspects of botany and related areas of knowledge.

Later in the eighteenth century a new type of botanical garden appeared, the applied botanical garden, and many such gardens were created in the colonies themselves. Applied botanical gardens were government institutions and elements in government policy to promote national and colonial economies. Little or no teaching took place within the confines of this type of garden; practical gardeners also assumed a greater importance in the applied botanical gardens of the later eighteenth century than men of science.[5]

France, Britain, and the Netherlands all established intercontinental systems of colonial botanical gardens and stations for the purposes of promoting economic botany in their respective spheres. The Dutch gardens at Capetown dated from 1694, and other Dutch stations existed in Ceylon and in Batavia on the island of Java in Indonesia. Fairly late in the eighteenth century the British established an extensive network of colonial gardens satellite to Kew Gardens in England. The English maintained gardens in Saint Vincent in the West Indies (1764), Jamaica (1775), Calcutta (1786), Sydney (1788), and Penang, Malaya (1800).[6]

Of the three nations, the network of French botanical gardens proved the most extensive and dynamic in the eighteenth century. The French established their first overseas botanical station in 1716 in Guadeloupe. The mature network included stations in Guadeloupe, Martinique, and Cayenne in the Western Hemisphere; several Indian Ocean outposts; a European link through the Jardin du Roi and other institutions in France; and, in chronological sequence, the government gardens in Saint Domingue.

While French colonial botanical gardens interacted among themselves, their connection to France and to French institutions was in many ways primary. On the one hand, they received administrative direction through the ministry of the navy and colonies. On the other hand, they linked organizationally and scientifically to the Jardin du Roi and the Paris Academy of Sciences. In this connection the botanical garden in Nantes, France, played a special role as an intermediary between Paris institutions and efforts at economic botany in the colonies.[7] Nantes was, of course, a major port. In 1688 an apothecary garden got started under the control of the professor of botany at the university in Nantes: a typical case of a medical botanical garden. Then, in 1709 the national authority of the king transformed the Nantes apothecary garden into a government garden for the purposes of acclimating plants from overseas. Royal letters patent confirmed this status in 1719 and again in 1726. Regulations tied the Nantes botanical garden to the Jardin du Roi in Paris, Nantes becoming in effect a dependency for colonial botany. The rules required ship captains to deposit seeds and plants gathered overseas in the Nantes garden, and they did so from all over. Government botanists in turn sent specimens from Nantes to the Jardin du Roi in Paris and to gardens in Montpellier and elsewhere in the French provinces.

In 1716 the Paris Academy of Sciences and its members at the Jardin du Roi took up a project to acclimate useful plants in the West Indian colonies. The Academy elected Michel Isambert as a *correspondant* and sent him to Guadeloupe with bees, silkworms, and coffee trees.[8] The abbé Bignon and the regent himself, the Duc d'Orléans, supported the project. Isambert died on arrival in Guadeloupe in 1716, but one J.-B. Lignon continued the project. Encouraged by Bignon, Lignon sent specimens back to the Jardin du Roi; the Paris Academy elected him a *correspondant,* and the king appointed him botaniste du roi at Guadeloupe.[9] The royal gardens in Guadeloupe were upgraded in 1775 with J.-A. Barbotteau appointed superintending botanist for the Windward Islands; the

Paris Academy elected Barbotteau a *correspondant* in 1776. J.-F. de Foulquier (1744–89), intendant first at Guadeloupe and then at Martinique,
seems to have played an important role in developing the gardens under
his jurisdiction late in the eighteenth century. Foulquier became a *correspondant* of the Paris Academy and the Cercle des Philadelphes, as did
several royal physicians serving in Martinique and Guadeloupe.[10] The
Antilles began to possess the infrastructure of applied botany.

But the first significant French efforts for economic botany took
place in the Indian Ocean island colonies of Île de France and Île Bourbon. These islands, today known as Mauritius and Reunion, form part of
the Mascarene chain off of Madagascar. The French East India Company
established the first garden on Île de France in 1735. The garden, called
Pamplemousses (Grapefruits), graced the grounds of the intendant. After
1750 colonial administrators upgraded and enlarged Pamplemousses,
which then functioned as an experimental station. A second government
garden on Île de France at Réduit began in 1748. Authorities established a
third official garden on Île de France at Palma in 1775. Two other botanical gardens took shape on Île Bourbon before 1767.[11]

This expansion of French Indian Ocean botanical stations is to be
understood in conjunction with the activities of one Pierre Poivre (1719–
86), whose name, "pepper" in English, only coincidentally relates to his
heroic work to secure spices and a spice trade for the French.[12] Poivre
made several daring voyages to the east to break Dutch monopolies on
pepper, cinnamon, nutmeg, and other valuable spices. He sailed first to
Indochina in 1749; on his second voyage in 1751 he traveled to Manilla;
the third trip went to the Moluccas in 1753 or 1755. On each of these
voyages Poivre bought or stole plants and seeds and smuggled them back
to Île de France.

Poivre returned to France in 1757, where he received a royal reward
of 20,000 livres tournois for his efforts. The French government took
over direct administration of the Mascarene colonies from the East India
Company in 1767 and redoubled the efforts begun by Poivre. Poivre
himself returned in 1770 as adjunct to the intendant for botanical development. Poivre organized two more expeditions to the Dutch Moluccas,
one in 1769 and another in 1771–72, the latter returning with four hundred nutmeg and seventy clove trees. Officials distributed these plants
among the botanical gardens of the Indian Ocean for further dissemination and, presumably, local production. They also transferred Indonesian
spices to colonial botanical gardens in the Americas. The Indian Ocean
stations formed an essential institutional basis for the emerging French
network of applied botanical gardens.[13]

The French also made efforts to develop the botanical resources of their possessions in Guiana and Cayenne and thereby to tie the South American colonies into the international framework of French economic botany then taking shape. More than elsewhere the French government used the position of botaniste du roi to accomplish these ends. Guiana and Cayenne were such inhospitable places that the state could not rely as much on a nascent medical corps or on missionary naturalists to perform the pathbreaking work of science and colonialism that needed to be done.[14] Many of the men involved scientifically in Guiana and Cayenne were physicians and served as official royal physicians, but they doubled in the role of royal botanists.

Pierre Barrère (1690–1755) served as botaniste du roi in Cayenne from 1722 to 1725, an early date. The post paid 2,000 livres. Barrère received instructions for his voyage from the Paris Academy, which made him a *correspondant*. In 1741 in France he published his modest but informative *Essai sur l'histoire naturelle de la France equinoxïale*.[15] J.-F. Artur (1708–79), who trained at the Jardin du Roi in Paris, served as médecin du roi in Cayenne from 1735 to 1770 with emoluments of 1,500 livres. Artur became a *correspondant* not only of the Paris Academy but also of the Jardin du Roi itself, as one in the network of correspondents of the royal Cabinet d'Histoire Naturelle organized by Buffon; Artur seems also to have been in contact with the Royal Society of London. Stimulated by La Condamine when he passed through Cayenne on his way back from Peru in 1743, Artur studied rubber and the rubber tree in close coordination with the botanical establishment in Paris. Rubber was an exotic substance, and the study mounted by Artur has been called a classic case of pure and applied research working hand in hand.[16]

J.-B.-C. Fusée Aublet served as botaniste du roi and director of the botanical garden in Cayenne from 1762 to 1764. Fusée Aublet transferred to South America from Île de France, where he had served from 1753 to 1761 as director of the garden at Réduit.[17] He produced a massive four-volume *Histoire des Plantes de la Guyane française* with four hundred plates in 1775. Louis XVI and the Paris Academy sent L.-C.-M. Richard to Guiana in 1781 as a government botanist explicitly to pursue applied botany. J.-B. Leblond, a well-traveled physician and member of several academies, received appointment as médecin botaniste du roi and was sent to Cayenne in 1786 to investigate quinine and develop French sources for quinine. The job was deemed so important it paid 6,000 livres.[18]

Other botanists and physicians served in the equatorial jungles of French Guiana and Cayenne in the eighteenth century, but the point should be clear that the South American colonies became the objects of considerable scientific investigation and of many practical steps in the pursuit of economic botany.[19] Botanical gardens and the infrastructure of applied botany in the South American colonies formed yet another element in the emerging global system of French colonial gardens. The stations in the Indian Ocean, in France, and in Martinique and Guadeloupe have been mentioned as components of that system. For a while in the 1750s it looked as if an experimental garden in Senegal, headed by Michel Adanson, might succeed and join in the larger system. Adanson managed to publish a *Histoire naturelle du Sénégal* in 1757, but the station itself failed.[20] Authorities invigorated and expanded the system of French colonial botanical gardens in the 1770s and 1780s, with the creation of botanical gardens and experimental stations in Saint Domingue the last element to fall into place.

Louis XVI created the Jardin Royal in Port-au-Prince in 1777. The Port-au-Prince garden and other botanical gardens in Saint Domingue in large measure grew out of the French colonial experiences just mentioned in matters concerning economic botany. By the same token, the botanical gardens in Saint Domingue developed as a response to a unique set of circumstances and a separate case of state-sponsored applied research. The case concerns the importation of the cochineal insect and the beginnings of cochineal dye production in Saint Domingue. Because so much of the story of economic botany in Saint Domingue revolves around the cochineal insect, a word should be said about the bug and its significance.

Related to mealybugs, cochineal (*Dactylopius coccus*) is an insect parasite of *Opuntia* and *Nopalea* cacti (Figure 17). The dried insect, 10 percent carminic acid, acts as a powerful red dye.[21] Pre-Columbian Aztecs *domesticated* the cochineal insect on cactus plantations and used the resulting dyestuff widely in the native American cloth industry. The domesticated form became known as *fina* cochineal; a wild form known as *silvestre* was also harvested, dried, and used as a dyestuff. Pre-Columbian production of cochineal amounted to something like 10,000 pounds a year.

The Spanish took over and expanded cochineal production in their empire and introduced the dye into Europe with great success. Oaxaca, Mexico, became the major center of Spanish cochineal husbandry. Plantations numbered upward of 60,000 cacti, and annual production in the period 1760–82 regularly topped 1 million pounds.[22] The Spanish annually sent a special cochineal fleet back to Spain; one of these went down off of Louisiana in 1766 with nearly 600,000 pounds of the product aboard. By the end of the sixteenth century, cochineal already stood equal in value to precious metals as an export from the Spanish possessions in America.[23] The Spanish completely monopolized the production and first stages in the distribution of cochineal through the 1770s, and they made huge profits in reexporting the commodity to the rest of Eu-

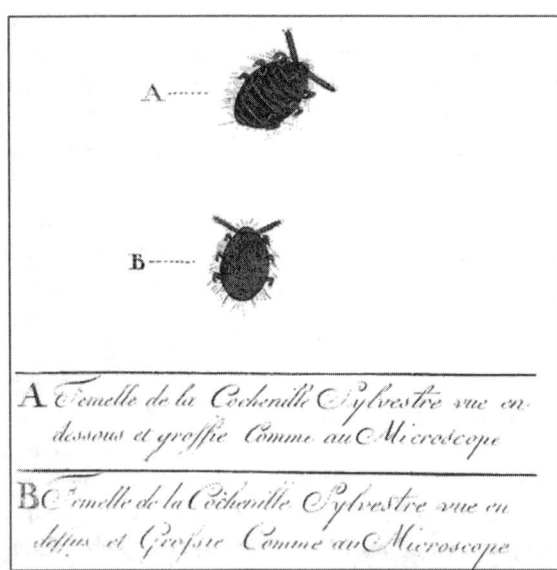

A ℱemelle de la Cochenille Sylvestre vue en
    dessous et grossie Comme au Microscope

BC ℱemelle de la Cochenille Sylvestre vue en
    dessus et Grossie Comme au Microscope

FIG. 17.
The Cochineal Insect
and Opuntia Cactus.
Hand-colored
drawings by the abbé
de la Haye in Thierry
de Menonville (1787).

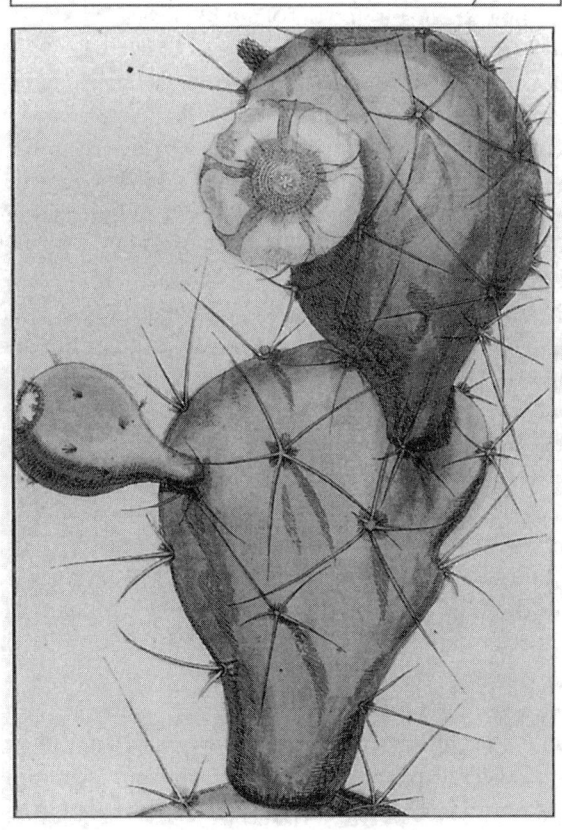

rope. The French paid the Spanish dearly for cochineal used at the royal tapestry works of the Gobelins.[24] World trade in cochineal was of non-trivial economic significance, and cochineal remained the major red dye in use in Europe until the advent of aniline dyes in the mid- to late-nineteenth century. The French had every reason to break the Spanish monopoly.

Joseph-Nicolas Thierry (or Thiery) de Menonville (1739–80), seeking to be the American counterpart of Pierre Poivre, undertook a clandestine expedition from Saint Domingue to Oaxaca in Mexico in 1777. (The Chamber of Agriculture of Cap François in 1765 first proposed the idea that cochineal production might be successfully introduced into Saint Domingue.)[25] Although trained as a lawyer, Thierry de Menonville had studied with Bernard de Jussieu at the Jardin du Roi in Paris. He established himself in Saint Domingue sometime before 1774, and thereafter, encouraged by Mme de Rozière, niece of the colony's governor, he conceived the idea of stealing cochineal from the Spanish. With the approval of the naval ministry and 6,000 livres for his expenses, Thierry set off in January 1777, as a "new argonaut" after a valuable fleece.[26]

Pretending to be an eccentric botanist and painter, Thierry de Menonville conned a passport out of a Spanish official, a former secretary to Don Ulloa. He spent six weeks at Vera Crux before he made a secret trip of over 500 miles to Oaxaca, where he bought samples of *fina* cochineal from unsuspecting Indians. In an adventure fraught with dangers, Thierry de Menonville succeeded in returning to Saint Domingue with his precious cargo in September 1777.[27]

Using funds left over from the operation, the administrators in Saint Domingue purchased a piece of property on the eastern edge of Port-au-Prince as a government botanical garden where Thierry de Menonville could cultivate his cochineal. Attached to the intendant's office, this Jardin Royal des Plantes, or simply the Jardin Royal, is to be distinguished from a later Jardin du Roi built elsewhere in Port-au-Prince. The government named Thierry de Menonville director of the Jardin Royal and botaniste du roi with appointments of 6,000 livres to run the garden.[28] (That figure equals two and one-half times the salary paid to first royal physicians in the colony!) Lacking Mexican *Opuntia*, the nopal cactus or "Spanish racquet," native to Saint Domingue, served as the host for the cochineal.[29]

Cochineal had great potential for France, including saving the "immense sums" spent on the product for the Gobelins works.[30] The cultivation of cochineal offered several advantages particular to Saint Domingue with its slave-based economy. Explicit cost analyses showed that the driest and most unproductive land in the colony could be used for cochineal. According to those analyses, French slaves were cheaper than the free labor of Mexican Indians; the weakest slaves could tend cochineal and the nopal cactus with comparative ease; in addition, the work was apparently seasonal, so slaves could be used elsewhere and still cultivate cochineal. The poorest immigrants to Saint Domingue would also find cochineal suitable for profitable cultivation.[31]

Experimental cochineal husbandry proceeded satisfactorily under Thierry de Menonville. P.-J. Macquer, professor of chemistry at the Jardin du Roi in Paris and *pensionnaire* of the Academy of Sciences there, apparently performed the first practical tests of Saint Domingue cochineal as a dyeing agent during Thierry's tenure.[32] But Thierry de Menonville died hardly two years later, in 1780. Joubert de la Motte, then serving as médecin du roi in Port-au-Prince, took over Thierry's appointment in 1781 as botaniste du roi and *médecin naturaliste du roi* especially to supervise the propagation of cochineal in the colony. The appointment brought 2,000 livres. Either the appointment came too late or Joubert was incompetent as a practical cultivator, for the Jardin Royal languished, and the *fina* cochineal secured at such great risk by Thierry de Menonville perished.[33]

Fortunately for them, colonists discovered the wild, *silvestre* form of the cochineal insect in Saint Domingue, and based on Thierry's work, Joubert and others continued efforts to develop cochineal husbandry in Saint Domingue. Joubert capitalized on having Thierry's private papers, and in 1783 he returned to France, leaving the Jardin Royal to deteriorate further under the supervision of an ex-soldier, one Lamotte. Joubert brought with him to France samples of the cochineal dyestuff and a plagiarized manuscript, "Histoire Abrégée de la cochenille de Saint-Domingue." Tests of Joubert's samples at the Gobelins factory proved highly successful. The king himself kept Joubert's memoir and encouraged his enterprise.[34]

Joubert returned to Saint Domingue with his royal stipend raised to 6,000 livres, but he returned to the enmity of those who had known Thierry de Menonville. Joubert declared Thierry's Jardin Royal no longer suitable as a botanical garden, and he began another, located on crown land behind the hospital in Port-au-Prince. Joubert's garden, called the Jardin Botanique du Roi or simply the Jardin du Roi, served as the major botanical station in Saint Domingue for the rest of the life of the colony.[35] The garden was poorly located and never entirely a success. The waterworks serving the Port-au-Prince hospital provided some irrigation for the Jardin du Roi, but the garden had stony soil and stood exposed to dry winds from the east. A partial wall, never completed, failed to protect the garden from the wind. The king's slaves worked in the garden, and the government once paid for 250 pots. Joubert himself died in 1787. One wonders whether he lived to see in print the revenge exacted on him by Arthaud and the Cercle des Philadelphes when they published Thierry de Menonville's *Traité de la Cochenille* that same year.

A private colonist, one A.-J. Brulley, initiated the major effort to sustain and promote cochineal production in Saint Domingue after Thierry. Brulley took wild cochineal from Thierry's Jardin Royal and used it to seed his own nopalry of four thousand cacti in the Marmelade parish in the north. Brulley got several harvests of cochineal, one of 5 pounds live weight. Ant attacks on the cactus hosts proved a problem, and Brulley developed ant repellents of unknown efficacy. In 1787 Brulley sent samples of his cochineal product to higher authority in the ministry of the navy. The Paris Academy of Sciences conducted tests and concluded that the samples very nearly approached the quality of Mexican *fina* cochineal. The king awarded Brulley a gratification of 3,000 livres.[36]

Brulley also provided the original cochineal stock for the botanical gardens of the Cercle des Philadelphes.[37] To anticipate, one of the earliest institutional acts of the Cercle des Philadelphes following its creation in 1784 was to inaugurate its own botanical garden and to involve itself in the cochineal effort. The Cercle likewise committed itself at an early date to publishing Thierry's *Traité*. By November 1785, the garden of the Cercle des Philadelphes had produced three harvests of cochineal. The Cercle sent samples of dyestuff from its gardens, perhaps along with Brulley's, to Paris in 1785 for tests. Swatches from these tests survive, still brilliantly scarlet after two hundred years.[38] Brulley noted that the garden of the Cercle des Philadelphes did not compare with his larger-scale, commercial undertaking, but he could hardly disagree with Arthaud when the latter wrote about the cochineal project: "The Cercle offers the colony the means not just to increase the number of its plantations, but also to open a new branch of commerce that can be as useful to the colony as to the State."[39]

The work of the Cercle des Philadelphes regarding cochineal and economic botany will be discussed further in a later chapter, but for the present its efforts and those of Brulley, Joubert, and Thierry de Menonville make clear the extent to which everyone saw cochineal husbandry as valuable for commercial development of the colony.[40] The beginnings of a cochineal industry arose in Saint Domingue just as revolution cut off the possibility. Colonial authorities had taken the necessary first steps, and in the process the institutions and infrastructures of eighteenth-century applied botany arrived in Saint Domingue.

Support for economic botany increased considerably when César-Henri, count De La Luzerne (1737–99) arrived in Saint Domingue as governor-general in April 1786. Not quite fifty at the time, La Luzerne was a nobleman and a career military officer, working his way up in the king's service. Before his appointment in Saint Domingue he served as lieutenant general of the army, and in December 1787 he ascended to the position of royal secretary of state for the navy and the colonies. La Luzerne is a major personage in this account. Not only did he do much to promote science and economic development while in the colony, he carried his ties in Saint Domingue to higher appointment in Paris, where he nurtured the Cercle des Philadelphes and ultimately became the protector of the Société Royale du Cap François.

Even before he set out for Saint Domingue La Luzerne made arrangements for a cargo of exotic plants to be sent to his future station from the government botanical stations on Île de France in the Indian Ocean. In Port-au-Prince La Luzerne daily attended to the botanical garden on the grounds of the governor's residence, the third botanical garden in Port-au-Prince. La Luzerne also introduced exotic animals into the colony. Peccaries from Cartagena and *agamis* from Cayenne were set loose on Gonave Island, and La Luzerne may have been responsible for importing sheep from the Cape of Good Hope.[41] The obvious goal in these endeavors was to develop new food sources and new branches of animal husbandry.

La Luzerne also pressed for new and more professional personnel to take charge of applied botany in Saint Domingue, and the circumstances under which this move occurred deserve to be noted. Soon after La Luzerne's arrival in Saint Domingue the naval ministry asked him and his coadministrator, Barbé de Marbois, about promoting silviculture in Saint Domingue as a resource for shipbuilding. According to ministry documents, "The Administrators respond[ed] that it would be extremely difficult, not to say impossible, to get colonists, who all desire to return to Europe, to grow trees whose only value is their wood and from which colonists could not see the fruits of their labors for a hundred years." La Luzerne and de Marbois went on to say, however, that cochineal production would take millions away from Spain. Seemingly oblivious to Joubert, they noted that no government naturalist had been appointed to Saint Domingue, and they requested that Richard be transferred from Cayenne.[42]

After Joubert de la Motte's death in 1787, the government did bring Hippolyte Nectoux from Cayenne as the new botaniste du roi for the government gardens in Port-au-Prince. After La Luzerne became minister of the navy in France, Nectoux took full charge as director of the

Port-au-Prince gardens with appointments of 4,800 livres.[43] An activist director, Nectoux took the helm in Port-au-Prince and joined in a notable series of plant exchanges with other government gardens in Cayenne, Île de France, and elsewhere. In this way Saint Domingue linked up with and completed the global system of eighteenth-century French colonial botanical gardens and contemporary French efforts at applied botany.

One shipment of exotic plants from the Indian Ocean arrived prior to the era of La Luzerne and Nectoux. That was in 1773, with plants brought aboard the *Artibonite* distributed to colonists and doubtless producing some local impact.[44] In 1786 La Luzerne and the intendant de Marbois sent the *Sincère* to Cayenne to pick up clove, cinnamon, nutmeg, and other spice trees and plants originally from Île de France. Richard sent the plants from Cayenne, but Nectoux, coming to take up his post, returned with them to Saint Domingue, after depositing a part of the cargo in Martinique. All but fifteen of four hundred plants survived on the voyage, and twenty-seven different types of plant made it to the Jardin du Roi in Port-au-Prince. The *Sincère* made another trip to Cayenne in early 1788. Richard sent clove trees and other economically useful plants back to Nectoux, who distributed them to the Cercle des Philadelphes and to private individuals for cultivation.[45] The system had begun to operate at full steam.

Cloves received special consideration as an economically valuable product capable of being introduced profitably into the French Caribbean colonies.[46] As Mozard commented in the *Affiches Américaines,* "The seedling clove trees [sent by Richard] are succeeding very well in the Port-au-Prince Jardin du Roi. One can conjecture that in a few years this precious spice will become a separate branch of colonial commerce." Mozard remarked on later occasions that a lot of money could be made in cloves and that "this branch of commerce is too interesting not to be assured the protection of the government."[47] The *Gazette de la Martinique* similarly spoke of "the great importance of clove to the State and the great riches to be had by the cultivator," and Versailles addressed a special memorandum to the administrators of the Greater and Lesser Antilles supporting clove development.[48] In Paris the great Lavoisier seems to have become involved in the chemical analysis of cloves.[49]

As a result of the arrangements made by La Luzerne before he departed France for Saint Domigue in 1786, a botanical shipment from the Indian Ocean arrived in Saint Domingue on July 15, 1788, aboard the slave ship *Alexandre.* Sent by Céré from the Île-de-France gardens and accompanied by a government botanist, Darras, the shipment included pepper plants, cinnamon trees, mango trees, mangosteen fruit, and a few breadfruit trees from Tahiti. Kept in the ship's hold, nearly all the plants died before arriving at Saint Domingue. But seventeen different types of tree and plant and sixteen different types of seed did survive, including the breadfruit tree.[50]

Colonial authorities recognized the value of breadfruit (*Artocarpus altilis*) as a food supplement especially suitable for the climate of their Caribbean possessions and for the hundreds of thousands of captive mouths that had to be fed there. Mozard wrote of the 1788 shipment from Île de France, "One must hope especially that the breadfruit succeeds; this will be of invaluable benefit to colonists."[51] Nectoux and others certainly wished the same in distributing breadfruit trees to twelve private and public stations around Saint Domingue, including the Jardin du Gouvernement in Port-au-Prince, the second garden of the Cercle des Philadelphes in Cap François, the gardens of the Charity Hospital in Cap François, and the private plantation garden of Paul Belin de Villeneuve in Limbé. Moreau de Saint-Méry reported that one could find mango, the Senegal date, Chinese mulberry, the Cape of Good Hope palm, and "the precious breadfruit" all in the gardens of the Charity Hospital.[52] The breadfruit plant arrived in good condition on Belin's plantation, and Moreau similarly reported that "everything promises that this very precious tree will have the most complete success"; Belin's description of the breadfruit tree followed in the *Affiches Américaines*.[53] The full effect of the breadfruit project remains unknown, but the instance illustrates again the mounting efforts to promote economic botany in Saint Domingue in the 1780s and the rationale behind those efforts.

Charles Arthaud was wrong in his view that the government would retreat from its commitment to intercolonial plant transfers because so many specimens had died on the shipment from Île de France in 1788. A year later, in June 1789, a second shipment from Île de France and the Indian Ocean arrived in Saint Domingue aboard the *Stanislas*. J. Martin accompanied the botanical cargo, and, although many plants again did not survive the long crossing, enough clove trees in particular did to make the voyage worthwhile.[54] No further opportunity for exchanges between Saint Domingue and the Indian Ocean occurred after 1789.

The separate global networks of botanical gardens created by the British and the French served competing colonial powers, which makes the cooperation and exchange instituted among French and British botanical gardens in the West Indies in the 1780s all the more remarkable. The French took the island of Saint Vincent from the British and held it from 1780 to 1783 during the American War. The occupying French general proved an unexpected benefactor of the large British botanical garden on Saint Vincent, and he helped to import exotic plants for Saint

Vincent from the French colonies. This spirit of fraternity continued and increased in the postwar period. The French sent cloves to Saint Vincent from Martinique in 1787, and in 1788 Nectoux made the first of two trips to Jamaica and the British colonial botanical garden there. From his first trip Nectoux returned aboard the *Gonave* with only seven plants, but with tea among them. In 1789, sailing aboard the *Arielle,* Nectoux apparently became friends with his professional counterpart, one Clarke, British crown botanist in Jamaica. This time Nectoux returned to Saint Domingue with twenty different types of tree and eleven different types of seed, many of them with potential medicinal value.[55] Recognizing the characteristic, Enlightenment spirit of the time, Mozard reported on these exchanges as follows:

> The trees and plants that the Administration recently received from Jamaica were accompanied by very friendly letters, wherein the Governor of that colony and the Director of the botanical garden promise to share everything valuable they receive. . . . The time has passed where nations try to monopolize certain of nature's riches. The flame of philosophy has dispelled the obscurantism that produces this antisocial system [of monopolies]. . . . Everything rare in the gardens in Port-au-Prince and the Cap will be speedily conveyed to our neighbors.[56]

While these exchanges proceeded on the periphery of the French colonial system, a feedback loop operated from Saint Domingue to Paris. In particular, Nectoux maintained a connection with André Thouin (1747–1824), chief gardener of the Paris Jardin du Roi and associate member in botany of the Paris Academy of Sciences. Nectoux sent seeds to Thouin, who sold duplicates to professional and amateur botanists in England, Holland, and Germany. (One does not know whether they split any profits.) Thouin and Nectoux also arranged for samples to be sent through official channels because transport by individual ships' captains proved so unreliable.[57]

The arrival of Palisot de Beauvois in Saint Domingue in 1788 strengthened the botanical connection to Paris. A naturalist-adventurer and a baron, Palisot de Beauvois (1752–1820) became a *correspondant* of the Paris Academy of Sciences in 1783, and in 1786 the government and the Academy sent him "to conduct research in natural history along the coasts of Africa and in America."[58] Palisot worked in Benin in Africa before danger and illness prompted him to seek a long-distance asylum in Saint Domingue. Once there and recovered, he resolved himself, metaphorically, to tilling the same garden Plumier had so profitably cultivated, and he joined the local colonial scientific community. He read a number of papers at the Cercle des Philadelphes, which elected him a French national associate. He botanized in the colony and became a close colleague of the local botanist, the abbé de la Haye. Palisot sent memoirs to the Paris Academy, and he and de la Haye sent at least six shipments of seeds to Thouin and the Jardin du Roi through December 1789. Palisot also made contact with A.-L. Jussieu over the publication of de la Haye's botanical manuscript, "Florindie ou histoire physico-économique des végétaux de la Torride."[59]

A major discovery with great potential for French applied botany occurred when someone found an indigenous form of Peruvian cinchona in Saint Domingue in 1785. The famous "Peruvian bark," taken from the cinchona tree (*Cinchona ledgeriana*), served as a specific against malaria and the malaria parasite in humans. Malaria was a deadly disease, and Peruvian bark (containing quinine) was the most wonderful of wonder drugs before penicillin. The secret of Peruvian bark in treating malaria spread in Europe after its introduction in 1640, but the Spanish controlled the Andean sources of the bark and profited greatly from their monopoly. For example, Louis XIV paid 2,000 *louis d'or* for the secret but not the resource in 1679. The Spanish monopoly of cinchona seems to have continued until 1859 when the English finally broke it; for a time they annually paid 53,000 English pounds for quinine for their troops in India.[60] The possibility discovered by the French in Saint Domingue of an alternative source of cinchona was one to be pursued with all vigor.

Botanists transferred the newly discovered cinchona plants to the Jardin du Roi in Port-au-Prince, where they prospered. Nectoux and Gauché prepared papers on the "Quinquina indigène," and the administrators asked Nectoux for samples to send to Paris for tests. According to a letter from Thouin to Nectoux, an unnamed but "famous" Parisian chemist performed the chemical analysis, and an "expert" physician conducted the clinical trials. Medical authorities passed the ambivalent word back that Saint Domingue quinine could replace genuine Peruvian bark in certain respects but that it proved inferior in others.[61]

The Cercle des Philadelphes became involved in an unexpected way with Saint Domingue quinine. Probably because of previous work on cochineal, the president of the Cercle in 1789, J.-B. Auvray, arranged for tests of quinine from Saint Domingue as dyeing agent. Dye masters performed these tests in Rouen, France, in cooperation with the Academy in Rouen.[62] Mozard does not report the results of these unusual tests, but they evidence again the desire to turn colonial science to every avenue with practical potential.

The French baron de Wimpffen, traveling in Saint Domingue in the period 1788–90, commented about the Jardin du Roi in Port-au-Prince that "it is as well furnished as could be reasonably expected from its infant state."[63] An inventory of the Jardin du Roi printed in March 1788 allows us to specify that state.[64] The nopalry consisted of four to five hundred cacti covered with cochineal provided by Brulley. Seventy-two different types of exotic trees and plants graced the botanical garden, including clove, cinnamon, ipecac, aloe, fig trees from various parts of the world, various types of palm tree, jasmine from the Cape of Good Hope, Madagascar indigo, Surinam mustard, green tea plants, litchi trees, Chinese rose bushes, Virginian chestnut trees, Louisiana wax trees, Cuban cedar trees, and other trees and plants from Egypt, Bengal, and India. Bamboo grew well in the nearby government garden; perhaps it did in the Jardin du Roi as well. Pecan trees from Mississippi probably grew in the Jardin du Roi, as the government had distributed them for propagation trials in 1787.[65]

In a way, the very success of the Jardin du Roi in Port-au-Prince prompted the effort to establish a fourth botanical garden in the capital, in addition to Thierry de Menonville's original but abandoned Jardin Royal, Joubert's Jardin du Roi, and the gardens on the grounds of le Gouvernement. Nectoux, La Luzerne (promoted minister), and the king himself agreed that the Jardin du Roi should be moved and a new and larger garden be created.[66] But La Luzerne's letter to Nectoux discussing this matter carries the date October 31, 1789, and by that time the events of the French Revolution began to overtake France and Saint Domingue.

The revolution ended plans for another botanical garden in Port-au-Prince. The authorities slashed the budget for the Jardin du Roi in September 1790, and by August 1791 the new botanical garden project was all but dead.[67] Some talk in France suggested that a proposal might succeed before the Legislative Assembly for extending the network of French agricultural societies and inaugurating commercial gardens and nurseries in Saint Domingue, but subsequent turmoil in France ruled that out. Still, as late as January 1792, the Provincial Assembly of Saint Domingue, itself a product of the unfolding revolution in both hemispheres, declared that a new botanical garden in Port-au-Prince was an institution necessary for the colony's welfare.[68] Given that civil war and a serious slave revolt raged in Saint Domingue at the time, this call by the Provincial Assembly for a new botanical garden indicates just how deeply useful economic botany was felt to be for colonial development.

❧ TEN ❧

Meteorology and
Popular Science

G IVEN THE HEALTHY state of meteorology and meteorological research in Europe in the 1770s and 1780s and given the maturation of
Saint Domingue in the same period, one would expect to find more than
a flourish of colonial activities to gather and process meteorological data.
One might well anticipate significant organized efforts in the colony to
observe and record the weather. In fact, however, colonists made comparatively few meteorological observations in eighteenth-century Saint
Domingue. No official or fully organized meteorological campaign ever
got off the ground, colonists reported only limited data, and reports appeared in scattered sources. The conclusion relative to meteorology in
the colony is one of surprising failure. A major reason for that failure—lack of cooperation from the Cercle des Philadelphes—is equally
surprising.

Meteorology proved a very popular science in France and Europe in
the 1770s and 1780s. Throughout Europe, perhaps a thousand individuals acting alone and for institutions regularly kept meteorological journals recording data on temperature, barometric pressure, precipitation,
and wind direction and conditions. The *Mémoires* of the Paris Academy
of Sciences, the *Philosophical Transactions* of the Royal Society of London,
and most other learned and scientific publications routinely published
weather data. Then, in 1778 the Société Royale de Médecine in Paris, as
part of its larger program of medical and epidemiological research, began
a systematic campaign to collect data and study the weather.[1] Largely
through the system of French intendants, but also in conjunction with
provincial academies of science, the Société Royale de Médecine estab-

lished a network of 150 or so meteorological correspondents in France and, to a limited extent, elsewhere. With royal support the noted French meteorologist, Father Louis Cotte, coordinated the effort and published the compiled data in the *Mémoires* of the Société Royale de Médecine. The program continued through 1793.

The Meteorological Society of Mannheim (Societas Meteorologicae Palatinae) launched an even more ambitious institutional project of coordinated meteorological observations in 1780.[2] The Mannheim Society made a point of engaging other scientific societies to join its effort, and while the number of stations reporting to Mannheim, thirty-seven, was smaller than the number of correspondents of the Société Royale de Médecine in Paris, the international scope of the Mannheim project was much greater. The Mannheim Society had observers throughout the German-speaking world, Italy, France, eastern Europe, Russia, and Scandinavia. One station operated in the Americas at Harvard College. Stations returned standardized reports, and the Society duly printed these in twelve massive volumes of now impenetrable *Ephemerides* appearing between 1781 and 1795.

Meteorology was an infant science in the eighteenth century, but everyone recognized its potential practical benefits for agriculture and the economy. Given the widespread commitment to a loose Baconianism characteristic of the period and given the technical capability, achieved only in the last decades of the eighteenth century, of producing precisely calibrated meteorological instruments, the projects of the French Royal Society of Medicine and the Meteorological Society of Mannheim would seem to follow inevitably. Meteorology also proved of larger popular and amateur appeal because taking weather data did not require a high degree of expertise, but brought with it a real scientific cachet.

Hence the surprise that European meteorology made so little impact in Saint Domingue. In 1764 a military engineer and geographer then in the colony, S. Calon, did propose a meteorological research project.[3] He offered to furnish barometers, thermometers, wind gauges, and standard tables gratis to those who would join him. "We shall examine together," he declared. Calon was possibly attached to the surveying team that perished in mapping the interior of Saint Domingue in 1764. In any event, nothing came of his project.

A few colonists kept private logbooks of weather data, and in 1767 some public discussion occurred of barometric variations.[4] But little or nothing developed by way of colonial meteorology over the next twenty years. Before he died in 1780, Thierry de Menonville sought to promote colonial meteorology by offering specimens of his Mexican cochineal in exchange for meteorological observations from correspondents.[5] But by

1784 Saint Domingue clearly lagged behind developments in Europe. In September of that year a colonist, one Baussan, wrote to Mozard and the *Affiches Américaines* about the matter.

> I am astounded that you are not receiving any colonial meteorological observations. We are living with an unpardonable carelessness in this regard. A series of these observations . . . [like] the ones published in the *Mémoires* of the Paris Academy and in the *Connaissance des Temps* . . . would be very useful.[6]

The charges raised by Baussan stung Mozard. In response, as the editor of the *Affiches Américaines,* Mozard took on the job of providing a central clearing house and immediately began to promote colonial meteorological investigations. Mozard invited Baussan and others to send him their reports for publication in the *Affiches Américaines,* and he got the intendant to provide a postal cover for the activity.[7]

Baussan followed up by sending his records of rainfall at Léogane covering the years 1761–84.[8] Baussan noted that he had seen meteorological observations from a place called "Tivoli" in Saint Domingue in the memoirs of the Paris Academy for 1781. Mozard replied that he never heard of Tivoli. That prompted Lefebvre-Deshayes to write in from Tivoli, his plantation outside Plymouth near Jérémie on the southern arm of the colony. Lefebvre-Deshayes applauded the new push for meteorological observations in the colony and said that he had been sending his observations to Paris for ten years and Baussan's for four years through J.-J. Lalande, so some private colonial logs did reach the mainstream, after all.[9] The observations of Lefebvre-Deshayes and Baussan had in turn been picked up by Cotte who published them in his own meteorological works and those of the Société Royale de Médecine. The practice continued into the later 1780s.[10]

Mozard published Baussan's, his own, and other observations, and he continued to press for their utility. For example, in 1787:

> We invite people who made meteorological observations last year to send us extracts for publication. It is the comparison of these observations that makes them interesting. . . . Agriculture, the physical sciences, medicine, and politics can make useful inductions from them. What would lovers of science not give to have a complete synopsis of meteorological observations for the colony over the last twenty-five years.[11]

The next year, on the occasion of a drought, Mozard complained:

> If meteorological observations had been made and published in Saint Domingue over the last twenty years, we would know if the

colony's temperature today is cooler or warmer, drier or wetter than it was formerly. This knowledge would be very useful to agriculturalists who work by reflection and not by routine.[12]

Along these lines, Mozard felt that local meteorological observations could be particularly useful for economic botany, especially in developing comparisons between the climates of Saint Domingue and the Indian Ocean colonies.[13]

Beginning in later 1784 meteorological observations did start to flow to Mozard and the *Affiches Américaines,* and one might say that a little meteorology got going in Saint Domingue through the agency of the colonial newspaper. Reports appeared regularly from Port-au-Prince, Cap François, Léogane, Les Cayes, and elsewhere in the colony.[14] Mozard's enthusiastic claim in 1787 that "meteorological observations are being made everywhere in the colony" exaggerated the situation, but his effort did meet with some success.[15] To guide meteorological work in Saint Domingue, in 1788 Mozard printed extracts from the work of the Italian theorist Toaldo.[16] In Port-au-Prince Mozard himself sold mercury and alcohol thermometers and blank tables for recording meteorological observations. The royal press in Cap François sold very good quality barometers.[17]

The Cercle des Philadelphes took form in August 1784, just as this flourish of colonial meteorological activity began. One would think that, as an activist organization with its own (indirect) ties to the Royal Society of Medicine in Paris, the Cercle des Philadelphes would have imitated Continental learned societies and seized the occasion to coordinate meteorological research in Saint Domingue. At least one would expect the active participation of the Cercle des Philadelphes in the meteorological effort growing up around Mozard. This expectation is so strong, in fact, that it has led one commentator to posit the existence of a missing volume of memoirs of the Cercle des Philadelphes understood to be devoted to meteorology.[18] The opposite is the case, however. The Cercle des Philadelphes expressed little interest in meteorology or meteorological research. Although a bit of meteorology occasionally did appear in its literature, the Cercle des Philadelphes did not participate in any of the meteorological programs going on in Europe or in Saint Domingue, and it prepared no volume of meteorological memoirs. Such a situation did not arise accidentally or inadvertently. Charles Arthaud and others in the Cercle des Philadelphes made a deliberate decision not to commit the institution to work in meteorology. In so doing, Arthaud and the Cercle des Philadelphes stood alone against all that was happening in Europe and Saint Domingue to question the utility of empirical studies of the weather. Their reasoning lies buried in one of several hundred questions posed by the Cercle des Philadelphes in its survey of Saint-Domingue agriculture published in 1787.

> Very precise meteorological observations have been made with very exact instruments, but the utility of these observations has not yet been sufficiently demonstrated. What is needed to extend the utility of meteorological observations is the ability to predict conditions and different revolutions in the atmosphere. Without that, one can only create a [crude] table juxtaposing weather observations with the events that seem to depend on them. [One never sees causes], one only sees outcomes. No power to control, no ability to prevent or to remedy anything results. This matter has seemed sufficiently interesting to the Cercle to be the subject of a question: The Cercle asks what can be the utility of meteorological observations for agriculture, for the health of men and animals, for the physical sciences in general, and for natural history.[19]

Mozard, a colonial associate of the Cercle and the promoter of meteorology in Port-au-Prince, clearly became aware of the attitude of the Cercle des Philadelphes in Cap François. He responded in March of 1786 in the *Affiches Américaines* under the rubric "Nouvelles Diverses":

> As for meteorological observations, it would take a long, detailed account to prove their utility, a utility some pretend to doubt. Let us say only, to justify giving space to meteorological reports, that the public papers in France and everywhere collect such observations with the greatest solicitude, that the Royal Academy of Sciences every year publishes results of observations, [and] that the Oratorian Father Cotte has been commissioned by the king to deal especially with this matter. One must believe that the learned would not waste their time in useless observations.[20]

The stance adopted by Arthaud and the Cercle des Philadelphes was obviously detrimental to the further pursuit of meteorology in Saint Domingue. Had things worked out otherwise, one can indeed imagine the Cercle operating a network of colonial stations connected to European projects. Baconian programs of widespread, coordinated, empirical observations of the weather—of the kind promoted in Europe by the Royal Society of Medicine and the Mannheim Meteorological Society and in Saint Domingue by Mozard—had to precede a deeper understanding of the forces of the atmosphere, much less how to predict or control them. Still, the Cercle des Philadelphes was the first and seemingly the only contemporary voice to point out the reality of the situation, that little of immediate practical value would result from these efforts.

That French science formed a strategic resource for the French government in colonizing Saint Domingue would seem unmistakable at this point. The initial investigations of government botanists and cartographers, the institutionalization of colonial medicine and applied botany, and the creation of a royal scientific society for Saint Domingue all support the conclusion. Even the relative failure of colonial meteorology took place primarily within the context of official institutions in Saint Domingue and in France concerned with meteorological research. Just the same, colonization involved more than strictly governmental efforts to develop Saint Domingue, and given the social impact of science in eighteenth-century France, one is not surprised to find that Saint Domingue experienced several popular fads for science and things scientific, notably in the 1780s. One of these fads was for balloons and ballooning.

Ballooning began in France in 1783. The Montgolfier brothers launched their first hot-air balloon in Annonay, France, on June 4, 1783. On August 27, J.-A.-C. Charles, professor of experimental physics, launched his hydrogen balloon in Paris. On September 19, Louis XVI and the court witnessed the release of a hot-air Montgolfier balloon at Versailles. The first human flight, by Pilâtre de Rozier and the marquis d'Arlandes, occurred aboard a Montgolfière on November 21, 1783, and on December 1, 1783, the first manned flight of a hydrogen Charlière balloon took place. For most of these early tests the Paris Academy of Sciences played a direct role in supervising test flights and in evaluating the practical potential of the new aerostats.[21]

Achieving human ascent into the air, a goal denied to Icarus, produced a tremendous impact on the popular imagination in the mid-1780s, as Charles Gillispie and others have shown. The excitement of the day's events might easily be underestimated by us two hundred years later when flight is common and when man has voyaged into space. A popular craze developed over ballooning: French peasants pitchforked alien invaders, while a more literate public became enthralled with the power of science to provoke such miracles, and all flocked to witness demonstrations of the new phenomenon.

Word of balloons quickly spread outward from Paris. The academy in Lyons launched its balloon on January 17, 1784; the "Académie de Dijon" flew on April 25, 1784; and other academies and learned societies took up ballooning later in 1784. No other French "province" took up balloons more quickly or more readily than did Saint Domingue. The first report of the Montgolfier trials in Paris appeared in the number of the *Affiches Américaines* for December 3, 1783, and another followed the next week. A longer report appeared in the number for February 4, 1784, and yet another followed on March 17. The latter article was run especially "for those who are working to build globes here," a clear indication that work on a colonial balloon was then underway. Further reports concerning Montgolfier and Charles balloons appeared in the *Affiches Américaines* later in March and into April of 1784.[22]

Efforts were indeed underway, and the first colonial aerostat took off from the Vaudreuil plantation outside of Cap François in March 31, 1784, just seven months after the original ascent in Annonay. The balloon released on the Vaudreuil plantation that day was the first lighter-than-air balloon flight in the New World, antedating the first flight in the United States by nearly nine years.[23] The colonials got their balloon off just two months after the Lyon Academy launched its balloon and a month *before* the balloon of the Dijon Academy. The first colonial balloon was small in comparison to those being built in France, only 20 feet tall and 24 feet in diameter. (The Lyon balloon, by contrast, measured 100 feet in diameter.) One Beccard, a government clerk in Fort Dauphin, built the first Saint Domingue balloon, and he remained the leading colonial promoter of ballooning while the fad lasted. The private, experimental, and unmanned launching on March 31 lasted seven and a half minutes. Mozard's remark about this "first aerostatic experiment to succeed in Saint Domingue" implies previous failures.

A second, much more public and ambitious flight took place in Saint Domingue ten days later, April 10, 1784, on the Galiffet plantation 2 leagues (5 miles) outside of Cap François.[24] Nine subscribers, most notably one Odelucq, underwrote the costs of the aerostat. Beccard again built the machine. The balloon stood 30 feet tall and 18 feet in diameter, and required 250 yards of taffeta to cover it. Four allegorical figures linked with painted garlands representing air, fire, chemistry, and physics decorated the balloon. In addition, the balloon sported the arms of the intendant, the governor, and Beccard himself, "whose friends insisted upon the honor for him." The governor of Saint Domingue at the time, Bellecombe, attended the launch. Operators put dry straw and charred wool into a brick furnace to rarefy the air, and after four and one half minutes the balloon strained for release. Cut loose at 7:40 A.M. the unmanned balloon "revolved slowly as it ascended," says Moreau, "giving spectators an opportunity to contemplate its embellishments." The balloon rose for five minutes to an altitude of 1,800 feet (300 *toises*); it remained stationary for three minutes, and it took five minutes to descend, landing 1,200 feet from its point of origin.

The balloon landed intact, and two further flights took place that day. They were visible in Cap François, and they doubtless caused a public commotion, but one knows not what to make of Moreau's remark that "black spectators did not allow themselves to cry out over the insatiable passion of man to submit nature to his power."[25] Twenty-four people signed the official report of the proceedings, including Bellecombe, Beccard, Mozard, and Arthaud's brother, Artaud. Said Mozard about the event, "We especially would have it that the most beautiful colony in the universe be the first to repeat these learned trials in the new world. . . . We are perhaps already more advanced here in aerostatics than many [French] towns with an academy, even though people in France generally believe that this country is plunged into barbarism."[26]

Shortly thereafter a third balloon flew from Cap François itself. A merchant in the town, one Benquet, let go a small Montgolfière from the Carénage neighborhood. Castonnet des Fosses claims that citizens decked out their houses in Cap François with flags and that a fancy ball followed to celebrate man's conquest of the air. The street from which the launch took place became known as the rue du Ballon.[27] Doubtless the occasion again proved spectacular, but the danger of fire presented by the event led the police to forbid further launches from town.

Balloon mania hit Les Cayes and the Caribbean coast of Saint Domingue in the later spring and summer of 1784. Letters from a colonist in remote Petit Trou, Jeanne-Eulalie Millet née Lebourg, to her sister in Nantes open a small window on events. By March 1784 Mme Millet had heard of the "elastic balloon of M. Montgolfier," and she asked her sister to keep her informed of the progress of the new machine, saying, "You see that novelty pleases women."[28] She mentioned two balloon flights in Les Cayes, one of which lasted twenty-two minutes and covered 18 leagues (over 40 miles). On September 2, 1784 she wrote of recent balloon flights in her own neighborhood.

> We have inflammable air [*sic*] here just as you do, but it does not raise as famous balloons for us, however. Three weeks ago I attended the flight of one twenty feet in diameter that raised itself up marvelously. It was made of paper decorated with painted garlands. . . . Our first balloon landed on a plantation eight leagues [twenty miles] away. The second burned five minutes after its launch. It was already very high.[29]

A wildfire movement, ballooning rapidly spread the aerostatic experience outward from Paris to all of France and its colonies. That Mme

Millet in the remote south of Saint Domingue could witness two ballooning trials by September 1784 vividly testifies to that fact. So does the
balloon flight that took place on Île de France in the Indian Ocean, also in
1784.[30] While hot-air balloons continued and continue to inspire wonder,
the popular fad did not last long in Europe or in Saint Domingue. By late
August 1784 Mozard could report that "the furor over balloons has diminished a little," and indeed only one other notice of a colonial balloon
launch appeared, that in 1788.[31] Still, in the heady days after the American
War, the new aerostats proclaimed the power of science to large audiences
in Saint Domingue as elsewhere, and it is not entirely a coincidence that
the Cercle des Philadelphes began meeting in Cap François in early August 1784 at just the time ballooning in the colony reached its height.

The contemporary world of science made itself felt in Saint Domingue
not just through such dramatic displays as ballooning. The culture of European science filtered into the general culture of the colony through a
variety of more mundane ways as well. The *Affiches Américaines,* for example, regularly printed notices of current scientific news from European sources, and, as has been seen, books, journals, libraries, and reading clubs represent other "institutions" through which colonists made
contact with the wider worlds of science and learning.

Professor Millon's public course in experimental physics provided
yet another such means. Millon first offered his *cours de physique* in Cap
François in May 1787. Supposedly to satisfy demand, he repeated the
course in March 1788, later in 1788, and again in early 1789. In May 1789
Millon took his show on the road and toured the other major towns
in the colony. Millon died shortly thereafter, but in 1791 one Primat,
"demonstrator in experimental physics and successor to the late
Mr. Millon," offered a similar course in Port-au-Prince. Millon gave his
lectures and demonstrations in his *cabinet des machines* in Cap François.
Eight sessions (held on Wednesdays, Fridays, and Saturdays) typically
made up the course, for which the fee was a hefty 66 colonial livres.[32]

Millon gave a good idea of himself and his course when he styled
himself, "Demonstrator of experimental physics, correspondent of several learned societies, and creator of diverse experiments concerning fire,
electricity, elastic fluids, the magnet, etc." In other announcements he
specified the particular topics covered in his course:

> Air properly called. Elastic fluids or aeriform matter designated by
> the names of fixed air, inflammable air, nitrous air, phlogiston (or
> pure fire free of any combination), and dephlogisticated air. Positive
> and negative electricity, medical electricity, and new experiments

on the electricity of clouds, the atmosphere, and the earth. The theory of lightning rods and especially the efficacy of lightning rods designed by the Professor. Earthquake prevention. Volcano prevention.[33]

A good part of Millon's course concerned chemistry and especially pneumatic chemistry. One can well imagine him producing the gases he names and demonstrating their properties with stands, beakers, troughs, tubes, and other apparatus of the trade. Millon's chemistry seems fairly advanced, what with his mention of dephlogisticated air, but Lavoisier's revolution either had not reached Millon by 1789 or he was not a partisan. A good part of Millon's course similarly concerned electricity and electrostatics, and one can imagine the professor demonstrating with appropriate instruments the generation of electrical charges, their storage, transmission, and other properties and effects. In this regard Millon operated within a recognized branch of contemporary science, experimental physics, and as a public demonstrator of the subject, Millon was a minor clone of the famous abbé Nollet, who pioneered the role and whose course in experimental physics was a standard fashion in Paris from the 1750s.[34]

That Millon and his course were socially acceptable is indicated by his election as a colonial associate of the Cercle des Philadelphes. Still, an element of showmanship cannot be denied to Millon. In one session, for example, the professor promised to demonstrate silk-screening with electricity and to produce a portrait of the king, and one wonders whether he tried out his nitrous air on his audience.[35] In this sense one must recognize a continuum from the respectable Millon through the more popular entertainments of magicians who wrapped themselves in the aura of science. One Pairet, for example, in 1774 demonstrated his perpetual-motion machine and "globe of the four elements," wholly new in the Americas and possessing curative powers. In 1784 the Italian "physicist and mechanician," Falconi, demonstrated his "sympathetic windmill" and "new discoveries in catoptrics" using his "little magic mirror" and "the magic lens or the incredible polemoscope." The "physicist" (*physicien*), Pinel, performed "physical experiments" with the head of "Theophrastus Paracelsus," which answered questions from the audience.[36] Distinctions were and are to be made between Millon and the magicians, but the similarities cannot be ignored.

Also notable in Millon's course was his concern with lightning rods as a practical extension of work in electricity. Benjamin Franklin, of course, had long since discovered the electrical nature of lightning, and lightning rods had become common in Europe at the time Millon began to promote them in Saint Domingue. Nevertheless, lightning rods first came to Saint Domingue through Millon and his course. A real danger in Saint Domingue, lightning killed people and damaged buildings regularly. Spurred by Millon, nine buildings in Cap François became fitted with lightning rods within a month of Millon's first offering in June 1787, with fourteen more in the process of receiving protection, including le Gouvernement, the powder magazine, and private houses.[37] Installing lightning rods must have provided the bulk of Millon's income at this time, for he charged a shocking 792 livres (12 "Portuguese") for each building he protected. Lightning rods began to be installed in Port-au-Prince in 1787 also, first by the local locksmith, Gaument, and then by Millon's successor, Primat, who charged 500 colonial livres a rod, an outrageous sum.[38] Colonial plantations also came to be equipped with lightning rods. A plantation owner on the northern plain, one Guillaudeau, became the veritable Nicola Tesla of old Saint Domingue; he constructed a tower 117 feet high and attached a lightning rod on the top to scrape the sky; Moreau de Saint-Méry later found the tower derelict.[39] Certainly unlike magic shows, lightning rods were of real practical value and another means by which science showed a genuine utility.

A handful of private *cabinets de physique* in the colony complemented Millon's. Several microscopes existed in Saint Domingue, but interest seems to have centered more on electrostatic generators and other electrical machines. Alexandre Dubourg, one of the founders of the Cercle des Philadelphes, owned a Ramsden 20-inch glass electrostatic generator, an electroscope, and a "magic table" used in electrical demonstrations. Baudry des Lozières, another founding member of the Cercle des Philadelphes, possessed the identical equipment. Charles Arthaud experimented with medical electricity and doubtless possessed some electrical apparatus. For a number of years one Thomin, head of the Chamber of Agriculture in Port-au-Prince, conducted a series of his own electrical experiments.[40]

This somewhat unexpected interest and research in electricity probably stemmed from the danger of earthquakes and the belief that electricity caused earthquakes. Citing the opinions of Buffon and the abbé Bertholon that "electricity is the principal cause of earthquakes," Mozard offered the following opinion of a colonial earthquake in 1788.

> The greater or lesser abundance of electrical matter is the probable cause. This conjecture does not seem too bold when one considers that our earthquakes are often felt at the same minute in all parts of the colony. Only the electrical fluid or light is capable of so stunning an action.[41]

However that may be, the tropics seemingly presented special electrostatic conditions, and the amateur electricians of Saint Domingue apparently experienced considerable difficulty in getting electrostatic equipment to work in a consistent fashion. The case makes evident in a small way the points explored by Steven Shapin and Simon Schaffer regarding the social construction of scientific knowledge and particularly the role of scientific communities in decoding facts of nature produced artificially by a machine. Moreau de Saint-Méry suggests the connection in his remarks about electrostatic machines in Saint Domingue:

> There is another meteorological [sic] instrument about which it is difficult to have a settled opinion in the colony. No one can deny that the air is charged with a great quantity of electricity, and especially in areas where thunder is so prevalent, the most skeptical is soon forced to admit the reality of an electrical fluid. But it is difficult to carry out experiments to test the absolute or relative intensity of this electrical fluid. Humidity is always more or less present in the air, and it hinders the proper working of the electrical machine. If the electrical machine is not operated by an experienced hand, and if the experiment is not conducted by someone highly knowledgeable of electrical theory, the results are null. I have known several motivated people who have had to give up their machines because they produced virtually nothing. The least change in the atmosphere, the nearness of someone's breath, [or] the inevitable perspiration of the operator suffices so that even the weakest sparks can no longer be produced.[42]

In the eighteenth century, as in centuries before and since, the question of the quadrature of the circle defined an interface between science and pseudoscience, and the issue of whether one can square the circle came up in Saint Domingue. A small flurry of notices about quadrature appeared in the *Affiches Américaines* in 1770. The editor, in response to a solution to the problem of quadrature offered by one Jacques Mailler, attacked Mailler and any notion of solving the problem.

> So often one hurries to announce futile and chimeric discoveries in the public papers, the weary reader is almost always disposed in advance not to believe those that could be useful and well established. So many solutions to the problem of longitude . . . so many new and useless solutions to the quadrature of the circle.[43]

The righteous dismissal of "circle-squarers" is one thing, but the reference in this passage to the problem of longitude is a surprise. (Wasn't the problem of longitude just then being solved by the French navy, which had tested chronometers in Saint Domingue?) Not to mistake the point, the editor later returned in his attacks on the underbelly of science explicitly to link the problem of longitude, the quadrature of the circle, and perpetual motion, all as fantasies that had misguided "the imaginations of geometers."[44]

In any event, solutions to the quadrature of the circle *and* to the problem of longitude continued to be proffered among the learned and semi-learned of Saint Domingue after 1770, and they make evident that the pseudosciences and concerns over the pseudosciences also showed themselves in Saint Domingue.[45] Along with Euler's announced "solution" to the (insoluble) three-body problem, this contemporary repudiation of the problem of longitude suggests that the line between the sciences and the pseudosciences is never so sharp as it later becomes.[46] A last notice from the newspaper heightens this sense of ambiguity over the social and intellectual boundaries of science. In 1777, apropos of the chemical cabinet of one of the famous Geoffroy family of apothecaries (associates of the Paris Academy of Sciences), the editor of the *Affiches Américaines* mentions three iron carriage nails that turned to silver. Since the transmutation seemed well authenticated and "since one must believe what one sees, we cite this fact as testimony of the possibility of the *philosopher's stone*."[47]

Ballooning was not the only popular craze to sweep Saint Domingue in the summer of 1784. Mesmerism and the fad for "animal magnetism" hit the colony then, too, and together the two fashionable passions drove the upper reaches of colonial society to a climax of enthusiasm for science and its potentialities. Indeed, when Mozard said that the furor over balloons had waned somewhat in late August 1784, he meant more that it had been superseded in the popular imagination by a like furor for mesmerism.

That furor erupted originally in France in 1778 when the Austrian physician, Franz Anton Mesmer, announced his discovery of a universal, superfine fluid that supported the effects of "animal magnetism."[48] Mesmer collected this etherial fluid in specially prepared tubs containing iron filings and bottles of mesmerized water. Ropes and iron bars connected to his tubs, and through them Mesmer could channel his fluid through the human body to provoke "crises" and to cure a multitude of ailments. It should be noted that mesmeric theory was not out of line with the mainstream of eighteenth-century scientific and physical theory that deployed a host of similar etherial fluids to account for a range of natural phenomena. In this respect Mesmer had the authority of Newton on his side.

Mesmer fell foul of the scientific and medical establishments in France and in Saint Domingue, however, because he insisted that his discoveries be kept secret among initiates and because he profited from an extravagant and monopolistic medical practice. After numerous run-ins between Mesmer and the Academy of Sciences, the Royal Society of Medicine, the Paris medical faculty, and the royal court itself, in 1784 two royal commissions investigated Mesmer's claims, one from the Academy of Sciences, another from the Royal Society of Medicine. With the illustrious Franklin as a member of both, these commissions wrote reports condemning Mesmer as a fraud and his fluid as nonexistent. These reports marked important steps in the official rejection and suppression of mesmerism, but at the time they appeared in later 1784 they stood in opposition to a great wave of popular excitement over Mesmer and the remarkable effects produced by his fluid. As Robert Darnton has shown, mesmerism became a cause célèbre with significant social and political implications, and the mesmerist movement persisted in various forms through the French Revolution.[49]

The mesmerist movement arrived in Saint Domingue in June 1784—at the very time the royal commissions conducted their investigations and while the mesmerist controversy peaked in France. That may not be unexpected, and one can even be surprised that news of Mesmer first arrived in the colony as late as 1784, Mesmer being in France from 1778. The unexpected lies in the spectacular manner in which mesmerism came to Saint Domingue, for when Count Anne Chasternet de Puységur arrived in the colony on June 8 aboard the *Frédéric-Guillaume* at the head of the cartographical mission described in a previous chapter, he brought with him not just word of mesmerism, he brought the thing itself. Anne Chastenet de Puységur, an ardent mesmerist who had studied with the master, was widely known, along with his elder brother, the marquis Maxime Chastenet de Puységur, for their use of "mesmeric somnambulism" (hypnotism). The two brothers mesmerized on a vast scale in the Bayonne area, and both published on the subject. The younger Count Puységur installed a mesmeric tub on his ship, and he used it to preserve the health and happiness of his crew for what turned out to be a four-month crossing to Saint Domingue in 1784. With his beautiful and equally partisan wife in tow, the thirty-two year old naval officer came to Saint Domingue as much to bring the benefits of mesmerism as to complete the cartographical duties assigned to him.[50] A dramatic event in the life of the colony already in the throes of ballooning, Puységur's visit provoked dramatic reactions.

Puységur immediately installed mesmeric tubs in the poorhouse in Cap François, and treatments began. Spurred by Puységur during the fourteen months he remained in Saint Domingue, the mesmerist movement spread rapidly throughout the colony, and a network of mesmerist practitioners sanctioned by the doctor in Paris expanded along with it.[51] By August 1784 mesmerism penetrated to the Caribbean and the southern department of Saint Domingue. In a letter dated August 25, 1784, Mme Millet, the correspondent encountered earlier who reported on balloon flights in the south, wrote to her sister about mesmerism.

> A magnetizer has been in the colony for a while now, and, following Mesmer's enlightened ideas, he causes in us effects that one feels without understanding them. We faint, we suffocate, we enter into truly dangerous frenzies that cause onlookers to worry. At the second trial of the tub a young lady, after having torn off nearly all her clothes, amorously attacked a young man on the scene. The two were so deeply intertwined that we despaired of detaching them, and she could be torn from his arms only after another dose of magnetism. You'll admit that such are ominous effects to which women should sooner not expose themselves. [Magnetism] produces a conflagration that consumes us, an excess of life that leads us to delirium. We will soon see a maltreated lover using it to his advantage.[52]

Independently, Moreau de Saint-Méry reported piously: "Magnetism had its disciples, its apostles, and consequently its miracles in the southern department. But it was also ridiculed, and it died. The miraculous was rejected by all faiths, except those that admit the Resurrection."[53] Still, at the time, Mme Millet experienced real and dangerous effects from mesmerism.

A report coming out of the central Artibonite Plain in September 1785 provides another glimpse of the Mesmer affair in Saint Domingue and some more sober reasons why mesmerism could seem attractive and worthwhile. The report in question is from the judicious Artibonite plantation owner, irrigation expert, and later member of the Cercle des Philadelphes, Jean Trembley, writing to his Swiss cousin, the famous naturalist Charles Bonnet.

> The great debates surrounding mesmerism hardly seem to be settled definitively by the very respectable report of the commissioners, academicians, doctors, and *physiciens* [in Paris] that attributes the effects only to the play of a biased imagination. Their decision does not prevent that there are still partisans. . . . Two mesmeric tubs in this colony were directed by Monseigneur, the Count de Puységur, officer of the royal navy, and by other adepts. Marvelous cures that could hardly be attributed to any play of the imagination have been

reported. A cripple brought from the plain to Cap François on a lit-
ter walked freely afterward. A female slave paralyzed for fourteen
years was entirely cured in a short time without her realizing that
she was being treated, etc. A plantation owner on this plain made a
big profit in magnetizing a consignment of cast-off slaves he bought
at a low price. Restoring them to good health by means of the tub,
he was able to lease them at prices paid for the best slaves. The rage
for magnetism has taken hold of everyone here. Mesmeric tubs are
everywhere. [But] today hardly anyone speaks of them any longer,
perhaps because of already having spoken too much about them.[54]

The enigmatic suggestion that mesmerism spread covertly among
slave owners in order to maximize profits is an astounding one. Similar
to inoculation, mesmerism exemplified just the kind of useful application
of science that plantation owners and managers might be expected to
have vigorously exploited. Mesmerism was a potential boon for the slave
system, and had mesmeric treatments really proved effective, slave owners certainly would have been allowed to treat slaves.[55] But when mesmerism passed out of the hands of whites and into the nonwhite population, one can be equally sure that colonial authority would crack down.
Creole mesmerism, so to speak, became a particular problem in the
mountainous Marmelade parish in the north, where a mulatto named
Jérome and his black assistant Télémaque promoted their brand of mesmeric and magical treatments.[56] The Conseil Supérieur of Cap François
first intervened in 1786 to curb the practice formally, and the next year it
condemned Jérome to the galleys and Télémaque to the public pillory for
continuing their activities.

Nothing has been said thus far of the resistance offered to Puységur
and to Mesmer by the medical and scientific communities in Saint Domingue. In a word, reaction was immediate and entirely negative. A long
skeptical notice about Mesmer and animal magnetism, possibly by
Arthaud, ran in the Cap François edition of the *Affiches Américaines* for
June 9, 1784, the day after Puységur arrived in town. Mozard followed
for the next few weeks with heated attacks on Puységur from Port-au-
Prince. These attacks picked up again in 1785 at the end of Puységur's
stay in the colony.[57] At one point Mozard attacked Puységur on the sticky
point of his mission:

> The apparatus he has had installed in the men's poorhouse gives him
> the means to administer magnetism to a large number of ill at once.
> But the operations with which he is charged by the minister [of the
> navy] do not allow him to consecrate but a few moments in provid-

ing solace to humanity. Perhaps he will be obliged to communicate his secret.[58]

Not surprisingly, Charles Arthaud led the medical and scientific effort to discredit Puységur and mesmerism. As first médecin du roi pro tem in Cap François and as a touchy medical practitioner whose professional views mesmerist doctrine challenged so strongly, Arthaud opposed mesmerism the moment it arrived in Cap François, and immediately in June 1784 he joined forces with two other individuals to undertake tests to disprove mesmeric claims. One of Arthaud's colleagues in these tests was Alexandre Dubourg, a local botanist and apothecary whose electrical equipment was enumerated previously. The other person was probably Arthaud's distinguished surgical counterpart, J. Cosme d'Angerville, chief royal surgeon in the northern department.[59] The three conducted their investigations in June and probably into July of 1784. In essence Arthaud and his two colleagues formed an ad hoc colonial committee on Mesmer wholly analogous to the official commissions then meeting in France. And indeed, when the reports of the French commissions reached Saint Domingue later in the fall of 1784, the scientific and medical establishment in Saint Domingue found itself completely in accord with its counterparts in France.

A close focus needs to be drawn to this committee of three that critically tested mesmerism in June and July of 1784. After a recess Arthaud, Dubourg, and the third member met again in early August, but this time for another purpose: to establish a colonial scientific society. They formed a planning committee, and on August 15, 1784, the Cercle des Philadelphes emerged with nine charter members. With ballooning, mesmerism, and everything else going on in Saint Domingue that summer, the time had come for a colonial institution like the Cercle des Philadelphes.

More will be said about various factors at play in the foundation of the Cercle des Philadelphes in Part III to follow, but no one disputes the role of Puységur and mesmerism in triggering the formation of the colony's scientific society.[60] In his manuscript history of the Cercle des Philadelphes penned in 1788, Charles Arthaud made no bones about the stimulus provided by the Puységur–Mesmer episode.

Thus we are brought to the establishment of the Cercle des Philadelphes. Doubtless a few partisans of science and letters had previously conceived the idea of this institution, but circumstances were not favorable to their aims, which were not pursued, and their plan, while well founded, was not fulfilled.

A singular event brought about the creation of the Cercle des Philadelphes. We are historians; we must be truthful. The large river whose beneficent waters irrigate the countryside and assure its fertility must not mistake its source. A tremor of the earth gave birth to this river, and weak and bent reeds seem to protect its native banks.

It was perhaps not a taste for science, a love of letters, [or] an enthusiasm for glory that led the founders of the Cercle first to unite. It was rather a contradictory curiosity, a contrary spirit, the desire to stop the progress of a mysterious and tyrannical doctrine [mesmerism], to put down a superstitious and rash enthusiasm, to tear off a veil of lies, to unmask error, and to see the triumph of truth that gave rise to the first assembly of the Cercle.[61]

Part III

The Cercle
des Philadelphes
(1784–1792)

The story of science and medicine in eighteenth-century Saint Domingue culminates in the establishment of a formal scientific society, the Cercle des Philadelphes, in 1784 (Figure 18). The creation of an official colonial institution overseas devoted explicitly to science and the practical arts marks something new and important in the history of science, medicine, and French colonization. The institutionalization of science in the colonies promised to regularize and extend the benefits of science and medicine that had already made such a difference in colonial development. The Cercle des Philadelphes has appeared many times in this narrative, and its significance as capping the history of colonial science in Saint Domingue should already be apparent.

FIG. 18. Silver coin. The *jeton* of the Cercle des Philadelphes. BN, Cabinet des Médailles.

ELEVEN

Origins: Science or
Freemasonry?

T HE CERCLE DES Philadelphes was born in the mad-for-science summer
of 1784. Peace was then firmly established after the American War of
Independence, the last spurt of growth in the colony was well under way,
and a rage for things broadly scientific then swept Saint Domingue. To recall, the year began with ballooning fever, the first colonial balloon being
launched on March 31, 1784. Mesmerism erupted onto the colonial scene
two months later in June, and later in September 1784 the effort to organize
colonial meteorological observations was launched. In light of and as a part
of all this coincident "scientific" activity, what more natural and necessary
than to organize a colonial scientific society? The inauguration of the Cercle
des Philadelphes on August 15, 1784, suggests the obvious answer.

The particular role of the Mesmer controversy in triggering the creation
of the Cercle des Philadelphes is well known and has been discussed. One
recalls that in June and July of 1784, after the arrival of the flamboyant mesmerist, Count Chastenet de Puységur, Charles Arthaud, then first royal
physician *per interim* in Cap François, formed an investigating committee
with his friend, the botanist Dubourg, and a third person (probably his surgical colleague, Cosme d'Angerville), to conduct experiments to discredit
Mesmer and Puységur. That job successfully completed, the "Mesmer committee" added six new associates and met again in early August 1784 with
the new aim of planning a scientific society. The Cercle des Philadelphes
emerged formally per se on August 15. The connection linking the Mesmer
controversy to the first assembly of the Cercle is thus close and recognized
by all who are familiar with the Cercle des Philadelphes.[1]

Were it not for a major historiographical complication, the story of the foundation of the Cercle des Philadelphes would be an entirely straightforward one, hinging on the colonial response to mesmerism as the triggering event for the creation of a colonial scientific society. Unfortunately, the small literature concerned with the Cercle des Philadelphes almost universally holds another factor also to have been of major importance in the creation of the new institution: the Masonic movement and a desire to extend Masonic institutions. The common view that the Cercle des Philadelphes emerged as part of the Masonic movement so popular in the eighteenth century is, however, wholly wrong, and it seriously distorts the deeper roots of the Cercle des Philadelphes as an institution of science. Because the false connection is made so often in the literature and because its implications are so misleading, the issue of the Cercle des Philadelphes as a Masonic association needs to be addressed directly and in detail.

In a short article about Masonic lodges in Saint Domingue published in 1927, Joannès Tramond first ascribed Masonic roots to the Cercle des Philadelphes, but the mistaken equation of the Cercle des Philadelphes with the Masonic movement owes more to an influential article first published in 1938 by Blanche Maurel. Maurel's was the first scholarly study of the Cercle des Philadelphes, and in her article she describes the Cercle des Philadelphes as a secret society evidencing the "characteristic role of Freemasonry." She claims that "the Masonic origins of the Cercle, its Masonic tendencies and activities are not to be doubted," and she concludes, "Masonic in its origins, its members and correspondents, the Cercle des Philadelphes breathed the purest air of Freemasonry. Its spirit was the spirit of Freemasonry."[2]

Maurel's misreading of the roots of the Cercle des Philadelphes is characteristic of a now antiquated bit of historiography current in the 1930s and 1940s that held the French Revolution to be the result of Masonic conspiracies. Based on earlier conspiracy theories, this view of Freemasonry, put forward most especially by Bernard Faÿ in 1932, held a notable corollary, that the learned scientific societies of eighteenth-century Europe were really Masonic organizations and perpetuators of the Masonic conspiracy. In other words, Faÿ conflated the different institutional, social, and intellectual movements giving rise to Masonic lodges and to contemporary scientific societies.[3] Maurel seems to have accepted the general interpretation from Faÿ and applied it to the Cercle des Philadelphes.

Maurel gave new expression, a wider circulation, and the semblance of historiographical currency to her 1938 view of the Cercle des Philadelphes as a Masonic organization when she republished it verbatim in 1961.[4] One way or another, the Maurel–Faÿ Masonic interpretation seems never to have been challenged, for virtually everyone who has written about the Cercle des Philadelphes has mentioned Masonic roots and rationales for the institution.[5] As recently as 1985, Pierre Pluchon forcefully restated the argument for the Cercle des Philadelphes as a Masonic organization. Pluchon, whose scholarly contributions are otherwise so notable and so relied upon in the present work, states that the Cercle des Philadelphes definitely constituted a "colonial institution emanating from Freemasonry, . . . a scientific and Masonic institution . . . taking its place in the current of Freemasonry that invaded France and the colonies after the Seven Years War." The Cercle "propagated the Masonic ideal"; the "Masonic spirit" of the institution is "transparent"; its "Masonic character . . . demonstrated." For Pluchon, as for others before him, the "Masonic cult of fraternity and earthly happiness" lay at the heart of the Cercle des Philadelphes.[6]

This view of the Cercle des Philadelphes as fundamentally a Masonic organization leads Pluchon (and others) further to attribute a significant and explicitly political agenda and role to the institution. In this view the Cercle des Philadelphes aligned itself with political forces promoting free trade, colonial independence, and a general American patriotism. The Cercle supposedly found its inspiration in Protestantism, the example of that great Mason, Benjamin Franklin, the (putatively Masonic) American Philosophical Society, and the politics of the American Revolution. As a political organization, the Masonic Cercle des Philadelphes, like Masonic lodges and scientific societies in France, purportedly had revolutionary aims and was party to political processes hastening the fall of the Old Regime in France and in Saint Domingue.[7]

This Masonic and political interpretation carries with it notions of a conspiracy at work in the foundation and activities of the Cercle des Philadelphes. Pluchon says explicitly that the Cercle des Philadelphes "camouflaged its political project in the garb of Science," and he states that "the official object of the Cercle des Philadelphes" was avowedly different from its hidden Masonic and political goals. Quoting statements of Charles Arthaud, Pluchon alludes to some "true object" and "secret plan" lurking behind the public face of the Cercle and its commitment to science—some "new order of things" coming through "revolution." Pluchon writes about "a more discreet design" to the institution and about a secret and disingenuous institutional diplomacy that guided the Cercle. Masonic "coincidences" and "machinations" are likewise invoked.[8]

All judgments concerning the Cercle des Philadelphes are seriously affected by whether (or to what degree) the common view holds true and the Cercle des Philadelphes really was a secret Masonic organization. Any appreciation of the fundamental nature and character of the institution hinges on the question, and any evaluation of the Cercle des Philadelphes, particularly as a scientific society supported by the government, would be tempered significantly if the institution were found to possess strong roots in eighteenth-century Freemasonry. At the least, that would be a highly interesting finding worthy of further exploration.

In the years since the 1930s responsible scholars have rejected the notion of a Masonic conspiracy behind the French Revolution. Fiscal crisis and the *réaction nobiliaire* have long since replaced Freemasonry as precipitating the French Revolution. In addition, Daniel Mornet, Daniel Roche, and other scholars writing on eighteenth-century scientific societies have shown Faÿ's conflation of the contemporary scientific society movement and the Masonic movement to be wholly unsound.[9] These historians see eighteenth-century learned societies not as Masonic associations, but as having roots more directly in the history of science and scientific institutions. Eighteenth-century scientific societies are now viewed as forming part of a distinct scientific-society movement stretching back to the seventeenth century and the creation of the Royal Society of London and the Royal Academy of Sciences in Paris and back to the academies of the Renaissance. Indeed, following the work of Margaret Jacob, the question these days is more of science's influence on contemporary Freemasonry rather than the other way around.[10]

Maurel and Pluchon do base their Masonic interpretations of the Cercle on some contemporary evidence, and that evidence will be dealt with later in this chapter, but by and large there is little in the contemporary record to support the notion of a Masonic charter to the Cercle des Philadelphes. Indeed, the Masonic interpretation obscures a superior alternative interpretation that would take the underlying purpose of the Cercle des Philadelphes simply to be a colonial academy of science, like the many provincial and national academies of science elsewhere, with the real aims of the institution being not secret but straightforwardly civic and scientific. This non-Masonic view sees the Cercle des Philadelphes as an outgrowth of the learned-society movement that developed so strongly everywhere in the eighteenth century, but nowhere more than in France, with more than two dozen officially sanctioned learned and scientific societies in the provinces, in addition to the major academies in Paris.[11] The Cercle des Philadelphes becomes, in this view, the tropical equivalent of a French provincial academy. The Cercle served the public good and utilitarian ends, and it worked with the financial and legal support of French colonial and national governments. Indeed, given the thesis here of government as the main backer of science and medicine in the colonies, one would expect to find a secular and scientific organization like the Cercle des Philadelphes rather than a Masonic one. The

Cercle des Philadelphes must be taken at its institutional word strictly as a scientific society, rather than as a political and moral conspiracy perpetrated by Masons. Furthermore, the evidence reveals that in some respects the founders launched the Cercle des Philadelphes in opposition to contemporary Freemasonry and the state of Masonic lodges in Saint Domingue.

Part of the evidence against seeing the Cercle des Philadelphes as a Masonic organization stems from the fact that by 1784 Freemasonry and Masonic lodges had penetrated Saint Domingue fairly deeply. As noted earlier, a total of twenty Masonic lodges and twice that many chapters seemingly saturated Saint Domingue before the revolution. While the number of lodges may have continued to grow elsewhere in the colony, in Cap François Freemasonry declined in the 1780s. In the 1770s Cap François supported two, vigorous lodges, Verité and Saint-Jean-de-Jérusalem Écossaise: a common meeting in 1777 brought together 106 Masons. Thereafter, however, membership in the Verité lodge dropped from 56 members in 1777 to 26 in 1780, and by the early 1780s, the Saint-Jean-de-Jérusalem lodge collapsed entirely and folded into the Verité lodge.[12] It seems contradictory to think that under these deteriorating circumstances another, highly successful Masonic organization would suddenly appear in Cap François in the guise of the Cercle des Philadelphes.

Furthermore, a thousand or so Masons saturated the colony in the 1780s, perhaps one out of three eligible white men, as already suggested.[13] In contrast, consider the number of Philadelphes. A total of only thirty-one of the Cercle's members resided in Cap François, and the total number of associates elsewhere in the colony did not reach seventy throughout the life of the institution. Allowing for Freemasonry as part of the common culture of later eighteenth-century France and Saint Domingue, the significant question in this connection, therefore, is not whether some or even the majority of the Cercle des Philadelphes were Masons, but why any of the many Masons were also Philadelphes.

Many Philadelphes were Masons, of course, and social links forged in one area no doubt reinforced themselves in others.[14] But those relationships become complex. Consider, for example, the trio of Moreau de Saint-Méry, Baudry des Lozières, and Charles Arthaud. Moreau and Baudry were certainly committed Masons. One of the Cercle's most prominent members, Moreau directed the Saint-Jean-de-Jérusalem lodge in Cap François for four years while in the colony in the 1770s, and he continued Masonic activity later back in France.[15] Baudry des Lozières had a less successful career as a military officer and lawyer, but he was a founding member of the Cercle des Philadelphes and an active participant in the life of the institution. At one point Baudry established his own Masonic lodge in the south of the colony, which he christened the lodge of the Philadelphes.[16] Charles Arthaud, the "father and animator" of the Cercle des Philadelphes, was also a Mason of unknown fervor.[17] But Moreau, Baudry, and Arthaud connected not only as Masons and as Philadelphes, but also as brothers-in-law, having married the three daughters of one Mme Milhet, widow of a rich Louisiana merchant.[18] A complex web of relations, indeed, and to make the story of the Cercle des Philadelphes merely an extension of colonial Freemasonry is to oversimplify at best.

Part of the evidence in favor of seeing the Cercle des Philadelphes unequivocally as a scientific society stems from the fact that at least two other proposals were made prior to 1784 to create non-Masonic learned societies in Saint Domingue. Both proposals failed, but they signal the Cercle's true institutional antecedents and reveal that Saint Domingue, too, felt the effects of the international learned-society movement before the successful creation of the Cercle des Philadelphes in 1784.

The first proposal dates from 1769 and was made by a plantation owner outside Cap François, one G. Lerond. Lerond's original proposal has disappeared, but to infer from the reaction it provoked, Lerond initially suggested a colonial academy of belles letters. Lerond's proposal precipitated an extraordinary exchange in the then new *Affiches Américaines,* an exchange that throws a clear light on the precise circumstances regarding organized learning in Saint Domingue in the year 1769.[19] In letters to the *Affiches Américaines,* an otherwise unidentified colonist, one Delile, responded to Lerond's notion of a colonial literary academy:

> I have read, Sir, with real pleasure M. Lerond's [circular?] letter. The points of view he expresses on each line announce a true Patriot and a friend of the Fine Arts. M. Lerond is a learned man and a beautiful soul. However, I will say that he seemed to have no other aim in mind than to awaken drowsy minds and to replace their lethargy with the taste for literature. However else does he imagine to be able to lead us to an Academy? While the establishment of an Academy would be infinitely glorious to the Colony, Lerond has neglected to consider the obstacles opposed to the execution of his project. . . . Many intelligent people might still be found in Saint Domingue, but I repeat that they will justly be too jealous of their time to attend literary conferences. I say literary, because our Chamber of Agriculture is very far from having the character of an academy. . . . However, says M. Lerond, La Rochelle, Angers, and

Montauban certainly all have academies. That's true, but they also
have colleges and universities. Let such schools be started in Saint
Domingue, and I'll shut my mouth. . . . The love of Letters reigns
in France. In Saint Domingue it's Commerce and Agriculture. Because people don't come here except to make their fortunes quickly,
and because they find a negotiable check more useful than the Moral
Tales of Marmontel, we'll have to await the day when a project for
an Academy is anything more than a beautiful dream.[20]

Delile's attack prompted a vigorous defense by Lerond.

I can state with certainty that there are many people here who are
worthy of forming an academical body, who are blessed with crea-
tivity, talent and good fortune, and who are not as anxious to fly
back to their home fires as M. Delile thinks. . . . Fifty years ago he
might have been right. But today the same colony has arrived at a
point of splendor that entirely annihilates M. Delile's rhetorical
apostrophe. All fashions are found in the colony today: plays, concerts, libraries, sumptuous parties where gaiety and wit oppose
irksome boredom. What an elegant reform for women! Pirates have
given way to dandies with embroidered velvet jackets, and fancy
dressing is so common it has passed to colored women[!]

A love of learning is joined to this love of luxury. Those who
previously couldn't read or write are today poets, orators, and scientists [physiciens]. The printing press, that useful institution and
source of national pride, crowns all this luster, and from there come
the public papers, factums, and memoirs. Our literary amateurs
worked, and a journal at last appeared. What did we read there? Dissertations on the natural history of the country, bits of the physical
sciences, analyses, reflections on Agriculture and Commerce, and
different genres of poetical production.

It's not far from this progression to an Academy. . . . The
Colony is capable of something more than keeping its accounts and
extending the cult of Sugar.[21]

The journal to which Lerond refers may have been the *Iris Améri-
caine,* a short-lived and exclusively literary journal appearing in either
1767 or 1769.[22] Lerond may also have been referring to the *Journal de
Saint-Domingue,* a periodical offspring of the then active Chamber of Agriculture that appeared for fifteen months between November 1765 and
January 1767 and contained literary and poetic works. It is noteworthy
both that this *Journal de Saint-Domingue* failed and that the work by "literary amateurs" so vaunted by Lerond appearing in the journal dealt pri-

marily with the natural sciences and utilitarian topics.[23] Perhaps sources of support existed for an academy, not of literature, but of science and agriculture.

A surprising letter followed in the *Affiches Américaines* in response to the Lerond proposal of 1769, supposedly from a slave, Toussaint, Lerond's maître-d'hôtel. The slave identity may have been faked as part of the ongoing exchange in the newspaper. Yet this Toussaint could have been real, and a remote possibility exists that he was even the Haitian hero, Toussaint Louverture.[24] If this letter is genuine, the voice of this enslaved Toussaint, albeit literate, is the only one of his class to speak directly in this story.

> I see in several of your past numbers a frontal attack launched against the proposition of my dear Master, whose plans are entirely patriotic, like those of the Abbé de St. Pierre. . . .[25]
>
> The *Journal de Saint-Domingue* was a useful and, in my view, essential work that could have become even more so, but it only lived for fifteen months. Many today deplore its fall, but few subscribed then. . . . The Chambers of Agriculture and Commerce were unable to maintain themselves, and what have they produced since being reconstituted as purely a Chamber of Agriculture? Nothing, or really very little, despite an enlightened membership. . . .[26]
>
> My Master believes that an Academy can be established in Saint Domingue, and he is right. M. Delile believes that no Academy can exist in Saint Domingue, and he is right. . . . We shall not want for Academicians, that is to say, talented people fit for literature, but we will never have an Academy. The spirit of emulation in Saint Domingue as a whole is not turned toward the Fine Arts, even though they are appreciated and cultivated by individuals. The position and the talents of the plantation owner from the [fashionable] Quartier Morin are more envied and aspired after than beautiful verses or the genius of the author of the *Iliad* and the *Odyssey*. The latter would starve and only be a fool in Saint Domingue, whereas the former . . . [sic], but I mustn't be mean, and I shut up for fear of being treated as a slave.[27]

Everyone seems to have recognized that in 1769 the colony possessed no basis for formally organizing literature, poetry, and the fine arts; if anything emerged out of this exchange over a literary academy, however, it was the notion that possibilities existed for organizing science and technology in the colony. The idea seems seconded by Delile himself, the man who initially objected to Lerond's proposal for an academy, for later in November of 1769, Delile inserted the following notice in the paper:

M. Delile proposes shortly to publish an Academical Discourse: The Sciences and the [mechanical] Arts, vast and interesting areas, are its subject. He offers subscriptions to Amateurs for the costs of printing at the Royal Presses at the Cap and in Port-au-Prince. He offers his work gratis to the public, hoping only for its approval.[28]

Maybe Delile was more right in his first opinion, after all, for nothing came of his proposed journal of academical discourses.

Much less is known about the project of 1776 to create an academy of medicine and allied sciences in Saint Domingue. But Charles Arthaud was the man behind the idea, a fact that makes his later association with the Cercle des Philadelphes even more natural and expected. Arthaud, a junior colonial physician in 1776, proposed an academy of medicine, surgery, and natural history, something like a colonial combination of the French Royal Society of Medicine, the Academy of Surgery, and, yes, the Academy of Sciences back in Paris. Arthaud also thought that his "academical establishment" should exercise regulatory and medical police functions, and his idea for a medical academy formed part of the ongoing struggle with the navy ministry over control of colonial medicine that was reaching a high point in Saint Domingue in 1776.[29]

Nothing came of Arthaud's scheme of 1776 for a medical academy, no doubt because war intervened and for lack of resources, but Moreau de Saint-Méry later explicitly linked the Cercle des Philadelphes to the medical controversies of the latter 1770s and perhaps to Arthaud's proposal:

Fortunately, the Society of Sciences and Arts of Cap François [the Cercle des Philadelphes] is suitable to undertake work that M. de l'Étang began [in 1777] and that demands the cooperation of several observers. An ample harvest awaits the medical doctor, the chemist, the *physicien,* the botanist, and the veterinarian in this country where almost everything remains to be done scientifically and where nature offers such rich resources. A society of industrious men, mutually encouraging each other, is naturally the point where research, observations, and results should be communicated and examined, with the desire to make of them some applications useful to all humanity.[30]

Moreau de Saint-Méry here depicts the Cercle des Philadelphes as a typical eighteenth-century scientific society with explicitly scientific interests and "patriotic" goals. The point need hardly be repeated that such an institution should not be confused with contemporary Masonic lodges or the Masonic movement. In another spot, Moreau states that the idea of a learned reunion of colonists had been "discussed a thousand times."[31] The straightforward story is that the Cercle des Philadelphes of 1784 grew out of the general learned- and scientific-society movement and the previous proposals of 1769 and 1776.

The eighteenth century was the golden age of learned and scientific societies, and the creation of the Cercle des Philadelphes and its elaboration into the Royal Society of Sciences and Arts of Cap François must be seen in a worldwide context and as part of an international institutional movement giving rise to such learned societies.[32] In this connection the proposal of 1771 to found a colonial scientific society on Île Bourbon (Reunion) in the Indian Ocean is certainly noteworthy, for it evidences the global scale on which the scientific-society movement developed in the eighteenth century, and it shows that Saint Domingue was not the only French colony to feel the impact of the European scientific-society movement. Seemingly nothing came of the proposal of 1771 for the Île Bourbon. Another proposal for an Indian Ocean science academy surfaced in 1786, this time for the Île de France (Mauritius), but this proposal took its inspiration from the Antilles and the Cercle des Philadelphes, which by 1786 had become so active and well established in Saint Domingue.[33]

Elsewhere in the Caribbean, organized science and the scientific-society movement likewise made themselves felt. Brooke Hindle points to what he calls "significant organizing activity in science" in the contemporary West Indies.[34] In this connection Hindle signals the establishment of British botanical gardens in Jamaica and St. Vincent. The equivalent French botanical gardens in Cayenne, Martinique, Guadeloupe, and Saint Domingue have been previously discussed here, and their existence strengthens Hindle's point. Hindle also mentions the Barbados Society for the Encouragement of Arts, Manufactures, and Commerce extant in the Barbados from about 1777 to about 1787. Modeled after the Royal Society of Arts, the noted "economic" and "technological" society in London, the Barbadian arts society makes evident that eighteenth-century learned societies institutionalized cultural activities of all sorts and that the learned-society movement had penetrated the outer confines of European civilization by the end of the century.

Hindle claims, however, that "no philosophical societies were established" in the West Indies. He may be technically correct in the narrow sense that no "society"-type learned societies like the Royal Society of London appeared there in the eighteenth century. But if Hindle means that no scientific society of any type arose in the Caribbean in the eighteenth century, he is simply and sorely mistaken.[35] The Cercle des Philadelphes—and the Royal Society of Cap François it became—rather self-evidently represents the crowning glory of organized science and the contemporary scientific-society movement, certainly in the Caribbean, if not in the whole of the Americas. In the end, the Cercle des Philadelphes had little to do with eighteenth-century Freemasonry but, on the contrary, everything to do with eighteenth-century science and the almost universal movement to incorporate scientific societies.

In a speech before the Cercle des Philadelphes sometime after its creation, Charles Arthaud placed his West Indian institution squarely in the tradition of eighteenth-century learned societies just mentioned. After indicating their historical roots in Plato's Academy, the Museum and Library at Alexandria, and learned circles around Charlemagne, Arthaud presents a standard and more or less accurate picture of contemporary academies and societies:

> The taste for academies developed in the Renaissance. Many towns in Italy created academies in their midst with fairly bizarre names. The Accademia della Crusca in Florence, the Platonic academy in the same town, the Accademia del Cimento, and the Institute of Bologna stand out.
>
> Letters patent in 1660 established the Royal Society of London, under Cromwell's administration [sic]. The government gave it no funds and it works for the honor of working. The splendid discoveries it has produced have almost given England the same superiority in the learned world it had so long in the political world.
>
> France's academies are famous today. Francis I loved the sciences and the arts even more than Charlemagne did, but it was reserved to the glorious reign of Louis XIV to create these institutions.
>
> Many of the provinces envied the capital the advantages it enjoyed. In several towns persons who cultivated the sciences and letters united and formed societies which by royal favor thereafter received permanent constitutions as academies. These institutions not only polish a nation, but add to its glory. They contribute to the spread of useful knowledge, and in molding wise men, they furnish useful examples of good morals and virtue.
>
> The usefulness of academies has been felt by all governments in Europe. Russia, which a century ago hardly had a political existence, [now] has a famous academy [the Imperial Academy of Sciences of Saint Petersburg]. The development of the sciences and the arts destroyed the barbarism of that savage country.
>
> Can an academical society be useful in Saint Domingue? A thousand ideas present themselves . . . and who will dare ask, after these remarks, if our circle can be of use.[36]

Maurel and Pluchon base their Masonic interpretation of the Cercle des Philadelphes primarily on remarks made by Charles Arthaud in his manuscript history of the Cercle des Philadelphes penned in 1788. For example, both Pluchon and Maurel read a lot into Arthaud's sentence that picks up from his remarks about mesmerism: "After having employed the arms of Reason against animal magnetism, the founders of the Cercle turned to the true object that was to unite them, and from that moment they were occupied with their plan alone and with the means that could assure its execution."[37]

The question is, of course, what did Arthaud mean by the Cercle's "true object" and its "plan"? The answer for Pluchon and Maurel, again, is Freemasonry and a revolutionary political program. Enough has been said already to suggest another interpretation that would see the "true object" and "plan" of the Cercle des Philadelphes as the promotion of science and public utility through the creation of a colonial scientific society. Arthaud, for example, in the several pages leading up to this key sentence, speaks at length about the "benefits the sciences and arts" can bring society. In particular, he mentions the scientific study of agriculture, manufactures, medicine, *physique,* and natural history, and he concludes, "This is what led to the establishment of the Cercle des Philadelphes."[38]

However that may be, Pluchon and Maurel similarly both make a great deal over a prayer and pledge of allegiance that Arthaud reports occurring at the first official meeting of the Cercle des Philadelphes on August 15, 1784. The passage from Arthaud's manuscript history of the Cercle is worth quoting in its entirety.

> The meeting opened with an invocation to the Supreme Being. The works of man are for naught once he forgets that his reason is a gift by which God distinguished him from all the other creatures emanating from His Omnipotence and Supreme Goodness:
>
> Invocation
> Principle of everything, in whose breast we are but atoms, illuminate our intelligence with another ray of your supreme intelligence. Enlighten and support our zeal and our small works and deeds. If we are unable to understand you, at least allow us to revere you and give homage to your power.
>
> After expressing this religious sentiment, we thought the following oath should be made out of respect for the civil order and to impress on ourselves the [social] obligations we contract.
>
> Oath
> We engage ourselves to undertake nothing contrary to the laws of the kingdom, to conform to the laws of society, and to contribute

as much as is in our power to strengthen the bonds that ought to unite us.[39]

One can ask why Arthaud felt it seemly to record this invocation and oath in his history of the Cercle des Philadelphes written four years after the event, but even if genuine, when presented in its entirety, the passage seems pretty tame and not particularly Masonic or conspiratorial. The deistic appeal to the Supreme Being is of obvious note, but unexceptional for the 1780s, as is the rather mundane social contract theory of the state expressed in the passage. But there is no Masonic mumbo jumbo here, and the oath of allegiance bespeaks anything but a political conspiracy of revolutionaries. Despite Pluchon's suggestion, furthermore, no evidence suggests that this invocation and oath were required of new members or ever repeated after the Cercle's initial meeting on August 15, 1784.[40]

The evidence discussed thus far and other evidence to follow make an overwhelming case that the founders of the Cercle des Philadelphes sought to create a typical (colonial/provincial) learned society devoted to science, medicine, and regional development. By the same token, admitting that point, one should not be blind to the fact that the Masonic movement made itself felt strongly in Saint Domingue and that aspects of contemporary Freemasonry did carry over into the Cercle des Philadelphes. In particular, the name Philadelphes carries strong (Masonic) overtones of fraternity and universal brotherhood. Similarly, the connection in the name to Philadelphia in Pennsylvania and to Benjamin Franklin, with all the political and Masonic overtones implied therein, was also real and made deliberately. Arthaud, speaking about the name in the spring of 1785, might intimate Masonic and political associations in the choice of the name, Philadelphes.

> The Academy of Philadelphes at Venice is known to you. You are also familiar with the famous town on the banks of the Delaware that the virtuous Penn destined to become the capital of a grand empire. This town is the home of the illustrious Franklin. You will doubtless applaud us, Sirs, for coming together under the invocation of the genius of this great man, adopting the name of his homeland, and taking a name that signifies the sentiment that must unite us.[41]

A Masonic conspiracy could be woven out of Arthaud's dissimulating reference to some "Academy of Philadelphes" in Venice, the notion of Philadelphia as the capital of a great empire, or the encomium to Franklin. The newly independent United States attracted great attention and admiration, and the commitment to brotherly love on the part of the

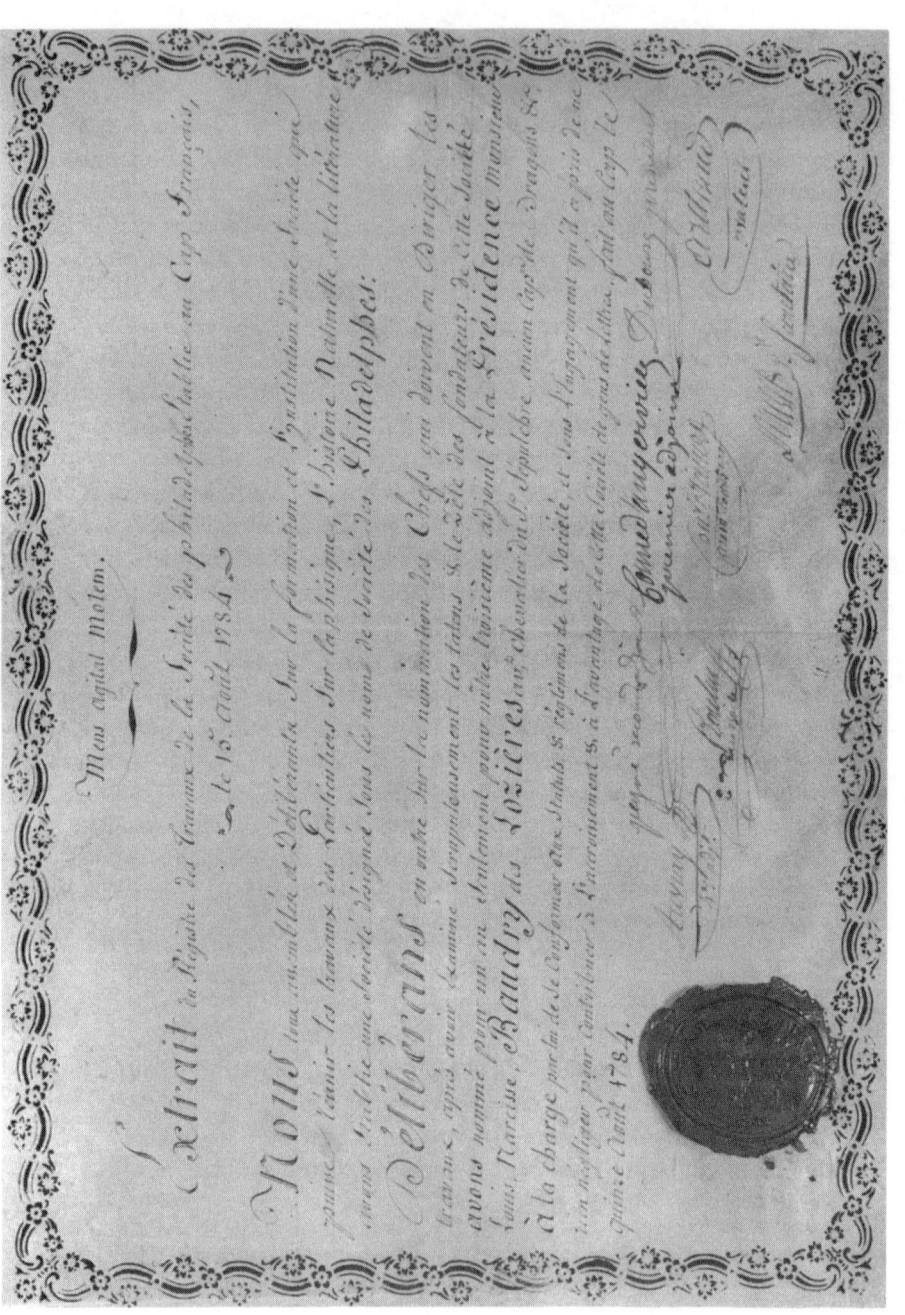

FIG. 19. Cercle des Philadelphes, membership certificate. APS, Manuscript Collections.

Cercle's founders seems to have been entirely genuine. No doubt these Masonic and political currents formed part of the flow of events, but they make the Philadelphes of Saint Domingue no less primarily a group interested in forming a colonial scientific society. For Franklin was also president of the American Philosophical Society and the leading man of science and learning in the Americas, and in placing themselves within the orb of Franklin's Philadelphia, the founders of the Cercle des Philadelphes in colonial Saint Domingue likewise paid homage to Franklin the man of science and a central figure in the frame of internationally organized science at the end of the eighteenth century. Certainly, Arthaud later engineered the successful approaches to Franklin and the American Philosophical Society on the basis of straightforward (and not conspiratorial) institutional and scientific commitments.

Masonic elements, insofar as factors at all, seem most influential at the Cercle's very outset. Indeed, several manuscript sources, drawn from the papers of Baudry des Lozières in the collections of the American Philosophical Society, suggest the point that before the founders mastered the details of organizing a scientific society, they initially imitated some forms and procedures they knew best as Masons. These sources shed a very particular light on the precise circumstances of the Cercle des Philadelphes at its very outset, and they permit a definitive answer to the question of whether the Cercle des Philadelphes was founded as an institution of Freemasonry. The first document is the parchment certificate of membership reproduced in Figure 19.[42]

The document is revealing in several ways. It carries the date, August 15, 1784, the very day of foundation of the Cercle des Philadelphes. Vouched for by the signatures, the authenticity of this document as a charter document of the Cercle des Philadelphes is unimpeachable.

The certificate is labeled, "Extract from the Register of the Société des Philadelphes Established at Cap Français." The use of the name "Société des Philadelphes" rather than "Cercle des Philadelphes" is curious, but on August 15, 1784 the founders had in mind the name "Société" des Philadelphes and not (the more Masonic) "Cercle" des Philadelphes. When they met again the next week (August 22), they changed the name from Société to Cercle des Philadelphes.[43] In a public speech the following spring Charles Arthaud explained why the original name was changed.

> We could have chosen another name. Literary Society or Academy would have flattered our vanity but alarmed our modesty. We examined our forces; we measured our zeal. We recognized that we are less Scientists than Lovers of Science. We thought that, in a country that so devours its resources, we should not pretend to equate

ourselves with Companies supported in every way to keep up emulation, to develop talent, and expeditiously to produce useful discoveries.[44]

Other curious elements of this certificate suggest an impromptu and tangentially Masonic beginning for the Cercle. For example, the motto that heads the document ("Mens agitat Molem"—"Minds Move Mountains") is not the motto the Cercle ultimately adopted ("Exercet sub sole labor"—"We Do Our Work beneath the Sun"), and it is never used again. Similarly, the wax seal with the first motto affixed to the certificate is nothing like the official seal of the Cercle and is also never used again.[45] Were this seal and motto simply handy? Something needed on the spot to make these documents official? Borrowed, perhaps, from a Mason?

However that may be, the document itself attests to the appointment of Baudry des Lozières as third adjunct to this society of Philadelphes, and it is signed by the eight other founding members of the Cercle, with each signer further indicating the position he held within the nascent organization. The signatures and the certificate thus make evident what happened: as each of the nine founders was also an officeholder, each founder received a similar certificate. Nine certificates were thus prepared and mutually signed by the other charter members in one (sacred?) act of institutional foundation.[46]

The offices the founders assigned themselves on the surviving Baudry certificate reveal an unusual, more Masonic initial organization to the Cercle des Philadelphes. (Table 3 lists those offices with other information.) The membership categories are not typical of an eighteenth-century scientific society. The position of orator (reader or speaker), in particular, would seem a carry-over from Masonic posts, as may be that of commissioner. Then, the position of adjunct is really the more egalitarian "adjunct to the presidency," as the certificate states, and not adjunct as in a typical academy. This egalitarian style of membership likewise characterized Masonic lodges more than learned societies.

The Cercle's founders initially decided upon these positions based on a proposal made by Baudry des Lozières on or just before August 15, 1784. A strong Mason himself, Baudry probably derived the offices from what he knew of Masonic practice.[47] From his descriptions these positions could have roots in contemporary Baconianism, but they were more likely carry-overs from Freemasonry. But at the very outset the founders were not masters of the standard organization and procedures of a scientific society, and to that extent they may have begun by imitating the formalities (rather than the substance) of Freemasonry.

Regardless, the initial organizational form of the Cercle des Phila-

TABLE 3. The Original Société des Philadelphes, August 15, 1784

Position Held	Founder's Name, Age, and Occupation
President	Alexandre Dubourg, 37, actor, merchant and amateur botanist
Orator	Charles Arthaud, 36, M.D., first royal physician (interim)
Secretary	Charles Poulet, merchant and shipowner, possibly a lawyer
1st Adjunct	Jean Cosme d'Angerville, master surgeon, chief royal surgeon
2nd Adjunct	Joseph-Benoit Peyré, private M.D., royal physician
3rd Adjunct	Louis-Narcisse Baudry des Lozières, 32, lawyer and military officer
4th Adjunct	Jean-Baptiste Auvray, large landowner and former merchant
1st Commissioner	Marc Couré, the younger; master surgeon, former surgeon in militia, later royal surgeon
2nd Commissioner	Barthélémy Roulin, surgeon, surgeon's-aide major, later royal surgeon major

delphes lasted for just over two months. When the Philadelphes put into effect a set of formal statutes in early November 1784, the organization assumed a very different character, one entirely typical of contemporary learned societies.[48] In November only the positions of president and secretary remained from the Cercle's organization in August. Otherwise, the categories of membership and office that the Cercle des Philadelphes came to incorporate (resident associate, colonial associate, honorary associate, etc.) were more or less standard for an eighteenth-century learned society. The Cercle quickly dropped any initial traces of Freemasonry, and its public character as a civic and scientific institution quickly settled into place.

Despite the several strange aspects of the Baudry membership certificate of August 15, 1784, therefore, the Masonic interpretation ultimately finds little support in it, for the document plainly states, "We have assembled and deliberated on the formation and institution of a Society that can bring together the work of individuals concerning *physique,* natural history, and literature." The reference to literature surprises a bit, and the language suggests that at its very outset the Cercle or proto–Cercle des Philadelphes was not so entirely directed to science, medicine, agriculture, and the practical arts as it subsequently became. But nothing of what is said about literature or the sciences in this certificate suggests anything further about Freemasonry.

We know the names and a bit about the founders, so it is possible to reconstruct somewhat the sociology and group dynamics of the original Cercle des Philadelphes. Before continuing the archival trail, we pause briefly to discuss the Cercle's founders.[49]

The first thing to remark: five of the nine original founders were physicians or surgeons. Indeed, they represent the official medical establishment in Cap François. In 1784 Charles Arthaud was interim first royal physician, and in early 1785 he used the launching of the Cercle des Philadelphes to secure permanent appointment as the chief government medical official in the colony.[50] Cosme d'Angerville, a master surgeon, held the post of chief royal surgeon for the northern department; he was Arthaud's official counterpart and coadministrator of matters surgical.[51] Previously a royal (staff) physician at the Môle outpost, Peyré had moved to Cap François to pursue private medical practice. Formerly a surgeon major in the militia, in 1784 Couré practiced as a private master surgeon, and he later became first royal surgeon in Fort-Dauphin. Roulin was also a surgeon, serving in 1784 as military surgeon's aide attached to the town of Cap François; he later became the royal surgeon major at the Cap.

The founders thus counted mostly medical men, and in the end medical personnel composed the largest single occupational category of all of the Cercle's members. In many respects, although not entirely, the Cercle and the Société Royale des Sciences functioned as a colonial academy of medicine. The medical—and not, once again, the Masonic—constituted more the central identity of the Cercle des Philadelphes.

The position of Charles Arthaud as founder and leader of the Cercle des Philadelphes is not entirely clear from the organization that came into being in August 1784. Possibly for reasons of deference, Alexandre Dubourg was named the first president of the Cercle. But the position of president did not suit Dubourg, and in the reorganization of the Cercle in November 1784, Arthaud took over that office.[52] Later (in 1786) Arthaud assumed the post of permanent secretary, the usual and natural position of institutional leadership in eighteenth-century scientific societies. Arthaud's role as administrative director no doubt reinforced the medical cast and orientation of the Cercle des Philadelphes.

Several notices about Alexandre Dubourg, the original president of the Cercle, list him as a merchant (négociant).[53] Two other original founders of the Cercle (Poulet, the first secretary, and Auvray) are to be associated with the merchant and commercial classes of the Cap, which makes "businessmen," so to speak, the second subgroup of founders after those in medicine. But labels deceive in this instance and hide the more complex story of the individuals involved.

Dubourg, for example, was actually the son of a poor tapestry worker from Caen.[54] Dubourg spent fourteen years as a soldier in the Périgord regiment, posted for a time in the Caribbean. But that is hardly all; Dubourg was also an actor. He turned to the theater in the army, and he continued as a comedic actor and theatrical producer in Saint Domingue after he left the army. Dubourg was well received socially and apparently was a very nice person. Once settled in Cap François Dubourg developed his skills and interests in botany, to which he seems genuinely to have taken. Dubourg assumed the directorship of the Cercle's botanical garden, and he later taught the botany course at the Cap sponsored by the Cercle des Philadelphes. We encountered Dubourg's electrical apparatus earlier, and his commitment to science seems active and sincere. He may have conducted some small "négoce" on the side, but Dubourg was above all a true and competent scientific amateur. He was apparently Arthaud's friend, and with his personality and language skills, Dubourg not incongruously became the man chosen to lead the Cercle as its president. The Cercle lost one of its stalwarts when Dubourg died just three years later in 1787, at the ripe old age of forty.

The connection of the Cercle to the business and commercial community through Charles Poulet would likewise seem to be weak. Poulet was a true merchant, and he became the Cercle's first secretary. But Poulet ended up with very little association with the Cap institution. The Cercle appointed an assistant secretary early in the fall of 1784, and Poulet either died or dropped out by the early spring of 1785.[55] That leaves J.-B. Auvray as the sole real representative from the merchant and landowning classes among the founders. Auvray had been a successful merchant and syndic of the Chamber of Commerce in the Cap, but he shifted his wealth into property and became a major landowner.[56] Auvray remained a strong backer of the Cercle and an active resident member, even when he might have become an honorary member. Auvray served as president of the Cercle on its becoming the Société Royale des Sciences et des Arts du Cap François in 1789.

In 1784 Baudry des Lozières, the remaining original founder, had let go of his military career and was making his way in the ranks of lawyers appearing before the Conseil Supérieur and other judicial bodies in and around the Cap. But, to repeat, Baudry was also Arthaud's brother-in-law, and that may explain a good part of Baudry's association with the Cercle. Certainly, his connection to the legal establishment in Cap François, while real, in no way constitutes much of an initial link between the Cercle des Philadelphes and robe circles in Saint Domingue.[57] The later development of strong connections with the colony's legal and judicial establishment represents an important and separate turn in the institutional growth of the Cercle des Philadelphes.

We know the ages of only three of the founders, and Dubourg at thirty-seven, Arthaud at thirty-six, and Baudry at thirty-two were probably among the oldest of the founders. Such a group in its mid-thirties might seem young, but was on the older side when compared with the general youth of the colony. Pierre Pluchon takes issue with Blanche Mauel's categorizing the Cercle's founders and members as primarily *grands blancs,* meaning planters and leisured plantation owners. But Maurel also took the term to include skilled professionals and other white males in the colonial ruling elite, and in this sense her classifying the Cercle des Philadelphes and its founders as *grands blancs* is certainly apt, as Table 3 confirms.[58]

Another remarkably pertinent document puts to rest entirely any notion that the Cercle des Philadelphes sprang from eighteenth-century Freemasonry or that the group of founders entertained some Masonic conspiracy when they joined to form the Cercle des Philadelphes. The document in question is an untitled and undated five-page manuscript of a speech by Baudry des Lozières. Baudry evidently delivered the speech to an organizational meeting of the proto–Cercle des Philadelphes at or just before the charter meeting of August 15, 1784.[59]

Baudry's address was the original formal proposal for the Cercle des Philadelphes, of great interest in its own right and for what it says explicitly about the Masons. After an awkward and stilted opening Baudry at last broached his topic.

> Our [previous] reunions, which were at first only for amusement, for friendly association, [and even] for dissipation in this unsociable country, can assume a more serious character, one more worthy of the select members who compose this assembly. One day this colony will be graced with a brilliant Society, a center of enlightenment, from which the colony will achieve a great reputation and the greatest gain. I could be mistaken, Sirs, and at each instance I ask for your indulgence.
>
> I will not hide from you, Sirs, that this idea of a future utility elevates my soul, electrifies it even[!], and the notion can only ensnare yours. From this point of view, here is what my sincere love for the public good and the high opinion I hold of your faculties inspire:
>
> Let us form a society for *physique* and literature, and let our se-

cret be only that of genius or of science. Let us not subject ourselves at all to formulas that impose impediments to Reason and that cannot fail to be tiresome to sensible people, such as those who compose this honest assembly.

Instead of these childish affectations that themselves in large measure have overthrown the temple of the Masons, our formulas must consist of plenty of politeness, respect, and decorum among ourselves. Let us pay little heed to the matter [?], let us put order in all that we do, but above all let us avoid anything that can bring ridicule upon us.

Let our assembly carry the name *Musæum Physico-Littéraire de St. Domingue* or the *Muséon Américain*.

Several things need to be said about Baudry's proposal, most particularly for present purposes his remarks about the Masons and their silly rituals. In proposing a scientific and literary society, Baudry rejected secrets, formulas, and tiresome impediments to reason and science. What he says explicitly about the "overthrow of the temple of the Masons" makes plain that, even though a fervent Mason, Baudry in the final analysis proposed his *Muséon Américain* in opposition to the Masonic movement and the state of Freemasonry in Cap François in the 1780s.

This "smoking gun" provided by Baudry des Lozières would seem to prove definitively that the Cercle des Philadelphes originated as a learned and scientific society and not as a Masonic organization.[60] But one need not dig so deeply into the archival record to arrive at the same conclusion. When one considers the many programmatic statements made in the name of the Cercle des Philadelphes, one repeatedly encounters the same commitment to science, medicine, and useful application. Not another word publicly or privately follows about Freemasonry. And then, the record of work of the Cercle des Philadelphes, as detailed in the next chapters, more than anything else shows the Cercle to be an institution entirely concerned with science, medicine, agriculture, and the arts and not at all with Freemasonry.

In the body of his speech Baudry went on to discuss various membership categories, public meetings, prize contests, members' dues, and other internal procedures proposed for the nascent learned society, including a "sober dinner" to be held every other month "to reinforce the bonds that unite us" and a special May Day celebration of "the renewing of nature." Baudry seems obsessed with preserving upright conduct to the point where, in addition to the usual academic requirements regarding morals, he urged the creation of a special committee of the Cercle to report "on the conduct of members and their reputations in the world."[61]

His sensitivities on this point suggest that some scandal, some previous excess on the part of members—very likely Baudry himself!—lurked behind the soon-to-be-respectable Philadelphes.[62] However that may be, Baudry concluded his speech to his fellow founders by recognizing the immediate next steps that needed to be taken.

> That posited, Sirs, we cannot afford to mistake the truth that no one can hold an assembly without the express permission of the Administrators. We must therefore immediately communicate with the Commandant at the Cap who represents the Governor General and with the Ordonnateur who represents the Intendant. We should offer them honorary seats and ask their protection and their aid in obtaining the consent of the Administrators.
>
> With the approval of these administrators we will then prepare and send a memoir to the minister in Paris, informing him of our patriotic designs and asking him for particular, but purely honorific privileges. We would have it understood that we are not asking for any exemptions, for we wish to be good citizens no less than well-informed ones. [That is, the Cercle should not cost the government anything at first.]
>
> I believe, Sirs, that in this manner we will acquire a solid form and an assured stability. Our statutes, well thought out and observed, can but improve us and bring the esteem and respect of our fellow citizens. Envy can hurt us, but only slightly, and the glory we will take from our patriotic works will be great compensation for these small annoyances.
>
> Such, Sirs, are my very succinct and, lacking time, slightly confused views about a truly important matter. A brilliant epoch may result and new growth occur in this superb colony!
>
> It is pleasant to think that we are the founders of so beautiful an edifice. However inadequately, I trust I have proved to you, Sirs, how much I desire to contribute with you to the public good and how much I have it at heart to justify your choice in me and in all those you accept.
>
> Let us collectively form a distinguished society, and by our activity and our beneficence become a noted star in the firmament of politics. And learning![63]

Yes, Baudry intended his institution to be committed to learning, but on the manuscript Baudry clearly penned the final phrase—"And learning!"—as an afterthought. Evidently, for Baudry (who was not very learned), the connections of his proposed museum to science and learning came second to the potential status of the institution in the "firmament of politics." Did Baudry mean by this expression that the Cercle was to carry out a political program, as the Masonic interpretation would have it, or was he thinking of the potential of the institution as a political and corporate entity among the others constituting colonial and metropolitan France at the time? The later was no doubt the case, and in what Baudry repeats about patriotism and the commonweal, the politics he espouses is, if anything, not that of revolution, but of public service and advancing colonial development (through science and learning). At the least, Baudry and his cofounders correctly understood the necessity of government approval and the political steps that needed to be taken to secure their newborn institution.

❧ TWELVE ❧

Milestones on the Road
to Recognition

THE SMALL GROUP of civic and scientifically minded founders who launched the Cercle des Philadelphes in August 1784 were hardly assured of the survival of their organization, much less could they anticipate the extraordinary successes it would achieve. The founders faced building an institution de novo. They understood the necessity of securing government support. They also committed themselves to a program of work and research, and that, too, had to be begun. Various aspects of institution building merged, as the nascent Cercle des Philadelphes at once began to formalize its own rules and procedures, to recruit a body of members, to fulfill its institutional agenda, and to take steps toward gaining official incorporation as a crown learned society. Following the Cercle's road to provisional recognition in 1786 and letters patent in 1789 provides an excellent way to trace the complex history of these developments.

In seeking letters patent, the Cercle des Philadelphes broke no new ground bureaucratically. The obscure protosociety in Saint Domingue followed well-established procedures whereby an originally private group like the Cercle des Philadelphes became an official French scientific society, and in being the last eighteenth-century French society to repeat this course, the Cercle des Philadelphes in many respects epitomized the process. Indeed, in taking less than five years to secure letters patent, the Cercle des Philadelphes outdistanced many of its provincial counterparts and demonstrated that the government considered the colonial institution on something of a fast track.[1] Because our sources are so good, we have a privileged insight into this process and the workings of the colonial bureaucracy of the Old Regime.

In promoting itself to government, the Cercle des Philadelphes undertook a twofold campaign: on the one hand, the institution repeatedly prodded officials for recognition and support; on the other hand, the Cercle undertook a series of scientific works, which, in part served to convince government of the worth of the institution. In all of its efforts, the Cercle sold itself on the basis of its utility and immediate practicality. The Cercle promised to be useful at every turn, and at no time did the Cercle or its spokesmen especially concern themselves with scientific theory or advancing knowledge for knowledge's sake. While not disavowing intellectual or long-term goals, the Cercle offered short-term rewards to its potential patrons, and the Cercle received support on that basis. The Cercle des Philadelphes was an institution of science, all right, but not of natural philosophy.

The founders understood that the authorities would not permit private or unauthorized assemblies of any sort in Saint Domingue, so on August 22, 1784, at the second meeting of the Cercle, twelve [sic] Philadelphes drafted a petition to the governor-general and the intendant requesting official permission to meet.[2] Properly submissive, the "supplicants" announced their project "to gather information on everything concerning the physical and natural history of Saint Domingue, to develop specific descriptions of each district in the colony with details about the soil, crops, rivers, mineral waters, and illnesses, and to unite everything influencing the health of colonists and slaves." The administrators responded favorably on September 16, 1784, saying not only that they saw no harm in the collective project to write a book [sic] about the physical and natural history of Saint Domingue, but that they were particularly pleased with the citizen-petitioners and could not praise their good intentions enough.[3]

Thus approved, the Cercle drafted its prospectus "to announce to the public the formation of the Cercle," and it sent this draft to the administrators along with a cover letter offering them the title of "Protectors of the Cercle des Philadelphes." The administrators "eagerly" granted permission to print the prospectus, but judiciously declined the honorific titles, saying the society had no need of protectors.[4]

The Cercle printed its four-page prospectus on the royal press at the Cap (Figure 20).[5] The prospectus began by noting that the principal towns in France had academies and that the colony in Saint Domingue could profit from a similar institution, especially as regards manufactures and the health of colonists. The prospectus aligned the Cercle with the

Nº 11

PROSPECTUS

DU

CERCLE DES PHILADELPHES,

ÉTABLI AU CAP.

Les principales Villes de France ont un dépôt des connoiſſances humaines, toutes ont des Académies, & l'on ſait combien l'émulation, que ces inſtitutions ont fait naître, a contribué à l'avancement des Sciences & à la gloire de la Nation.

La Colonie de Saint-Domingue, digne de l'attention du Gouvernement, par ſon étendue, par ſes productions, par ſon Commerce, par ſa population, eſt vraiment ſuſceptible d'un pareil établiſſement; elle a eu beſoin des Arts & des Sciences pour connoître les moyens néceſſaires à ſes Manufactures & à la conſervation de ſes Habitans, & on ne peut trop s'occuper à perfectionner les connoiſſances qui ſe rapportent à ces deux objets.

D'ailleurs, l'Hiſtoire phyſique, naturelle & morale de la Colonie,

FIG. 20. *Prospectus du Cercle des Philadelphes* (1784).

tradition of physicians and naturalists who had worked in Saint Domingue, and after briefly describing the organization of the Cercle, it went on to detail the "vast" institutional plan and program the Cercle had established for itself.

> We aspire to having a general description of the Colony. We are asking for specific descriptions of different districts. We would like observations on the soil, on minerals found there, on trees and plants that grow there, [and] on agriculture and manufactures undertaken there. The history of insects will be interesting and quite valuable to us, as will histories of birds and shellfish.
>
> We seek astronomical and meteorological observations. We will be obliged by research on the constitution of the air, on temperature, on the winds, and on the quality of the water, sweet or mineral. [We seek information] on reigning illnesses and on diseases particular to each district. We would like philosophical observations on the constitution and habits of people born in the colony [and] on the changes in temperament and in physical and moral constitution experienced by Europeans [in coming to the colony. We desire information] on the character, talents, and mores of slaves, and on ways of improving their lot, without harming the interests of colonists[!] Finally, we wish to have observations concerning diseases affecting livestock, ways of treating these diseases, and especially of preventing them.

The prospectus noted that since the government seemed about to promote cochineal husbandry, the Cercle would also investigate the valuable red bug and its host nopal cactus. The Cercle further proposed forming a (public) library, a cabinet for chemistry, *physique,* and natural history, and a botanical garden. The Cercle welcomed poetical and literary contributions "in order to foster good taste and to introduce variety into its works." The Cercle invited everyone who shared their "zeal for the public good and for the advancement of science" to contribute to the support and improvement of the institution. "Such is our goal," the prospectus concluded. "Perhaps we are forming an enterprise that is beyond our forces. But if success does not follow our hopes, at least we will have done all we could to deserve success."

Thus, from its beginning the Cercle des Philadelphes articulated a multifaceted but unified program of scientific inquiry and useful application. The prospectus of 1784 outlined that inquiry, and over the next years other versions of the "mission statement" repeated the same items and areas of interest: comprehensive economic and natural historical surveys, studies of colonial medicine and diseases, examination of colonial agriculture and manufactures, and a pragmatic concern for slaves.[6] The approach was decidedly Baconian and empirical. The Société Royale de Médecine in Paris and its program of medical and scientific research were influential in shaping the agenda of the Cercle des Philadelphes. Likewise, Moreau de Saint-Méry's own long-standing project to compile a complete encyclopedia of the colony also figured in the formulation of the Cercle's plans and goals.[7] The state would obviously be interested. The Cercle des Philadelphes began with an explicit mission to investigate the colony and be useful. It stuck to its commitments over the next years, and to a large extent successfully carried out its scientific program.

Wrote Arthaud, "The Cercle made it a point to send its Prospectus to all officials, to all notables, and to all those who had the reputation of loving science and the study of nature." Governor-General Coustard graciously replied to the Cercle, "Certainly nothing is more worthy of praise than the enterprise you have formed. . . . If I am unable to contribute in terms of talent, at least I offer you my zeal and my offices on all occasions for which the latter may be useful."[8]

The first public notice of the Cercle des Philadelphes appeared in the *Affiches Américaines* for November 27, 1784. Charles Mozard, the editor, based his opening remarks directly on the prospectus. He, too, placed the Cercle in the tradition of the academies and even claimed that the Cercle des Philadelphes was the first such learned association in the Americas.[9] In the rest of his notice, Mozard gave his own views of the limitations and the promise of the new group:

> The members of the Cercle des Philadelphes do not hide from the many obstacles they must overcome before their Assemblies will produce all the utility the colony will one day receive. Everything new excites scornful smiles and the jeers of useless beings. But soon appear men truly inflamed with the love of good. They triumph over adversity. They scorn banter, and their assiduous works bring rich harvests to a land that had been judged incapable of producing anything. Thus they obtain the respect and the gratitude of their fellow citizens.

Mozard makes clear that certain segments of colonial society initially derided the Cercle des Philadelphes. Arthaud himself later wrote about the "spite, ingratitude, contrariety, and satire" that greeted the Cercle.[10] To the colonial and frontier town of Cap François—still rough around the edges if no longer at heart—the idea of a learned assembly of locals at first must have appeared pretentious and absurd.

The Cercle next created a set of statutes to specify the institution's form and procedures. The Cercle appointed a committee on October 24,

1784. The committee reported on November 14, at which time the Cercle approved its new statutes and sent them to the administrators for their approbation and an imprimatur.[11] The governor-general and the intendant replied cordially from Port-au-Prince on December 9, and the twenty-one page pamphlet, *Statuts du Cercle des Philadelphes,* issued from the royal press at Cap François in early January 1785.

The statutes themselves are unexceptional in setting forth a typical eighteenth-century learned society.[12] The statutes dictated three classes of members: resident associates (members living within 5 miles of Cap François), colonial associates (members resident in the colony and elsewhere in the West Indies), and national and foreign associates (members living abroad). The initial statutes did not include honoraries, but that class developed very early on, as we will see. The statutes set initial limits at twenty members in each category, but the numbers proved flexible. A president and a permanent secretary constituted the Cercle's effective officers, although provisions were eventually fulfilled for a vice-president and an assistant secretary. Other society positions included director of the botanical garden, guardian of the *cabinet de physique,* and librarian.

The statutes set forth the organization and operation of the Cercle in great detail. The Cercle held ordinary meetings on the second and fourth Sunday of the month from 10:00 A.M. until noon, members not signing in by 11:00 being considered late.[13] The president occupied the place of honor, with members seated by seniority. (An egalitarian circle the Cercle was not.) The rules required appropriate morals for members and potential members. At first candidates faced literal blackballing, with a single negative vote preventing their election, but the Cercle later amended this rule so that only three-quarters of the balls had to be white. The colony's administrators had to approve all new members, and the statutes themselves were null without the approval of the administrators. The Cercle had to pay for itself, at least at the outset, and resident associates assessed themselves 396 colonial livres (6 "Portuguese") in reception dues. Because of increased expenses and delinquency, however, at the outset members in fact each had to pay over 1,000 colonial livres annually to keep the Cercle going.[14]

Eighteen men signed the Cercle's statutes, which means that six new associates joined the Cercle in September, October, and November of 1784—two a month or one every meeting.[15] The inauguration of the Cercle's new statutes on November 14, 1784, entailed a reorganization of the group and a turning away from the initially egalitarian and Masonic character of the Cercle at its foundation in August.[16] This reorganization apparently provoked antagonism within the Cercle. Arthaud exacerbated the situation, perhaps, by presenting a rigid and hierarchical seating plan, with himself as president at the center, and he may have anticipated trouble, for he pleaded for peace in presenting his plan. Later, at the meeting on November 28, 1784, in his speech accepting the presidency of the Cercle, Arthaud alluded to internal conflict, and went on to cite Fontenelle on the harmony that should reign in an academy.[17]

Arthaud's presidential acceptance speech of November 28, 1784, is interesting on several further counts. He spoke at length about opposition and the hostile reception encountered by the new Cercle des Philadelphes. He positively overflowed with notions of public utility, and he pointed with pride to the warm reception given the Cercle by the colony's administrators: "They have perfectly well sensed the potential utility of our society," he said. Arthaud praised the administrators for following the intelligent policy (*sage politique*) of support for the sciences and the arts begun by Colbert. Later in his speech, in discussing medical research, Arthaud displayed his political loyalties further when he said bluntly, "Occupying ourselves to this end would serve the colony, but it would serve the state better," a curious distinction we have encountered before. In outlining the Cercle's scientific program Arthaud suggested that the Cercle investigate the colony's mineral resources, and he proposed developing a comparative table of seasonal variations for Saint Domingue and Europe. Arthaud recommended that the Cercle sponsor prize competitions for research into these and other areas. In this way also the Cercle des Philadelphes strove to imitate European academies of science, provincial and otherwise. Arthaud ended his maiden speech as president by exhorting his confreres: "If we need protection to sustain ourselves, we have to endeavor to merit that protection, and we can only do that by constantly demonstrating the importance of our creation and the utility of our existence. We have the desire to do good, but to succeed, we must multiply our efforts."[18]

If the Cercle des Philadelphes took the fall of 1784 to establish itself, it gave over the winter and spring of 1785 to promoting itself to larger circles of colonial and metropolitan society and officialdom. Most notably, armed with its prospectus and new statutes, the Cercle des Philadelphes rather naively set out to get its letters patent from the king. On February 21, 1785, Cercle associates wrote directly to the minister of the navy and the colonies in Paris, de Castries, and asked quite directly for "letters patent and the title of academy."[19] The Cercle des Philadelphes wrote to the minister a second time on April 1, 1785, again asking straightforwardly for a "genuine confirmation." A third letter and a third copy of their prospectus and statutes followed on May 12, 1785, this time noting that under de Castries's ministry the Cercle could become "the first academy to exist in the French colonies."[20]

De Castries in Versailles had never heard of the Cercle des Philadelphes in Cap François, so on July 7, 1785, on the advice of subordinates he sent the prospectus and statutes to the administrators in Saint Domingue, demanding a report on the Cercle. The colonial administrators responded to the minister with a seemingly favorable report five months later on December 1, 1785. They saw no drawback to the Cercle, they recommended the talents and zeal of its members, and they thought the institution might offer some real advantages, regarding cochineal production, for example.[21] The local administrators saw no inherent problem with letters patent, but because they knew the Cercle would ask for money, they opposed granting such letters until the institution had "displayed somewhat greater continuity." This was not the last time the Cercle was "put on hold."

The direct assault on the ministry at Versailles was not the only front on which the Cercle des Philadelphes fought to establish itself in early 1785, and while the preceding letters made their rounds within the colonial bureaucracy, the Cercle des Philadelphes simultaneously reached out to other constituents and potential patrons.

The judicial establishment in Cap François, for example, was not well represented in the initial Cercle des Philadelphes. Only three of the first eighteen members of the Cercle were lawyers, and only one (Deschamps) held a formal position in the court system, the comparatively minor one of second public prosecutor. In recruiting the colony's higher robe establishment, the Cercle des Philadelphes took a major step toward solidifying its position, and it achieved this end by electing men as honorary associates. "Having soon felt the necessity," as Arthaud put it, the Cercle created the class of honorary associates on January 9, 1785, and on February 13 it elected as its first honorary the president of the Conseil Supérieur at the Cap, one De Trouillet. The Cercle selected N.-L. François de Neufchateau, royal attorney general and chief public prosecutor, as an honorary member on May 1, 1785.[22] The election of two other honoraries from the Cap judiciary followed on May 30: Justice Routte and the ennobled F.-E.-F. Le Gras père, former chief attorney and rich property owner.[23] The Cercle thus quickly cemented personal and institutional relations with the important Conseil Supérieur at the Cap.

With the election of the chevalier L.-L. D. de P. Dugrès, the military commandant at Cap François, the effort to expand the Cercle through honorary association extended to the military establishment. The Cercle was not about to overlook the centers of power represented by Dugrès,

who might have done more for the Cercle had he not died shortly thereafter, in November 1785. The election of Dugrès inaugurated a small but
notable tradition of the Cercle's electing military men, and especially the
military leadership of the Cap area.

Similarly the Cercle chose Foulquier de Bastides, the intendant at
Martinique, on March 21, 1785. Possibly after conferring with his brother
administrators in Saint Domingue, the Martinique intendant accepted
honorary membership in the Cercle, and his acceptance not only gave a
political stamp of approval to the Cercle from another quarter of colonial
administration, but it carried with it something of a scientific cachet, in
that this Foulquier was also a *correspondant* of the Paris Academy of Sciences.[24] Over the next two years, in addition to Foulquier in Martinique,
the baron de Clugny (governor-general of Guadeloupe), the baron de la
Borie (governor-general of Saint Lucia), and F.-J. Lequoy de Mongiraud
(another administrator at Saint Lucia) in turn became honoraries of the
Cercle des Philadelphes, and the Cercle set another important pattern of
recruiting colonial administrators from elsewhere in the West Indies.

In their letter to Foulquier de Bastides and his coadministrator in
Martinique, the Philadelphes wrote of their "desire to have the torch of
science burn brightly in these secondary states."[25] The Philadelphes were
not above stretching their credentials, claiming to be "newly established
in Saint Domingue under the protection of government," as if the limited
set of permissions granted the Cercle to that point constituted government recognition or protection. On the basis of that rhetorical ploy, the
Philadelphes then asked the enlightened administrators in Martinique to
follow the example of their no-less enlightened counterparts in Saint
Domingue and honor the Cercle with their "approbation." A nice piece
of institutional maneuvering, that, but in writing to the administrators in
Martinique, the Cercle really hoped to get help in publicizing itself to
interested colonists in Martinique and to promote the work of the Cercle
in "*physique,* botany, meteorology, mineralogy, astronomy, and the
other essential parts of the sciences." The administrators obliged by publishing the Cercle's letter along with a description of the institution drawn
from the prospectus and statutes. Such publicity must have had its effect,
for over the next years the Cercle elected eighteen associates from the
other French Caribbean colonies of Martinique, Guadeloupe, and Saint
Lucia. The Cercle elected the majority of these new members as colonial
associates, putting them on par with nonresident members of the Cercle
in Saint Domingue.

In March 1785 the Cercle also contacted a Spanish count in Havana
named de Galvez about spreading word of the Cercle des Philadelphes to
New Spain. People in Saint Domingue knew de Galvez, a Spanish gen-

eral, from the time of the American War, which may explain the contact, but de Galvez was just then leaving Cuba to take up the post of viceroy in Mexico. The Spaniard promised to communicate anything he might discover in Mexico and to alert "persons whose intelligence and application could be useful to you." Nothing resulted from this overture to the realm of Spanish America, for de Galvez died soon thereafter, but a Spanish colonial contact continued in the person of Martin de Navarre, "intendant at Louisiana," whom the Cercle elected an honorary associate in November 1786.[26]

By the time of the first public meeting of the Cercle des Philadelphes on May 11, 1785, the roster of members had grown to thirty-five.[27] Only three new resident associates joined the active membership in Cap François since the previous November, bringing the total number of resident members to seventeen.[28] The Cercle des Philadelphes also selected a baker's dozen of its first colonial associates in the period through May 1785, and these associates living elsewhere in Saint Domingue and the West Indies formed an important new group in the overall makeup of the Cercle. Four were physicians, three of them government physicians (médecins du roi). Joubert de la Motte was the most notable; he served, one may recall, as royal botanist and naturalist in charge of the botanical garden in Port-au-Prince.[29] The Cercle similarly picked as colonial associates two other men in royal employ: Joseph Gauché, soon to be appointed director of the mineral spa at the Eaux de Boynes, and J.-J. Bertrand de Saint-Ouen, royal engineer in charge of the irrigation of the Artibonite.

Two important new colonial associates, Étienne Lefebvre-Deshayes and Paul Belin de Villeneuve, were (lesser) nobles and rich plantation owners. Lefebvre-Deshayes, a notable amateur naturalist and colonist from the Jérémie area, also acted as Buffon's correspondent through the *cabinet du roi* in Paris.[30] Belin de Villeneuve was a member of the Cap Chamber of Agriculture and a technologist known for his researches into sugar production.[31] The other colonial associates elected at this time include Charles Mozard, editor of the *Affiches Américaines;* de Vicendon-Dutour, a lawyer practicing before the Port-au-Prince Conseil; Marrier de Chanteloup, lawyer and naval commissioner; Tanguy de la Brossière, a colonist at Les Cayes; and the abbé de la Haye, another colonial naturalist we have seen, curate of remote Dondon.

The spring campaign and membership drive of 1785 culminated in the Cercle's first public meeting on May 11. According to Arthaud, many people from all walks of life supposedly attended this "first colonial tribute to the sciences and letters," but the record makes no mention of government representatives.[32] Arthaud opened the meeting with a discourse on the formation of the Cercle and its utility to the colony.[33] Arthaud again laid out the goals of the organization, concluding that "our end, therefore, is to observe and collect everything relating to *physique*, astronomy, navigation, agriculture, manufactures, and medicine and to devote ourselves to public utility." Arthaud carefully noted that only arts of first necessity should be cultivated in Saint Domingue, and on this occasion he proposed a special role for the Cercle as a possible center for medical correspondence and medical training. Arthaud underscored the political acceptability of science and the Cercle, saying, "No one doubts that the pursuit of the sciences makes man better suited to the civil state and more docile before the laws to which he must submit himself," and Arthaud finished this part of his presentation with an elaborate statement of political loyalty, wherein he invoked the names and interests of Louis XVI, the colony's administrators, and the Conseil Supérieur at the Cap.

More in his capacity as first royal physician, Arthaud then read his description of the town of Cap François, a prolegomenon to a complete medical and epidemiological history of the colony's leading urban center. Cosme d'Angerville, chief royal surgeon at the Cap, followed with memoirs on spinal bifida and the formation of kidney stones, and to inaugurate its cabinet of curiosities, d'Angerville presented to the Cercle his preserved specimens of a case of spinal bifida, a hydrocephalic specimen, a two-headed sheep, and kidney stones found in the late Count D'Argout. Alexandre Dubourg followed d'Angerville with a botanical discourse concerning a new species of palm tree that Dubourg named after Madame Bellecombe, wife of the governor-general of the colony. (Unfortunately, it turned out that the species had already been discovered, and so the denomination proved invalid, but the sentiment was clearly correct.) Then came J.-B. Peyré reading a paper on the use of quinine as a treatment for intermittent fevers, a topic proposed by the Société Royale de Médecine in Paris. Finally, Baudry des Lozières read a commemorative *éloge* of Messrs. Larnage and Maillart, coadministrators of the colony in the 1730s, noted for both their public spiritedness and the unusual harmony in which they governed the colony together.[34]

The Cercle had planned to present a further half dozen papers, but time ran out, as, no doubt, did the patience of the audience.[35] The meeting concluded with the presentation of the Cercle's program of prizes for the upcoming years 1786 and 1787. The details of the Cercle's first series of prizes will be examined next, but about the public meeting of May 11, 1785, as a whole, Mozard commented in part as follows in the *Affiches Américaines*:

Everyone can see from the simple account of this meeting we have just given how valuable the establishment of the Cercle des Philadelphes can become to Saint Domingue. One will easily perceive that its members are all animated by this useful emulation that will rapidly accelerate the progress of knowledge.[36]

The *form* of the Cercle's prize contest was entirely typical of the hundreds, indeed thousands of such competitions sponsored by eighteenth-century learned and scientific societies in France and elsewhere.[37] In most respects, too, the *content* of the Cercle's first prize program was unexceptional and straightforward. For 1786, for example, the Cercle des Philadelphes solicited essays dealing with the causes of convulsions in infants (tetanus), preferred fertilizers for particular soils, and "the best means of constructing buildings required for colonial agriculture and manufacture, from a simple slave's hut to the most complicated mill." Also for 1786 the Cercle solicited *éloges* of the founders of the men's and women's poorhouses in Cap François, and for 1787 it asked about scurvy in Saint Domingue and its treatment. Gold medals awaited the winners. In addition, the Cercle indicated that it would make nominal awards of silver medals at anytime to anyone submitting a satisfactory memoir as part of the Cercle's campaign to gather information about the colony, and, in this connection, the program posed questions concerning meteorological and hydrological conditions in the colony and illnesses affecting whites, new arrivals, and slaves [sic].

The "extraordinary prize" for 1787, however, on the best means of preserving paper from the ravages of insects, deserves particular attention. François de Neufchateau, the royal prosecutor before the Cap Conseil and newly elected honorary of the Cercle des Philadelphes, sponsored the contest and underwrote the prize, a generous 25 "Portuguese" (1,650 colonial livres or 1,100 livres tournois).[38] Insect attacks on paper threatened the very basis of colonialism, for in the tropical conditions of Saint Domingue, anything written or printed on paper soon became a meal for various bugs, worms and larvae. Not only could colonists hardly keep libraries—an obstacle to the pursuit of science and learning recognized by all concerned—but, more seriously, public, legal, and commercial documents similarly succumbed within a short period, with dire consequences for the disposition of property and title. Moreau, among others, provides a vivid description of this serious problem:

One can imagine the ravages caused by insects that devour paper in the Colony. Old registers and acts are rendered absolutely illegible.

Insects convert the paper they eat into a type of earthy gluten, and this gluten hardens, causing the sheets of paper to stick together and to tear if separated. . . . The age of the pieces is not always a guide to the damage, for the nature of the paper and the humidity of the place contribute to accelerate the destruction. In 1783 at the registry of the Sénéchaussée court in Cap François I saw parish records dating from 1774 that were already partly illegible.[39]

Other courts and bureaus were affected. (See Figure 15.) The Conseil Supérieur at Port-au-Prince, for example, at one point had its entire records copied on account of the problem with insects. In part because of the susceptibility of colonial records to insect attacks, the national government created a separate colonial archive in France in 1776 and ordered that officials keep duplicate colonial records.[40] Naturally, various remedies appeared, including the potion sold by Mozard for 33 colonial livres and good for a hundred books: "A halfway intelligent slave can apply it," advertised Mozard, and, in fact, colonists tried to combat the problem by having slaves manually pick out bugs and worms from among books and papers.[41]

The problem of insect attacks on paper thus proved a nontrivial one to which someone in the judicial system would be especially sensitive. The solution encouraged so directly and munificently by Neufchateau promised significant practical benefits, and the project was perfect for the new Cercle des Philadelphes to prove its social utility. Unfortunately, even though the Cercle and its correspondents worked diligently on the problem over the next several years, no satisfactory solution was forthcoming. The Cercle received numerous samples, and all but one succumbed to insects. The sole sample not to be eaten had been treated with arsenic and was deemed too dangerous for actual use. The Cercle remanded the Neufchateau prize in 1786 and 1787. The Cercle wrote up and published its disappointing results in 1788 as its *Dissertation sur le papier,* and proclaimed the Neufchateau prize yet again, this time on indefinite basis.[42]

Almost exactly nine months elapsed between the foundation of the Cercle des Philadelphes on August 15, 1784, and its first public meeting on May 11, 1785. The period was one of gestation, and the institution displaying itself at the public meeting in May 1785 emerged anxious and at last ready for real work and the fulfillment of its civic and scientific program.

The Cercle's modest effort to develop an insect-proof paper is one small entry in the substantial record of work and accomplishment achieved by the Cercle des Philadelphes over the next five years. In the summer of 1785 the Cercle undertook two other small projects, similarly connected with François de Neufchateau. The first involved contact with the Académie Royale des Sciences et Belles-Lettres of Bordeaux, which had approached Neufchateau and the Cercle about a transatlantic test of Parmentier's potato biscuits *cum* sea biscuits. The Cercle named Arthaud, Peyré, and Ducatel to examine and report on the biscuits sent from Bordeaux. Arthaud wrote the report, and in due course the Cercle received the official thanks of the Bordeaux Academy for its "interesting and very well conducted experiments."[43]

In the second matter begun that summer, Neufchateau, as royal attorney general, asked the Cercle to conduct scientific tests of inks used in keeping public documents and commercial records. Arthaud and Ducatel conducted the experiments, and they concluded that "red ink is subject to decomposition by fixed air conducted by humidity," and on that basis and the recommendation of the governor-general, the Conseil Supérieur at Cap François forbade merchants and public officials from using red or "Campeche" ink.[44] The incident is a wholly minor one, but it illustrates well the authority and social role of science in the eighteenth century, especially when institutionalized in a learned society, even one as green as the Cercle des Philadelphes.

The summer of 1785 saw the Cercle establish a formal "literary correspondence" with the Paris Musée, the famous free school and scientifically inclined public association sanctioned in part by the Paris Academy of Sciences.[45] Moreau de Saint-Méry presided over the Paris Musée in 1785, and he initiated the link with the Cercle. Over the next four years the Cercle elected as national associates in France five other members and officers of the Paris Musée in addition to Moreau, ranking the Paris Musée just behind the Royal Society of Medicine in Paris in terms of strength of institutional association forged through common members. If not much more came of the formal tie to the Paris Musée, like the new contact with the Bordeaux Academy, it speaks of the Cercle beginning to make its way among the learned corporations of the day.

The most important project begun by the Cercle des Philadelphes in the summer of 1785 concerned organizing its botanical garden. The Cercle's "sacred commitment" to a botanical garden and to promoting cochineal husbandry dated back to 1784, but only in May 1785 could the Cercle take concrete steps.[46] The Cercle secured a private piece of land with indigenous cactus, and, using a handful of cochineal insects obtained from the colonist Brulley, it sowed its first experimental garden and nopalry on June 12, 1785.[47] But Brulley's resources were apparently limited, so the Cercle approached the colony's administrators about further supplies from Thierry de Menonville's old Jardin Royal in Port-au-Prince. The administrators replied that they were delighted with the Cercle's efforts, but that, while they would have the royal botanist (Joubert) look into the matter, they feared little would be forthcoming from the ruins of Thierry's garden.[48]

The Cercle's first experimental plot proved very successful, with the first harvest of cochineal insects being made in only two months.[49] The Cercle forwarded samples of its cochineal to the administrators, and the institution had no hesitancy in then approaching the administrators about a lease on public land for a bigger garden. The administrators responded positively, and official permission for the concession came through a fortnight later.[50]

For Arthaud, the grant of public land to the Cercle des Philadelphes rightly marked "a new epoch of government favor," and at the first anniversary meeting of the Cercle held on August 15, 1785, he ranked it proudly as first among the "flattering approbations" the Cercle had received from government in the year since its foundation. The original nine founders and the other members had every reason to feel proud of their institution and its accomplishments over the previous year, as a review of this chapter would indicate. They had indeed come a long way toward "erecting a haven for science" in the colony, and to celebrate they capped the anniversary meeting with a "philosophical banquet."[51] Following the customary calendar of learned societies in France, the Cercle des Philadelphes then adjourned for a month.

In the fall of 1785 the Cercle des Philadelphes began a new project related to its campaign to promote cochineal: the publication of Thierry de Menonville's *Traité de la Culture du Nopal et de l'Éducation de la Cochenille*. Arthaud began the effort by delivering a panegyrical address for Thierry de Menonville when the Cercle rejoined on September 19.[52] The Cercle then approached the colony's administrators about printing Thierry's manuscript, and they in turn granted permission both for a prospectus and for the work itself.[53] The prospectus noted proudly that "the Cercle offers the Colony not only the means to increase the number of its enterprises, but it is introducing a [whole] new branch of commerce that can be as useful to the Colony as to the State."[54] In reporting on the project in the *Affiches Américaines,* Mozard remarked "The Cercle is doing the colony an inestimable service that will enamor posterity to the memory of those who conceived and executed the project to form this *Société Académique,* which that much more deserves official protection."[55]

The Cercle des Philadelphes decided formally to proceed with publication of Thierry's *Traité* on November 28, 1785, but nearly two years elapsed before the volumes finally appeared in the fall of 1787. At least a year of that delay was taken up because the work was printed in France, and so became subjected to the baroque formalities of official censorship there.[56] The Cercle may have economized by having Thierry's *Traité* printed in France, but the complicated publishing process presented an evident obstacle to the ready dissemination of useful knowledge in Saint Domingue. All the subsequent volumes produced by the Cercle des Philadelphes received approval much more expeditiously—at a stroke—in the colony and were printed on local presses.[57]

The Cercle's edition of Thierry's *Traité* finally did appear in 1787 (Figure 21) in a handsome two-volume octavo set of 436 pages with four colored plates provided by the abbé de la Haye. The first volume was given over to Thierry's picaresque account of his voyage to Mexico and to sundry related pieces, and the administrators were probably right to doubt the utility of this material.[58] The second volume, however, was a natural-historical and practical manual devoted to the nopal cactus and to *fina* and *sylvestre* cochineal, and it provided the details for nopal farming and cochineal husbandry. In his notice in the *Affiches Américaines* Mozard strongly praised the work and the Cercle des Philadelphes for publishing it, although he disagreed that cochineal husbandry might someday rival the importance of coffee in the economy of the colony.[59]

With the effort to publish Thierry's *Traité* underway in the fall of 1785, the Cercle des Philadelphes continued with its own horticultural trials. The Cercle's original nopalry produced a third harvest of cochineal in November 1785, and other possibilities for applied botany were in the air.[60] By the same token, little progress occurred regarding land for an official botanical garden promised by the administrators the previous August. The Cercle had to renew its petition, and not until November 21, 1785, did the authorities finally deed over to the Cercle two *carreaux* on the Morne du Cap.[61]

Once invested with this new property, however, Dubourg and Peyré, the director and assistant director of the Cercle's gardens, stood ready with a public course on botany. They offered a twenty-session introduction to the subject, based, notably, on the Linnean system, with sessions scheduled for Tuesdays and Saturdays from 2:00 to 4:00 P.M. in the "Jardin du Cercle."[62] In their prospectus the instructors took care to insist about botany: "There is nothing abstract about this science. It is not at all based on algebraic calculations. Its subject is not a million miles from us, but right beneath our eyes." Their ardor may have been naive ("We are burning with the desire to be useful"), but their sincerity was unimpeachable, for nothing indicates that they charged a fee for the course. The botany course ran from January to April 1786, and it seems to have been genuinely well received.[63] The record gives no indication

TRAITÉ
DE LA CULTURE
DU NOPAL,
ET DE L'ÉDUCATION
DE LA COCHENILLE

Dans les Colonies Françaises de l'Amérique;

PRÉCÉDÉ D'UN

VOYAGE A GUAXACA,

PAR M. THIERY DE MENONVILLE, Avocat en Parlement, Botaniste de Sa Majesté Très-Chrétienne.

Auquel on a ajouté une Préface, des Notes & des Observations relatives à la culture de la Cochenille, avec des figures coloriées.

LE tout recueilli & publié par le Cercle des Philadelphes établi au Cap-Français, isle & côte St. Domingue.

━━━━━━━━━

AU CAP-FRANÇAIS,

Chez la veuve HERBAULT, Libraire de Monseigneur le Général, & du Cercle des Philadelphes.

à PARIS,

Chez DELALAIN, le jeune, Libraire, rue St. Jacques.

& à BORDEAUX,

Chez BERGERET, Libraire, rue de la Chapelle St. Jean.

━━━━━━━━━

MDCCLXXXVII.

FIG. 21. Thierry de Menonville, *Traité . . . de la Cochenille* (1787), title page.

that Dubourg or Peyré repeated the course, however. By itself, the Cercle's botany course is not especially significant, but it indicates an expanding range of activities promoted by the Cercle, and it testifies to the continued success of the institution during its second year.[64]

As institutional activity increased, the Cercle des Philadelphes and its leaders did not lose sight of the goal of letters patent and formal incorporation, and they used the institution's nascent record of civic and scientific accomplishment to pursue the matter of official recognition and support. A second round of overtures began in October 1785 when the Cercle des Philadelphes asked the minister in Paris, the Maréchal de Castries, to accept the dedication of their edition of Thierry de Menonville's *Traité*.[65] While the minister's office facilitated the publication of the Thierry volume, de Castries refused the dedication, notably on the grounds that the Cercle des Philadelphes lacked corporate status and royal approval. The Cercle wrote back, saying they understood about the dedication, but that they remained enthusiastic and hopeful of favorable reports from the local administrators.[66]

As another step in wooing government officials, the Cercle des Philadelphes entertained François Barbé de Marbois, the new intendant in Saint Domingue, at a meeting on October 31, 1785.[67] Barbé had just stepped off the boat, and he stayed as intendant in the colony through the fall of 1789. Barbé de Marbois was the very model of the rationalist state administrator, and, while not a special friend of the Cercle des Philadelphes, the Cercle won its laurels under his intendancy, at the very peak of the Old Regime in Saint Domingue. In welcoming Barbé to the precincts of the Cercle des Philadelphes in Cap François, Charles Arthaud began by reaffirming the institution's commitment to useful knowledge. He then bluntly asked for Barbé's patronage and support.

> We implore you to accord us your protection and to help us procure the sanction of our sovereign and the legal existence we fervently wish. We have no desire to complain about our [financial] sacrifices. We know that all the [learned] societies in France made similar ones. But we know that the government, sensing the importance of cultivating the sciences, gave to these societies the means to sustain themselves and continue their works.[68]

Following Arthaud's pitch, Dubourg read a paper on cochineal; Verret reported on an irrigation project; Peyré presented a memoir on the Eaux de Boynes; and Dutrosne discussed distilling schnapps from sugar cane.

Arthaud later recorded that, in response to all this, Barbé thanked the Cercle des Philadelphes and promised to do everything in his power to secure support for the Cercle, not the least being a positive report to the minister in Paris.[69] One may doubt Arthaud's account or Barbé's honesty, for on December 1, 1785, hardly a month after the meeting held in his honor, Barbé de Marbois and the governor-general, Coustard, reported on the Cercle des Philadelphes to the minister in France in a way that squelched any immediate prospects for letters patent.[70] Although the administrators in Saint Domingue thought the Cercle meritorious in many ways, they concluded that "His Lordship will doubtless judge it apropos to have the Cercle des Philadelphes wait some time further before granting it the letters patent it has solicited," and, accordingly, a cold bureaucratic hand writes "Wait" (*Attendre*) atop the ministerial memorandum concerning this matter.[71]

César-Henri, the count of La Luzerne, arrived in Saint Domingue to assume the office of governor-general in April 1786, and immediately he received the written compliments of the Cercle des Philadelphes.[72] Thus began an important relation, wherein ultimately La Luzerne, as state minister, would award the Cercle its letters patent and assume the role of protector of the institution. In a letter effusive even by the standards of the time, La Luzerne responded to the Cercle's greetings, saying that no one desired the success of the Cercle more than he did, that he well recognized the government's obligation to support societies devoted to public utility, and that they could count on him for real proofs of his convictions.[73] Given such encouragement, the Cercle des Philadelphes responded with a detailed memoir further outlining the institution and its goals. In forwarding this memoir, Arthaud appealed for La Luzerne's support, complaining that, while relations with the intendant, Barbé de Marbois, were (supposedly) satisfactory, he had heard through Moreau de Saint-Méry that the minister was not well disposed toward the Cercle.[74] When deeds came to be compared with words, however, La Luzerne had nothing to offer and could only don a conciliatory mask.

> In the Republic of the Sciences and Letters, many institutions that are very famous today suffered the same fate in their infancy as your emerging society. For many years they were simply assemblies of learned men, almost ignored, approved at most, and only very rarely supported by the Government. The emulation found there, the discoveries coming from their research, the utility that resulted, and the popular support that in the end became clamorous slowly but surely consolidated those institutions meriting it, and the sovereign found himself, so to speak, forced to give them a legal existence.
>
> These thoughts should ward off any discouragement and should stimulate the Cercle des Philadelphes further, especially as the cen-

turies are becoming more enlightened. . . . The Cercle will not, I say, have to depend on me alone to obtain its desire. It is easy to foresee that all of my successors will make it not only their duty but an honor to push for and encourage a society that is only concerned with truly useful objectives for the Colony.[75]

The notion of La Luzerne's successors or of more enlightened centuries to come could not have been very encouraging to the Cercle des Philadelphes as it prepared to celebrate itself in a second public meeting.

The Cercle's public meeting for 1786 took place on June 20.[76] The Cercle held this meeting in the new rooms it had rented on the rue Vaudreuil, at the corner of the rue Sainte-Marie, one block west of the place d'Armes.[77] The main room on the second floor was large (60 feet long). A sizable meeting table and the rudiments of the Cercle's library occupied one end of the room.[78] The Cercle displayed its cabinet along the walls, and at the far end of the room work space existed for chemical demonstrations.[79] Arthaud's private apartments led off from the Cercle's quarters, and he naturally superintended the locale and the Cercle's growing number of possessions.

The administrators did not attend the 1786 public meeting, but sent their regrets, read as the first order of business.[80] The familiar series of papers and reports followed. Baudry des Lozières began his *éloge* of Le Gras père with the scheduled compliment to Governor-General La Luzerne, and he proposed that the Cercle publish Le Gras's manuscript on the physical and moral training of slaves.[81] Arthaud followed with the sad *éloge* of Lefebvre-Deshayes, the Cercle's notable associate and correspondent of Buffon and Lalande, who had traveled to Cap François from the south of Saint Domingue the previous winter especially to visit the Cercle des Philadelphes, only to die while staying with Arthaud.[82] The Cercle heard proposals to publish the abbé de la Haye's tropical botany and a manual for testing mineral waters. Arthaud took the podium again to read his "Plan for Public Education in the Colony of Saint Domingue," a speech that probably stepped over the line politically, for two years later, when he wrote about this meeting, Arthaud claimed he spoke on the safe topic of the aboriginal Indians. Millot then presented his meteorological record for the twelve months preceding March 1786, but time ran out before the count d'Ingrandes could read his ode to Mlle le Masson le Golft, the Cercle's newly elected national associate from Le Havre.

The Cercle's public ceremony for 1786 closed with a presentation by Prévost of a new prize program and a report on the contests dating from 1785.[83] The results on this head were disappointing. The Cercle received only two entries for its question on infantile convulsions (tetanus), and it rejected both as too theoretical ("systématique" [sic]). The Cercle felt committed to its selections, so it repeated for the years 1787 and 1788 the entire slate of questions originally posed for 1786 and 1787. At the same time the Cercle reiterated its offer of a silver medal for any useful or interesting memoir. The Cercle printed its revised prize program and sent copies to the administrators.[84]

Even though the response to its prize contests was scant, the Cercle des Philadelphes was nonetheless the hub of a good deal of "scientific" activity on the part of its associates, conceivably motivated by that promised piece of silver. In particular, colonists sent over 250 papers to the Cercle des Philadelphes in the two years from mid-1784 to mid-1786. In part to pacify authors, the Cercle regularly published lists of memoirs it received in the *Affiches Américaines,* and these lists not only indicate this substantial level of scientific activity, but they also allow a study of the areas in which the Philadelphes and their friends worked.[85] The results are noteworthy, but not surprising. Thirty-one percent of the papers dealt with a medical topic of one sort or another, mostly case reports.[86] Twenty-two percent concerned agriculture, rural economy, topographical descriptions, and meteorology. Botany and natural history papers, in some instances connected to issues of rural economy, represented 15 percent of the papers, followed closely by the other physical sciences, excluding meteorology, at 14 percent. The remaining 18 percent of contributions included literature, poetry, history, economy, and ethnography. This distribution is entirely characteristic of work taken up by the Cercle, as is the almost complete lack of theoretical concerns.

A major boost to the Cercle's fortunes occurred in the summer of 1786, when it successfully recruited Benjamin Franklin as an honorary associate. On behalf of the Cercle des Philadelphes, Arthaud wrote to Franklin from Cap François on March 15, 1786, saying, "We have the honor to send to your excellency the first works of a society formed to cultivate those arts and sciences befitting a colony like Saint Domingue. Accept our homage to the founder of the sciences and of liberty in the New World, and permit us to place your illustrious name among those of our associates."[87] The great man replied from Philadelphia on July 9, 1786.

> Sir, I received the Letter you did the honour of writing to me the
> 15th of March last together with the printed discourse that accompanied it. It gave me great Pleasure to find that the Improvement of Science is attended to in a Country where the Climate was suppos'd naturally to occasion Indolence, and an Unwillingness to take Pains except for immediate Profit. I am very sensible of the great Honour

done me by the Society of Philadelphians, in naming me among their Associates; and I beg they would accept my thankful Acknowledgements, together with the second Volume of the Transactions of our [American Philosophical] Society here. I am much oblig'd by the favourable mention you were pleased to make of me in your excellent Discourse at the first Opening of your Assemblies. Your Account of the Cape, contains a Variety of knowledge respecting it that we had not before, and many Particular observations for preserving Health, that may be useful to our Northern People who visit your Island. Wishing success of the Labour of the Society, I have the honour to be, Sir, Your most obedient & most humble Servant, B. Franklin.[88]

As all are aware, Franklin possessed the highest international renown in the contemporary worlds of science and politics. Franklin loved to add to the number of his learned society memberships, in part because of his role as president of the American Philosophical Society in Philadelphia. But, even so, Franklin's formal association with the Cercle des Philadelphes stood strongly to the credit of the French society in Saint Domingue.[89] More than that, it established a link between the Cercle des Philadelphes in Cap François and the American Philosophical Society in Philadelphia. With his letter Franklin was able to send the second volume of *Transactions* from the American Philosophical Society, but mostly the exchange went the other way, with the more productive Cercle des Philadelphes sending one or another of its works off to Philadelphia as they appeared over the next years.[90] Franklin himself wrote a second letter to Arthaud on December 11, 1787, especially to thank him for "the Books and Pamphlets sent by your respected Society to your Brothers here," but the Cercle des Philadelphes also maintained contact with the American Philosophical Society through Benjamin Rush and, later, Samuel Vaughan, Jr., both of whom the Cercle elected national cum foreign associates. In 1789 the American Philosophical Society reciprocally elected Charles Arthaud and Moreau de Saint-Méry.[91]

Arthaud tells us he was able to deliver the good news of Dr. Franklin's acceptance at the second anniversary meeting of the Cercle des Philadelphes on August 15, 1786.[92] No other indication exists that such a meeting actually took place, but, regardless, the news about Franklin no doubt lifted the spirits of the Cercle des Philadelphes as the group completed two years of effort and began a third.

Like its connection to Franklin, the Cercle's project, launched in July 1786, to publish a volume on tetanus stood strongly in its stead. The project began after Arthaud received a circular on tetanus published by the Société Royale de Médecine in Paris and forwarded by the local administrators. The Société Royale along with the medical branch of the navy ministry had undertaken a major study of the dreaded disease of tetanus (*le spasme*) and was soliciting information from all quarters.[93] The authorities approached Arthaud more as the colony's first royal physician than as president of the Cercle des Philadelphes, but Arthaud seized the opportunity for the Cercle. He compiled the papers the Cercle had already received on the subject, adding his own observations and a summary overview. The Cercle approved the assembled manuscript on September 21, 1786. The administrators added their approbation and permission to print the volume in the colony, and the Cercle's 104-page *Dissertation et Observations sur le Tétanos* appeared with a 1786 imprint (Figure 22). The Cercle distributed copies to the Royal Society of Medicine in Paris, to the navy ministry, and to the American Philosophical Society.[94] This expeditiously produced volume, essentially a first volume of memoirs for the Cercle (since Thierry's *Traité* still languished in the publishing process in France), provided considerable additional visibility for the Cercle, at least in the matter of tetanus, and no doubt increased the reputation of the institution as the fall of 1786 progressed.

Still and all, while it had taken giant strides toward establishing itself as an institution and while it had accomplished a considerable amount in two brief years, the Cercle des Philadelphes had not made much progress toward securing official recognition and support: it remained a private organization with the merest formalities of government approval. In the early part of 1786, the Cercle seemed almost to move backward in its efforts to secure a formal place for itself. Consider, for example, the problems that arose over its botanical garden. The first concession of land, so warmly granted to the Cercle by the administrators back in August 1785, turned out to be held in private hands, and, unfortunately, after having paid to improve the property, the Cercle found itself forced further to pay to rent it. The Cercle approached the administrators about another grant of land, but on April 6, 1786, the administrators flatly rejected any concession to the Cercle des Philadelphes until the organization received letters patent.[95] The administrators followed with a second harsh letter dated April 20, 1786, saying that "since it is doubtful [the Cercle] will be accorded letters patent," its members should consult a lawyer about a mechanism to secure some sort of legal status for the organization. The administrators repudiated the previous concession, and they reiterated their view that "an association not recognized by the Sovereign, even though approved by the administrators, has no legal existence in the social order and is not capable of ownership."[96]

DISSERTATION

ET

OBSERVATIONS SUR LE TÉTANOS,

Publiées par le Cercle des Philadelphes au Cap-François.

> La Médecine a pris naiffance de l'obfervation : c'eft l'obfervation qui la conduit au degré de perfection, & c'eft par le défaut d'obfervation qu'elle n'eft quelquefois qu'un verbiage vide de fens.
>
> *Traité de l'Expérience*, par Zimmermann, *Liv. III, chap. III.*

AÜ CAP-FRANÇOIS,

Chez Dufour de Rians, imprimeur breveté du Roi.

M. DCC. LXXXVI.

Avec Approbation et Permission.

FIG. 22. Cercle des Philadelphes, *Dissertation . . . sur le Tétanos* (1786), title page.

Then, in the fall of 1786, perhaps because La Luzerne took over as governor-general, perhaps on account of Franklin's association with the Cercle, perhaps because of the tetanus volume, or perhaps, as Moreau de Saint-Méry claims, because of pressure from J.-B. G. de Vaivre, a former intendant of Saint Domingue and then head of the colonial department within the ministry of the navy, government authorities changed their attitude toward the Cercle des Philadelphes and decided to accept and promote the institution.[97] In early October 1786, for example, the administrators in Saint Domingue granted the Cercle's request for a new concession of land for a botanical garden that could be truly theirs.[98] Then in December 1786, they signaled a new acceptance of the Cercle in deciding to attend a special public meeting of the institution. A new relationship began between the Cercle des Philadelphes and the colonial and national government.

The Cercle held this extraordinary, formal meeting on December 11, 1786, largely to honor its distinguished guests, the administrators. But the Cercle had a second purpose, to unveil the bust of Louis XVI commissioned by the Cercle des Philadelphes.[99] Thus, as representatives of the king, the administrators came to witness and to lend an extra air of official authority to what was a very solemn act of political loyalty.

The meeting opened with the presentation of Louis's bust, which if true to its fleshy counterpart must have been something to see.[100] It stood on a special square pedestal, 4 feet high, with a bas-relief carved on each side. On one side were His Majesty's royal arms, surrounded by the attributes of fame and maritime commerce, along with the inscription, "Erected to Louis XVI, protector of liberty at sea, the sciences, and the arts, by the Cercle des Philadelphes, November 21, 1786, in the administration of Messrs. de La Luzerne and de Marbois." The figure of Mercury adorned the second side of the pedestal with the attributes of Commerce, Science, and the Arts. On the third side the bust's pedestal depicted an embodied Fortune along with the attributes of the Sciences, a juxtaposition Arthaud felt called on to explain by saying, "The association of wealth with scientific talent is very rare, but this emblem shows that it must always exist."[101] On the fourth side Concord personified held the Cercle's cartouche atop the inscription, "The Cercle des Philadelphes established at Cap François, August 15, 1784, in the administration of Messrs. de Bellecombe and de Bongars."

While arcane to us, the Cercle's bust of Louis XVI represents an elaborate political symbol of the Cercle's loyalty and of the place the Cercle saw for itself in a world governed by Louis Bourbon. If the iconography was not already plain enough, Arthaud drove home the point when he avowed: "The sacred person of the sovereign never leaves the center of government, but his name is cherished in all parts of his empire, and if we are anxious to possess his image it is to express to him the sentiments in the heart of all Frenchmen and to offer him the respect he holds from the love of his people."[102] Not too many years later, both in France and in Saint Domingue, such deeply royalist political sentiments became capital offenses.

In any event, Arthaud next addressed the audience on the question "Can an academy be useful in Saint Domingue?"[103] He answered affirmatively, of course, but the body of his address dealt with the charge that, lacking learned men (*savants*), Saint Domingue could not support a learned society. In response, Arthaud emphasized the importance of observation in science and the necessity of developing a cadre of observers to complement men of genius. Arthaud suggested a role for the Cercle des Philadelphes as a center for scientific correspondence. He went on to speak of the positive "influence of cultivating the mind on the police of subjects," and he envisioned a less selfish and indolent colonist attracted to a "new order of things" in Saint Domingue.[104] Arthaud praised the administrators as men of letters and applauded their desire "to extend the empire of the sciences here and to erect for them a sanctuary that in posterity will attest to the wisdom of an enlightened administration."

Half a dozen other papers and reports followed, but time again ran out before poor Baudry des Lozières could read his *éloge* of Christopher Columbus. The ceremonies concluded with a tour of the "cabinet of natural history and mineralogy," during which the administrators seemed to express a particular satisfaction with the Cercle's project to survey mining resources and to form a complete colonial mineralogical collection. All then stepped down into the rue Vaudreuil, the special meeting with the administrators at an end. The day represented an important turning point in the fortunes of the Cercle des Philadelphes, especially in cementing relations with colonial government authority at the local level. Much else lay in store, but the meeting of December 11, 1786, affirmed the place of the Cercle des Philadelphes among French institutions governing colonial Saint Domingue.

Unbeknownst to the parties meeting in Saint Domingue in December 1786, another major development unfolded simultaneously in France: the granting of royal approbation and provisional royal recognition to the Cercle des Philadelphes. The minister himself, the maréchal de Castries, finally became won over to the Cercle des Philadelphes, and he moved independently to recommend that Louis XVI grant provisional royal approval and recognition to the organization. De Castries's letter of December 29, 1786, from Versailles to the administrators in Saint Domingue carried the good news.

> The society formed at Cap François under the name of the Cercle des Philadelphes has undertaken useful work that can encourage people to improve production in Saint Domingue and that can even speed progress to other colonies. I proposed to His Majesty to authorize this society provisionally in its present form. . . . You will convey to the Cercle des Philadelphes this mark of the King's goodwill, which it should consider as a very glorious recompense for its first efforts and as a powerful spur in endeavoring it to realize the hopes that it has raised. The Cercle des Philadelphes can hope to acquire an even higher regard by publishing interesting memoirs and by devoting itself to objectives of recognized utility.[105]

In December 1786, while it paid tribute to the sculpted image of the king and entertained his highest representatives in Saint Domingue, the Cercle des Philadelphes, of course, had no idea that at the same time the king was granting provisional royal authorization at Versailles. The Cercle learned of its esteemed new status four months later. The Cercle was far from having letters patent, but with royal recognition and an official status the Cercle des Philadelphes had passed a major institutional milestone.[106]

THIRTEEN

On to Letters Patent

P ROVISIONAL ROYAL recognition accorded the Cercle des Philadelphes of Cap François in December 1786 was doubtless "a very glorious recompense" for the labors of the Philadelphes over the two and one-half years since the Cercle's foundation in 1784. But provisional royal recognition was not the same as official letters patent and a formal legal charter of incorporation, and the successes achieved through December 1786 in no way took away from the ultimate goal and determination of the Cercle des Philadelphes to secure such letters patent. From 1787 through 1789 the Cercle continued actively to build its record of scientific work and accomplishment and to use this record to press the government further for letters patent and formal establishment.

Provisional royal authorization signaled the end of standoffish relations between the Cercle des Philadelphes and government authorities. The Cercle remained on probation of sorts, but after 1786, rather than having to prove itself as a bona fide institution, the Cercle des Philadelphes began working closely with government. A minor indication is the project begun by the Cercle des Philadelphes with government support in early 1787 to inoculate mules against glanders. Various epizootic diseases (epidemics affecting animals) posed real economic difficulties whenever they broke out. For example, in one case a single well-tended plantation lost two hundred mules, and in another outbreak authorities sought the advice of the royal veterinary school at Alfort in France. Arthaud summed up the situation when he noted: "Animal diseases cause losses that are harmful to the colonist, to commerce, and to the state. The

Cercle could not occupy itself with a topic more useful to the colony, nor could it take a more solid step toward earning public recognition."[1]

Jean Gelin, royal veterinarian in Saint Domingue, graduate of the Alfort veterinary school, and resident associate, launched the Cercle's glanders project with a memoir distributed throughout the administrative and medical bureaucracy. Based on Gelin's memoir, Arthaud suggested an inoculation against glanders similar to that against smallpox. The administrators responded with 1,000 livres and two mules for clinical trials, planned in conjunction with the commandant and ordonnateur (the top military and naval officers) in Cap François. The trials failed, as Arthaud and his surgical colleague, Roulin, no doubt succeeded in merely infecting the hapless mules with the offending bacteria, but government interest in the project remained high, and further trials followed.[2]

As part of this campaign, the Cercle solicited reports on animal diseases from colonial observers. (Such a solicitation was already in characteristic institutional style.) The Cercle then completed the project by publishing the assembled collection along with its own experimental results in 1788 as the *Recherches, Mémoires et Observations sur les Maladies Épizootiques de Saint-Domingue* (Figure 23). The third produced by the Cercle, this volume ran to 250 pages and contained fifty-four papers; that several articles dealt with slaves may or may not have blurred the distinction with epizootics.[3] Charles Mozard, reviewing the work in the *Affiches Américaines,* saw great value in it, especially since veterinary expertise supposedly did not exist outside of Cap François and Port–au–Prince. "The Cercle's goal in publishing this volume," he added, "fits perfectly with the end of the institution: public utility."[4]

The Cercle des Philadelphes and the entire northern department of Saint Domingue suffered a setback in 1787 when Louis XVI's government suppressed the Conseil Supérieur in Cap François and merged it with the Conseil Supérieur in Port–au–Prince. A unified Conseil Supérieur for Saint Domingue made sense for France, but an independent Conseil Supérieur had operated in Cap François since 1701, and its suppression was an "irreparable loss" that reduced the northern region to a secondary position compared with Port–au–Prince and the surrounding Cul-de-Sac area.[5] Arthaud claims that the Cercle des Philadelphes survived this "revolution" by dint of continued support from its judicial associates transplanted to Port–au–Prince. The court in Cap François was a major pillar on which the Cercle had come to rest, and Arthaud lamented for the Cercle: "It covered its losses without recouping them, but the Cercle will never forget the regrets this has left."[6]

The third anniversary meeting of the Cercle des Philadelphes went ahead as scheduled on August 15, 1787. The abbé Dicquemare—astrono-

RECHERCHES,

MÉMOIRES ET OBSERVATIONS

SUR

LES MALADIES ÉPIZOOTIQUES

DE SAINT-DOMINGUE,

Recueillis & publiés par le Cercle des Philadelphes
du Cap-François.

> Nous ferions trop heureux, fi nous
> avions rempli dignement les vues du
> Gouvernement : nous le ferions encore
> plus, fi cet Ouvrage peut contribuer à
> l'utilité publique pour laquelle il a été
> uniquement fait.
> *Rech. hift. phif. fur les Mal. épizoot.*
> *par M. Paulet, D. M. P. M. T. II,*
> *page 477.*

AU CAP-FRANÇOIS,

DE L'IMPRIMERIE ROYALE.

M. DCC. LXXXVIII.

AVEC APPROBATION ET PERMISSION.

FIG. 23. Cercle des Philadelphes, *Recherches . . . sur les Maladies Épizootiques*
(1788), title page.

mer, naturalist, and recently elected national associate of the Cercle from Le Havre in France—sent his own bust, which the Cercle unveiled on the occasion.[7] Someone donated a portrait of the late colonial associate, Lefebvre-Deshayes, along with books and papers from Deshayes's estate.[8] Arthaud read the *éloge* of Cosme d'Angerville, chief royal surgeon and a founder of the Cercle, who unfortunately also had died during the year. The Musée in Bordeaux sent the Cercle des Philadelphes a formal "Diplôme d'association" uniting the two academical societies, and this presentation must have been a highlight of the meeting and a significant source of institutional pride for the Cercle.

At the third anniversary meeting in 1787 the Cercle des Philadelphes for the first time awarded a prize. The prize (presumably the promised gold medal) went to the Montpellier-trained physician from Nîmes, Baumès, for his memoir on infantile convulsions (tetanus). The Cercle had originally proposed the question in 1785 and received at least seven memoirs in response.[9] However, the Cercle could only be disappointed in being able to award only one prize after two complete cycles of its prize competitions, and it made no further prize proposals in 1787 for the years upcoming.

The Cercle des Philadelphes did present a new roster of members— the first since 1785—at the anniversary meeting in 1787, a roster subsequently published in the official "green book" of the colony, the *Almanach de Saint-Domingue*.[10] A check of the roster shows the Cercle's membership to have grown from thirty-five to ninety active members in the two-year interim. This considerable expansion brought its membership up to the numbers the institution had established for itself. Six new resident associates appeared on the list: three *habitants*, two lawyers, and one apothecary.[11] With Franklin leading the list, the Cercle likewise added a half dozen new honoraries. Thirty new colonial associates joined making a total of thirty-eight active colonial associates. These included thirteen medical professionals, thirteen plantation owners or managers, three surveyors, three military officers, two priests, and one *démonstrateur de physique,* Professor Millon.

The largest change in the Cercle's membership occurred in the category of national (foreign) associate, with that category growing from two to twenty-two members. Except for two men—the Philadelphia physician, Benjamin Rush, and Maërter, imperial botanist from Vienna—active national associates in 1787 all lived and worked in France, but only half were in Paris. Among the national associates not yet mentioned, Le Brasseur, the naval commissioner (ordonnateur) for the colonies in the naval ministry, was certainly an important new member at this time. The Cercle elected four officers and associates of the Paris Musée as national associates, indicating the deepening affiliation with that institution. The election of Dubois de Fosseux reveals another notable institutional connection for the Cercle. Dubois de Fosseux was not only the permanent secretary of the Royal Academy of Belles-Lettres in Arras, but he also directed a nationwide, government-subsidized communication network in France that united most of the academies and the learned of the French provinces.[12] The Cercle also chose M.-P.-G. de Chabanon of the Académie française as a new national associate. Chabanon was born in Saint Domingue, which explains the connection. Chabanon accepted his election, but he understood colonial conditions well enough to write back, "advising the Society to devote itself to useful matters rather than those of pleasure or adornment."[13]

Mademoiselle Marie le Masson le Golft became a new national associate from Le Havre and the Cercle's only female member. Coming to the Cercle's attention through Lefebvre-Deshayes and the abbé Dicquemare, Mlle le Golft was an amateur naturalist, and it was appropriate—but still characteristically sexist—that she should join and adorn the Cercle des Philadelphes.[14] Mlle le Golft was the author of the *Balance de la Nature* (1784), wherein she apparently "weighed" the "qualities" of various animals. She wrote other papers on natural history, on education, and on human ethnic and racial diversity, all of which she sent to the Cercle des Philadelphes after her election in 1785.[15] Mozard reports on her paper concerning the "toilette" of flies, wherein she concluded that flies are not ruminants but merely fastidious and clean, and about which Mozard concluded (facetiously?): "One sees that Mlle Masson le Golft does not spare her sex. There is spirit in this piece, but its ideas are more superficial than sustained. Those who know natural history will judge the learning of this author, who doubtless only wanted to let it be suspected."[16] This provincial mademoiselle of science provoked notably warmer reactions from others of her new colonial associates in Saint Domingue. Lefebvre-Deshayes offered his new species of sea anemone to be weighed in Mlle le Golft's "balance"; then, the count d'Ingrandes composed an ode on the conditions of women, dedicating it to Mlle le Golft.[17] About this ode and the question at hand, Arthaud added the surprisingly liberal comment: "This is a stale topic perhaps, but it is always interesting to repeat to women that it's more the fault of governments and of custom than of nature if men are guardians of the empire of the sciences."[18]

The Cercle des Philadelphes launched a major institutional undertaking in the fall of 1787: its general survey of colonial agriculture, slavery, and rural economy. Of self-evident utility, the project began formally only after La Luzerne, then governor-general, forwarded an agricultural questionnaire prepared by the abbé Tessier, a royal agronomist in France. Arthaud noted somewhat testily that "the Cercle planned from the outset to examine all parts of agriculture and rural economy in Saint Domingue," and indeed the Cercle had announced such a goal in its prospectus of 1784.[19] The Cercle responded with its own extensive survey based on Tessier, its *Questions relatives à l'Agriculture de Saint-Domingue*.[20]

According to Arthaud, "the questionnaire could be regarded as the prospectus to a work on colonial agriculture," and with five thousand copies printed it received wide circulation. Mozard announced the Cercle's survey in the *Affiches Américaines,* saying that while the circular could not be abstracted, the Cercle had charged him to notify interested colonists to contact Arthaud in Cap François.[21] La Luzerne, who had indirectly precipitated the matter, responded to the Cercle's strong initiative by promising to distribute its questionnaire to all concerned. "I find it very interesting," he added, "that this colony could study itself, that it would develop the ambition, and that enlightenment gathered here could be communicated to metropolitan France and all peoples."[22]

Divided into three parts, the questionnaire itself set out 250 questions organized into sixty-nine articles. The first and largest part (forty-nine articles) concerned colonial agriculture per se. The aim was a comprehensive description of colonial agriculture, but the questions were very specific, with a clear eye on the practical and on improving agricultural production. The details of growing cane, coffee, indigo, and cotton figure prominently. For example, the twelfth article asked how the soil is prepared for each type of cultivation, to what depth, and with what instruments. The thirteenth article inquired about the number of cane holes to be dug per acre and how many canes planted per hole. Other questions concerned pests, diseases, fertilizing, irrigation, and meteorological conditions.[23]

The second part of the survey (seven articles) concerned slaves, and the third part (thirteen articles) dealt with animals. More will be said shortly about the Cercle's attitudes toward slaves and the juxtaposition of slaves and animals, but the points emerge amply in the Cercle's questionnaire. Regarding slaves, the Cercle wanted to know about illnesses, mortality, acclimatization, food, housing (the cheapest), and manpower requirements for each sort of plantation. The fourth article asked about which Africans were the easiest to discipline. The third part on livestock similarly inquired about supply, use, maintenance, and cost. The Cercle here expressed a special interest in the raising of sheep. In both these parts of the questionnaire, the practical again shines through.

The Cercle des Philadelphes manifested a strong institutional commitment to agriculture and issues of rural economy, and thus functioned as a colonial agricultural society, akin to the many such societies flourishing elsewhere in the Americas and in Europe. In this respect the Cercle's active involvement contrasts markedly with the colony's essentially moribund Chambers of Agriculture.

The Cercle printed a list of thirty-four responses to its agricultural questionnaire in the *Affiches Américaines* for October 17, 1789.[24] The Cercle promised to publish these memoirs, but by October 1789 the shock wave of the French Revolution had reached Saint Domingue, and no volume of agricultural memoirs ever appeared. The project seems to have proved barren. But not entirely, for Moreau de Saint-Méry probably folded the various reports from the Cercle into his massive *Description de la Partie Française de Saint-Domingue*. Georges Anglade makes this point, and he shows that the Cercle's effort in fact began as an adjunct to Moreau's larger campaign to gather information about Saint Domingue.[25] The Cercle's agricultural survey thus had a fruitful outcome, after all.

The survey of 1787 represents the Cercle's deepest institutional involvement in colonial slavery. The questions posed about slaves make clear that, despite an aura of liberalism, the Cercle des Philadelphes was a profoundly racist institution and an unquestioning supporter of the slave society out of which it emerged. As a patriotic and activist organization, the Cercle energetically investigated slavery in Saint Domingue, and it was ever anxious to exploit science, medicine, and expert knowledge to bolster and improve the colony's slave system. Although not highest on the Cercle's explicit institutional agenda, issues involving slavery formed not a negligible part of its mission. The Cercle's involvement with slavery clearly reveals an institution committed to employing its resources and the resources of science to conserve and enhance rather than to challenge the existing colonial system in prerevolutionary Saint Domingue. In these matters the Cercle's goals were always practical: to increase efficiency, productivity, and profit, and the Cercle never deviated from the conservative and clear principle enunciated in its initial prospectus: "to moderate the lot [of slaves], without harming the interests of colonists."

In response to the Cercle's questionnaire, various individuals submitted memoirs dealing with various aspects of slavery.[26] J.-F.-de-S. Oulry, the count d'Ingrandes, a colonial associate from Fond-des-Nègres, took a special interest in food for slaves and warding off famine. He produced memoirs on the potato, the custard apple, and the conservation of grain, the latter memoir Arthaud calling "very interesting for the subsistence of slaves and poultry."[27] Arthaud painted a sympathetic picture of d'Ingrandes:

> An attentive and enlightened cultivator, M. d'Ingrandes knew that a colonist ought to treat slaves as men, that he concern himself with their conservation both for his own interest and out of a sense of humanity. . . . He hoped that decent lodging would be given to slaves, that one take care to clothe them, and that they be treated with moderation. He observed quite correctly that excessive punishments brutalize, harden the cruel, and excite an indignation and a resentment, the effects of which are only to be feared.[28]

By this account d'Ingrandes appears liberal and farsighted, but elsewhere Arthaud staked out the true boundaries of liberality and racial tolerance.

> Much has been written about the origin of slavery. Much has been mouthed against the ills and abuses of servitude without examining closely the relation of master and slave and without appreciating that those who command are often enslaved to those who take orders. One has even dared to discuss the emancipation of slaves in the colonies. If the specious reasoning brought in favor of this philosophical system could seduce our kings and rulers, the result would be that we would have to give up the colonies for want of the manpower to develop them, because their products would no longer balance the expenses necessary to keep them up, [and] because of the insubordination of the freed slaves. Their license and their efforts to gain a living will cause disorder and disastrous troubles for those whom nature, even more than the law, seems to have destined to rule.[29]

For Arthaud and others, the colonial system in Saint Domingue depended on slavery; the idea of emancipation was inadmissible heresy and the superiority of the white man a scientific fact of nature. Still, colonials who wrote about slavery strove to depict conditions of slaves in a positive light, particularly in comparing slaves in the colonies with peasants in Europe. The marquis du Puget, for example, elaborated at least one side of this comparison.

> I do not believe it possible to hold two different opinions on the fate of slaves when one has witnessed without prejudice the manner in which they are treated in the great number of plantations in all the

colonies through which I have traveled, and when one has seen as close as I have the shocking misery of the peasants of France. This word, liberty, is illusory, because without an independent fortune, as soon as you must work to live, you are the slave of those who would employ you. And if a personal interest does not immediately attach the [working] slave to his master of the day, he is inhumanely abandoned as soon as illness or old age prevent him from doing the work from which he obtains his feeble wages.[30]

By and large the association of the Cercle des Philadelphes shared a view with the rest of the colonial elite, considering slaves as a degraded species of humanity along a continuum with other animals.[31] On occasion, members of the Cercle des Philadelphes concerned themselves with the theoretical causes of racial differences. For Arthaud and (presumably) the colleagues he mentions in his article on albinos (Vatable, Gauché, Lefebvre-Deshayes), climate and environment play a direct role in creating different species (*espèces*) of men:

> If we judge by analogy, we would say that there are many species of men. It seems that nature varied her designs to establish harmonies between climates and organized bodies. . . . It is probable that the different species of men were formed through the necessary and reciprocal interactions that must coexist between men and the climates of the regions they inhabit. We cannot believe that black men who inhabit low-lying coastal areas at the equator are the same species of men who inhabit the elevated and mountainous interior of Africa.[32]

Arthaud thus saw more "degeneration" and blending among human types than Buffon and Maupertuis, with whom he contrasted his view. Historians of French racial attitudes document several conflicting tendencies among eighteenth-century theorists: monogenism and the church's view of the essential unity of mankind; polygenists, like Voltaire, who would make Africans an entirely separate species; the notion of the Noble Savage that tended to elevate non-Europeans; the idea of the Great Chain of Being that put Africans between man and animals; and environmentalists (like Montesquieu and Arthaud) who saw social and cultural differences among human groups to be more a function of ecological circumstances.[33] Despite these differences and a certain lack of theoretical clarity and consistency among their exponents, scientific consensus was virtually universal among European and colonial intellectuals as to the biological inferiority of blacks and the superiority of whites. The point certainly holds true for the Cercle des Philadelphes and their associates in

Saint Domingue, and no clearer instance could be found of theoretical science reflecting and serving the interests of colonialism and the forces of oppression.

The Cercle's edition of Thierry de Menonville's *Traité du Nopal et de la Cochenille* finally appeared in the fall of 1787, and the Cercle used the occasion to press again for letters patent. The Cercle des Philadelphes approached La Luzerne in September 1787 while he was still governor in the colony with the offer to exchange Thierry's *Traité* for La Luzerne's translation of Xenophon's *Retreat of the Ten Thousand*. Also in September the Cercle sent copies of Thierry's *Traité* and its agricultural survey to the minister in France (de Castries), asking him "to make these works known to His Majesty and to present him their desires to obtain the confirmation of their establishment." By the time their package reached France and was processed within the ministry, however, La Luzerne had quit Saint Domingue and replaced de Castries as state minister for the navy and the colonies. In January La Luzerne's subordinates passed on the copy of Thierry's *Traité* sent by the Cercle to de Castries and an internal memo relating to the Cercle's further request for letters patent.[34]

La Luzerne rejected the Cercle's request for the time being. On the memo just mentioned he penned: "Give them hope for a permanent constitution if they continue to merit one through their zeal and their works." The official letter from Versailles gave the proper encouragement along with the implicit rejection of letters patent: "I told the King of the utility of your works, and he was satisfied. His Majesty will willingly grant you a permanent constitution if you continue to merit this favor through your zeal to promote and diffuse useful works."[35]

On another level the Cercle's fortunes progressed nicely and with considerable support from La Luzerne. In October 1787 the administrators inquired about the Cercle's finances and its expenditures to date. Arthaud reported 40,000 colonial livres as the Cercle's total expenses from August 1784, a figure that accords well with 1,000 livres annual dues assessed resident associates. These amounts represent a significant investment in the Cercle by its active associates, clearly an affluent group. But since much of the capital costs of their society had been met, the Cercle requested an annual subvention from the government of only 3,000 livres. Officials approved the subvention, and the Cercle des Philadelphes began to draw monies from the general fund (*caisse générale*) of the intendant. The Cercle likewise received the government's postal cover as part of these new arrangements.[36]

Sometime in late 1787 and early 1788 the Cercle des Philadelphes took money from its accounts and sent off to France to have struck a commemorative medal or *jeton*. Figure 19 depicts the *jeton,* engraved by the king's royal engraver, Simon. On the front of the coin, in positively grotesque relief (he was then thirty-four years old) is Louis XVI, king of France and of Navarre, and the ominous yet quintessential date of the Old Regime, 1788. On the back, the insignia and legends of the Cercle des Philadelphes: a beehive and hovering bees beneath the rays of a meridian sun encompassed with the motto, "Exercet sub sole labor" ('We Do Our Work beneath the Sun'). At the bottom, the inscription, "CERCLE DES PHILADELPHES, ETABLI AU CAP, 1784."[37] Étienne Taillemite notes a strong resemblance with the emblem of the Musée of Court de Gébelin in France, with which Moreau de Saint-Méry was associated, and both Taillemite and Maurel see in the Cercle's iconography clear evidence of its Freemasonry.[38] Still, the less complex imagery of busy bees performing their communal labors beneath the tropical sun fits and befits the Cercle des Philadelphes just as well. Each associate of the Cercle received a silver *jeton,* and the Cercle also sent coins to the colony's administrators and to their wives.[39] Silver coins possessed monetary value, of course, but the payoff to members of the Cercle for their financial sacrifices over the years was literally and figuratively token.

The Cercle des Philadelphes wrote to La Luzerne in April 1788, overflowing with gratitude and enthusiasm. They thanked the minister for showing their work to the king in January (even if letters patent were again not forthcoming), and they thanked him for the 3,000 livres they received from the intendant. The Philadelphians then beseeched La Luzerne to assume the title of protector of the Cercle, and they begged for his portrait for "the first temple erected to the Sciences in the colonies of France."[40] La Luzerne wrote back, refusing his portrait and the title of protector, but approving the grant of 3,000 livres from the intendant.[41] La Luzerne wrote to Barbé de Marbois the same day, saying, "I told His Majesty, who approved and confirmed in advance [!] the arrangements you have made in this regard," but Barbé de Marbois replied with surprise from Port-au-Prince, reminding La Luzerne that he, La Luzerne, had approved the expenditure for the Cercle des Philadelphes while he was governor in Saint Domingue.[42] But the sly ways of Old Regime bureaucrats should not obscure the fact that in the process the Cercle des Philadelphes had fully become a government institution.

While pursuing their political contacts through 1788, the Cercle des Philadelphes remained no less active in carrying out its program of work and research. The Cercle's investigation of the mineral waters of Saint Domingue is another example of its continuing scientific and practical efforts. Arthaud knew that the early work of the Paris Academy of Sciences in the 1660s concerned mineral-water analyses, and he noted that "this subject, interesting for natural history and for human health, has merited government attention in all our countries." Gauché explicitly tied the Cercle's mineral water project to the work of the Paris Academy and to more recent calls by the Royal Society of Medicine in Paris for chemists everywhere to apply themselves to the subject. For Gauché also, the project should "prove to government that the Cercle des Philadelphes can, as much as any other society, take on the motto: We are born, not for ourselves, but for public service."[43] The Cercle published the resulting collection of papers at government expense in 1788 as the *Mémoires du Cercle des Philadelphes, Tome Premier* (Figure 24).

The Cercle's first volume of memoirs per se, this new tome was actually the Cercle's fourth book-length publication in as many years. The volume of 264 pages incorporated twenty-four articles, fourteen of which presented analyses of various hydrological sources in Saint Domingue. The Cercle des Philadelphes supplied subventions for some of the analyses. Joseph Gauché wrote seven of the memoirs; Arthaud wrote five, and the connection between Gauché and Arthaud, as director and medical inspector respectively of the Eaux de Boynes facility, should not be overlooked in this connection.[44] The mineral-waters project of the Cercle des Philadelphes promoted both the Cercle and the Eaux de Boynes to the authorities on the appealing basis of chemical and medical science. Reviewing its volume of *Mémoires* in the *Affiches Américaines,* Mozard praised the Cercle for its patriotic aims and for following the example set by the Paris Academy of Sciences. About the Cercle itself Mozard noted pointedly: "Their work no longer allows one to doubt the utility of this academical society."[45]

J.-L. Jauvin, naval ordonnateur in Cap François and new honorary associate of the Cercle des Philadelphes, initiated the project to improve methods for storing and preserving flour. One of Jauvin's jobs was to superintend the provisioning of the colony, and he became particularly sensitive to the problems of flour stores. Arthaud vividly described the issues involved.

> Because of alteration by fermentation and prompt corruption by infections of insects, the difficulties involved in preserving flours in the royal storerooms cause extra expenses and often reduce the colonial administration to furnishing very bad flour for the subsistence of the troops. The general public suffers when town bakers buy old flour from the storerooms, in order to mix it with fresher flour, a practice that gives a disagreeable color to the bread, a bad taste and

MÉMOIRES

D U

CERCLE DES PHILADELPHES.

TOME PREMIER.

* *
* * *
* *
* * *
* *
* *

AU PORT-AU-PRINCE,

de l'Imprimerie de M O Z A R D,

Associé du Cercle des Philadelphes & Rédacteur de la Gazette
de Saint-Domingue.

M. DCC. LXXXVIII.

AVEC APPROBATION ET PERMISSION.

FIG. 24. Cercle des Philadelphes, *Mémoires du Cercle des Philadelphes, Tome Premier* (1788), title page.

unhealthy qualities. Because of the difficulties and uncertainties of provisioning the colony with flour, especially in times of war, flour is kept in the royal storerooms as long as possible.[46]

Jauvin petitioned the Cercle des Philadelphes to research the best and cheapest means of preserving flour in the king's storerooms, and Jauvin's office underwrote the expenses. Ducatel and Arthaud constituted the committee, and they performed a series of inconclusive experiments that could not guarantee the safe preservation of flour in the colony for a year.[47] Sanguine about the outcome, Arthaud said that only through a whole series of works, including dead ends, could a few truths be discovered. The Cercle did not desert the flour-preservation project, but posed the problem as a prize contest later in 1788. In 1789 the Cercle proposed further experiments on preserving flour and again invited colonists to submit memoirs, most especially practical ones. The Cercle ultimately awarded an honorable mention to one Cassan, first royal physician in Saint Lucia, for his work in this area. What Cassan proposed exactly is not known.[48]

Along these lines, "wishing to fulfill the desires of the administration," in 1788 the Cercle distributed the clove trees it received from the ordonnateur Jauvin and from the royal botanists in Saint Domingue and Cayenne, Nectoux and Richard.[49] At this time also, the Cercle gave away grass seed intended for eroded soils and additional forage for animals. J.-B. Auvray, a founding member of the Cercle, had conducted trials with the grass in question (known as "La Luzerne"!), and he provided the seed. Pound and half-pound packages were available free at Ducatel's pharmacy in Cap François.[50]

In another notable project, in March 1788 the Cercle decided to publish the abbé de la Haye's botanical work, *Florindie, ou histoire physico-économique des végétaux de la Torride,* and the abbé in turn dedicated the work to the Cercle des Philadelphes. Curate in the remote, mountainous parish of Dondon, de la Haye was a devoted botanist and botanical artist.[51] The Cercle planned to publish de la Haye's *Florindie* (for *Flore Indienne*) by subscription, and to this end the Cercle printed a separate, four-page prospectus. The prospectus promised a two-volume, in-quarto illustrated natural history of Saint Domingue along Linnean lines for 66 livres per volume.[52] Probably because of the requirement to have plates engraved, the Cercle sent de la Haye's manuscript to be published in France and enlisted A.-L. de Jussieu to see the volumes through the press.[53] Had de la Haye's book appeared as anticipated, it would have been the most significant work on colonial botany in the second half of the eighteenth century and another feather in the cap of the Cercle des Philadelphes. Unfortunately, other events of 1789 forestalled this publication.

All of official Cap François turned out for the fourth anniversary meeting of the Cercle des Philadelphes held on August 15, 1788. As Mozard reported it, the meeting began at 4:00 P.M. in the Cercle's quarters on the rue Vaudreuil, "in the presence of Messieurs the army Commandant, the Ordonnateur, the officers of the seneschal court, military officers, members of the Chamber of Commerce, and a large number of other people distinguished in various stations."[54] Society wives of these men no doubt also embellished the occasion.[55] The establishment in Cap François (as elsewhere) had accepted the Cercle des Philadelphes, and the public meeting of the Cercle in turn served to legitimate the existing power structure by providing an occasion for it to display and renew itself. In this sense, the public meeting of the town's scientific society became an updated version of the medieval parade.

Arthaud opened the meeting with necrological notices of Philadelphians who had passed away since 1786. He presented, but could hardly have read his eighty-page manuscript history of the Cercle from 1784. The Cercle heard the prospectus for de la Haye's *Florindie,* along with the published report on the Neufchateau prize and the project to preserve paper from insects that dated from 1785. Several medical and natural historical papers followed. One Ycard, the Cercle's librarian, expounded upon the true sense of the word *philosophy;* artillery captain and resident associate Levavasseur recited a bit of poetry.

The busy anniversary meeting of 1788 concluded with the presentation of the Cercle's prize program for 1789, its first prize announcement since 1786.[56] The Cercle remanded the previously announced prize on the chemistry of sugar production, even though it awarded a silver medal to Quenet Duhamel, a colonial associate, for his paper on the subject. The Cercle likewise continued the Neufchateau paper prize. The Cercle announced the government-sponsored prize on preserving flour and a special prize sponsored by Siméon Worlock, a resident associate, concerning manpower requirements for sugar plantations. Finally, the Cercle posed a question in forensic medicine, asking about causes of death produced by worms and by different types of poison. Worlock's prize and the poison prize evidence again the Cercle's continuing concerns for rural economy and the police of slaves.

One of the Cercle's more notable resident associates, Jean Barré de Saint-Venant, was conspicuous by his absence at the anniversary meeting in August 1788. Barré de Saint-Venant, then fifty-one, was a highly successful manager and owner of sugar plantations. Essentially a founding member of the Cercle des Philadelphes, he had served as president of the society from 1786.[57] But Barré had left Saint Domingue for France earlier in 1788, in part on assignment from the Cercle des Philadelphes to lead a special delegation for letters patent before state minister La Luzerne in person. Prior to his departure, Barré wrote to the intendant, Barbé de Marbois, asking for a joint letter from the colony's administrators in support of the Cercle's petitions before the minister.[58] The Cercle des Philadelphes wrote a formal letter to La Luzerne, indicating that Barré would be the bearer of their requests before his office. Arthaud followed up in August and November of 1788 with packages to the minister wherein he repeated how productive the Cercle had been, how hard they were working for an official establishment, and how much they desired that he speak to the king on their behalf about letters patent.[59]

Barré de Saint-Venant, Moreau de Saint-Méry, and the marquis du Puget constituted the delegation that met with La Luzerne at Versailles sometime in the fall of 1788. Barré brought with him strong colonial credentials, as did Moreau, the youngest of the three, at thirty-eight. Du Puget, forty-six, a nobleman and military officer, was elected a colonial associate of the Cercle in 1786 while serving in Saint Domingue as inspector of colonial artillery. Du Puget subsequently returned to France, and at the time of the meeting with La Luzerne he was well connected at court as tutor to the Dauphin. Although no record exists of this meeting, one can easily imagine the delegation from Saint Domingue being received by La Luzerne in an office suite at Versailles. There, briefly amid the rest of the king's business, the veterans of Saint Domingue sat down to discuss the Cercle des Philadelphes. Doubtless at this point the minister consented to letters patent for the Cercle. La Luzerne assigned Moreau the task of drawing up the prospective document.[60]

Meanwhile, back in Cap François the Cercle des Philadelphes drafted its own statutes to serve as the basis for the letters patent. Arthaud sent the proposed new statutes and the desiderata of the Cercle regarding the details of its incorporation to the administrators in Port-au-Prince in September 1788. The administrators in turn forwarded the Cercle's materials to the ministry at Versailles in November, along with a strong letter in support of letters patent and increased funding for the Cercle. The Cercle sent a letter of its own alerting the minister that the administrators in Saint Domingue would be transmitting the Cercle's proposed new statutes, and they repeated their appeals that he put the matter before his majesty.[61]

In the meantime still, Arthaud's August letter, sending La Luzerne the Cercle's volume on epizootics and entreating his offices before the king, reached France and percolated upward through the ministry. An assistant prepared a summary document for the minister's attention on December 11, 1788. ₊The document reminded La Luzerne that he had promised the Cercle a "permanent constitution" if the Philadelphes continued their good work and that the Cercle was already receiving 3,000 livres. At the bottom of the document, the businesslike notation, finally: "Give them letters patent." [62]

La Luzerne granted clearance, but the letters patent themselves and several other points of institution building remained to be negotiated. Two draft versions existed of the letters patent, one (in thirty-five articles) written in Paris by Moreau de Saint-Méry, another (in seventy-one articles) sent via the administrators in Saint Domingue, and it fell to Moreau to fold the two versions together. [63] The administrators in Saint Domingue and then La Luzerne in France established the fine points of the new institutional charter in dealing with nine formal requests presented to them by the Cercle des Philadelphes. The authorities granted the request that funding for the Cercle be raised to 10,000 livres, and they approved the Cercle's legal right to receive additional gifts and to sponsor privately funded prize competitions. The government would not stand in the way of land for a new botanical garden, but it would not buy any land for the Cercle. *Jetons* were permitted the new society, but, as La Luzerne insisted in the margin, they had to be struck in France. La Luzerne and the administrators rejected the requests that the Cercle be allotted four of the king's slaves for the botanical garden and that its members be exempt from military service. They also rejected a franking privilege for the new society, presumably because the Cercle was already operating under the postal cover of the ministry. The internal ministry memorandum covering these points is dated April 17, 1789. [64]

The final version of the letters patent cleared La Luzerne's desk on April 30, 1789. The papers were shown to the king for his approval on May 13, 1789, and parchment letters patent were prepared. Louis XVI signed the documents creating the Société Royale des Sciences et des Arts du Cap François on May 17, 1789. [65] One can easily imagine this scene, too. It is a somewhat disturbing one. The setting is again Versailles, at a morning meeting, perhaps, and with a word or two of introduction La Luzerne presents the formal document incorporating the scientific society in Saint Domingue to his majesty, who signs his usual, "Louis." Other officials sign and endorse, and other seals and signatures are affixed then and later. Various liveried secretaries shuffle the papers. Perhaps the king said something; perhaps he merely turned to other business. The disturbing element enters when one considers that at that very moment elsewhere on the grounds at Versailles the Estates General of France—called for the first time since 1614—were meeting and had been meeting for close to two weeks. On May 17, 1789, the entire process of the Estates General stood at an impasse over verification of deputies. Louis XVI was deeply involved in these affairs. The declaration of a National Assembly in France and the swearing of the Tennis Court Oath were a month away, the storming of the Bastille just eight weeks away. The signing of letters patent for the Cercle des Philadelphes took place on the very threshold of revolution in France, and the moment, as historical vignette, epitomizes both the Old Regime and the history of science in the Old Regime.

The deed done, La Luzerne informed the administrators in Saint Domingue of his majesty's actions, and through the administrators he transmitted its official letters patent to the new Royal Society in Cap François. Arthaud replied warmly to La Luzerne from Cap François on August 22, 1789. For their part the administrators wrote back to La Luzerne, notifying him that the letters patent had been registered at the Conseil Supérieur in Port-au-Prince and that they, the administrators, would superintend the 10,000 livres allocated to the Société Royale in Cap François.[66]

The most remarkable feature of the letters patent was the new name, Royal Society of Sciences and Arts. "Cercle des Philadelphes" was too modest and, yes, too Masonic for a learned society incorporated by the king and the king's government. The legalistic preamble, in the king's voice, rambled on about the Society's foundation in 1784, its provisional recognition in 1786, and the Society's work, most recently its volume of memoirs. For these reasons the king judged that the time had come for a legal constitution, and he afforded his royal protection to the work of all members of the Society.

The first of the twenty-two articles of the document charged the Society "to make its principal occupation everything pertaining to the physical and natural history of the colonies and everything that might perfect farming, running plantations, the sciences and arts relative to manufactures, and the extension of commerce."[67] The scale of the new organization was larger than the preceding Cercle des Philadelphes, with forty resident associates envisioned over the previous limit of twenty. The letters patent reasserted the principle of government approval of candidates, but other provisions carried the preceding Cercle des Philadelphes intact into the new Société Royale, and otherwise left the renewed institution to govern itself, as before.

In 1789, then, Cap François was resplendent with a royal scientific society, the third chartered learned society in the Western Hemisphere, after the American Philosophical Society in Philadelphia and the American Academy of Arts and Sciences in Boston, and the one with the best support. The model of the French learned academy overreached the Atlantic and became successfully transplanted to the tropics. There, the new royal institution of science gave promise of even greater utility in the further development of colonial Saint Domingue.

As one of its first acts, the new Royal Society of Cap François wrote to the colonial administrators in Port-au-Prince and asked them to become honorary associates of the Society. They also renewed their requests that the minister assume the title of protector of the Society. The administrators in Saint Domingue wrote to the minister in France, saying that they had refused honorary membership because the Cap Society might conceivably differ with official policy. Still, seeing that the king had favored the Society with letters patent, they asked the minister's advice. La Luzerne replied to his subordinates in January 1790 that he had decided to accept the role of protector of the Royal Society in Cap François and that they should accept the posts of honorary associates. So they all were listed in the colonial almanacs that followed for 1790 and 1791.[68] A process of institution building that began with the founders' first meeting in August 1784 drew to a natural close in January 1790, when La Luzerne became the official protector of the legally chartered Société Royale. It seems unlikely, however, that the cherished portrait of La Luzerne ever adorned the Society's meeting hall on the rue Vaudreuil in Cap François.

Winning letters patent from the crown was one of two grand institutional successes achieved in 1789 by the Cercle des Philadelphes turned Société Royale. Likewise in 1789, coincident with the final push for letters patent, the colonial society in Saint Domingue also forged an extraordinary formal affiliation with the Royal Academy of Sciences in Paris. The additional backing from the world's leading scientific center added considerably to the support already forthcoming from the highest levels of government.

The idea of an overture to the Paris Academy of Sciences may have arisen with Barré de Saint-Venant as part of his mission to France on behalf of the Cercle des Philadelphes. The initiative might also have come from state minister La Luzerne himself, elected an honorary member of the Academy of Sciences in September 1788. La Luzerne was vice-president of the Academy for 1789, and he went on to succeed as president of the Academy of Sciences in 1790.[69] As former governor-general of Saint Domingue, it certainly flattered La Luzerne's career to see an association between the colonial science society of which he was soon to become protector and the mother of all such institutions in Paris in which he served as a ranking member and officer.

However the occasion was arranged, Barré de Saint-Venant headed a delegation from the Cercle des Philadelphes that appeared before the Academy within the royal precincts of the Louvre in Paris on Wednesday, February 25, 1789.[70] Moreau de Saint-Méry again accompanied Barré; the third member of the party was Barré de Saint-Leu, a naval officer and French national associate of the Cercle. The party presented several of the Cercle's printed works on this occasion, as well as an updated version of its statutes. Barré de Saint-Venant then addressed the assembled luminaries of French science on behalf of his associates in Cap François. His manuscript speech of fourteen pages is preserved. It may have taken half an hour to read aloud.[71]

Barré probably thought his audience not that familiar with Saint Domingue, so, after an opening in praise of the Academy of Sciences, he described the lush Caribbean paradise whence he came. As Barré would have it, for example, one had only to scratch the ground for a quarter-hour a day to secure a more than wholesome living, after which one could lend oneself to "the arts, the sciences and the crafts."[72] Representing planter interests before the Estates-General, Barré adopted a strongly political tone to his presentation to the Academy of Sciences:

> The Cercle des Philadelphes has witnessed with great gratitude and patriotic interest the moment where a prince, the likes of which heaven offers but rarely, proposes to regenerate the nation and consult his subjects on the means of securing their happiness.
>
> The Cercle des Philadelphes will prove its zeal by developing knowledge that can be offered to the sovereign and the nation regarding the common interest and by proving that, despite foreign rivals, our position is such that we can get along without the principle of the trade "exclusive" applied to colonial commerce.

Lest the learned gentlemen of the Academy think he overstepped his bounds in making clear the political agenda that brought him to France, Barré made the point that "political considerations regarding commerce have never been foreign to the sciences, and how could the Cercle des Philadelphes not be concerned with them, when its compatriots owe their existence to the continued exports of colonial produce."[73]

But Barré did not lose sight of his audience, and as much as possible he kept the focus of his presentation on points of scientific interest. For example, he remarked on an unusual diurnal cycle of barometric variations evidenced in the colony and, relative to geology, Barré went on at length about measuring the erosion of mountains through the silt content of rivers in order to establish a baseline for geologic time.[74]

But most of all Barré wanted to present the Cercle des Philadelphes to the Parisian academicians. Barré mentioned the Mesmer controversy and the origination of the Cercle in 1784, and he characterized the group as passionately committed to promoting the common good. He noted the Cercle's achievements and the recognition it had received from government to date. "Although just beginning and the only one of its type established at the end of the world, the Society seems to have been in existence for a long time." Barré let on that letters patent and 10,000 livres had just been approved for the Cercle, but he continued, "All these happy preliminaries will not satisfy us; we see here only the dawn of our existence and of our prosperity."

Barré elaborated the now familiar program of the institution he represented, emphasizing the useful at every turn. The practical benefits forthcoming from medical research received considerable attention, especially that regarding tetanus and venereal diseases. Botanical research promised to yield medically useful plants and possibly a substitute for quinine. The cultivation of spices in the colony was potentially a very fruitful area of further institutional activity. Research regarding the mineral waters of the colony would surely prove useful in medical practice, said Barré, and veterinary medicine had already seen a considerable advance through the work of the Cercle des Philadelphes on epizootic diseases. Barré envisioned a large field for the practical application of scientific principles in agriculture, manufacture, commerce, and the useful arts, and he emphasized the theme that science and enlightenment could overturn "dull routine and thoughtless prejudice."

> Such are the objects of the Cercle's work and energy. The tasks the Cercle has imposed on itself would exceed its forces if it could not hope that the learned of Europe and especially the most illustrious men of science of this realm would favor its efforts and encourage it through their counsel. . . . Our society is worthwhile, therefore, and if it did not exist, the Academy would doubtless want to create one like it.

Only at the very end of his speech did Barré make clear that in coming before the Academy of Sciences the Cercle des Philadelphes (and probably La Luzerne) had something more in mind than simple institutional goodwill or receiving the blessings of a senior society, or even establishing an ordinary institutional "correspondence." "In seeking support within the maternal embrace of the Academy of Sciences," Barré de Saint-Venant et al. proposed something more formal and specific: an actual union of the institutions, akin to (and even beyond) the special affiliation that from 1706 linked the Paris Academy and the Société Royale des Sciences of Montpellier.[75] Only the Royal Marine Academy at Brest enjoyed anything like Montpellier's special institutional association with the Paris Academy. Barré pleaded his case for a similarly special relationship with his society in Cap François.

> We await the moment when the Royal Academy of Sciences will give light to our eyes in granting us a favor it did not refuse the Society in Montpellier.
>
> Doubtless the academy [in Montpellier] possessed glorious titles that we cannot have yet, but given our distance, we need your help even more. Without it we cannot prosper. Without it interesting discoveries will remain in the shadows. The Academy knows all too well how important it is to enlighten this part of the world, and it will not refuse to adopt the Cercle des Philadelphes and invigorate it with its counsel . . . [for which the Paris Academy will earn] the gratitude, affection, love, and respect that a weak and poor child owes to the kind mother who adopts it.

Barré concluded his presentation by asking the Academy, regardless of the decision on affiliation, to accept a *jeton* of the Cercle des Philadelphes, and he ventured to request one of the Academy's in return. "May this exchange and mutual gift put a seal on the union we desire."

The assembled academicians no doubt applauded this spirited representation from Saint Domingue, and warm words were certainly forthcoming from the chair—conceivably occupied by La Luzerne himself, sitting as vice-president of the Academy, and other remarks may have been offered by B.-G. Sage, *pensionnaire* chemist and academy director for 1789.[76] Out of civility if nothing else, the Parisian academicians doubtless honored Barré's request for a commemorative coin from the Academy on the spot, but his other request for an institutional union clearly required separate consideration by the Academy.

The composition of the seven-man committee appointed to report on the proposed affiliation with the Cercle des Philadelphes reveals a harmony of minds among the leadership of the Academy, for all of its officers served on the committee, including La Luzerne, the marquis de Condorcet (permanent secretary), and the director, subdirector, and treasurer of the Academy.[77] A.-L. de Jussieu, forty-one years of age, *pensionnaire* botanist (concurrently attending to the publication of the Cercle's edition of de la Haye's *Florindie*), and the venerable J.-B. Le Roy, seventy, *pensionnaire* in general physics, rounded out the committee.

Le Roy wrote the committee report and presented it to the Academy a month after Barré de Saint-Venant and the delegation from Saint Domingue appeared before the Academy. For the most part the report merely summarized Barré's speech and information about the Cercle gleaned from the statutes left at the Academy by Moreau de Saint-Méry. The laudatory report noted candidly that "it is not hard to form a clear idea of the Society of Philadelphes. In effect they all want to apply knowledge they acquire to perfect agriculture and augment the productions of the colony." The report did not reject the Cercle on that account but, rather, it underscored the scientific character of the institution, for *"physique,* medicine, chemistry, botany, agriculture, meteorology, and mechanics form the bases of its activities."[78]

"Regarding the most ardent wishes of the gentlemen of the Cercle des Philadelphes that the Company grant the honor of an affiliation like that with the Royal Society of Montpellier," the committee *rapporteurs* pretended not to have an opinion, "it being up to the Academy to decide." But they went on to observe that the Academy had long since recognized the advantages of having representatives on the spot over relying on travelers or emissaries. More than that, the Cercle des Philadelphes would become an actual agent of the Paris Academy, ready and (with its own institutional resources) able to undertake scientific commissions in Saint Domingue assigned by the Academy in Paris.

> By favoring this nascent society, the Academy will align itself with a number of well-informed people who will make the most of the opportunity. The Academy will accelerate the progress of useful knowledge on the island of Saint Domingue in a marked manner, and thereby render an essential service to the colony. Finally, it is important to remark that the Society of Philadelphes established at Cap François on the island of Saint Domingue in the Torrid Zone is such a special case that the Academy should not fear that any other society might impose on the Company for similar treatment.[79]

That some lesser provincial society might claim the same privileges represented the major obstacle to any formal association with the Cercle des Philadelphes. But other problems existed, too, over exactly what was entailed by an affiliation like that with the Royal Society in Montpellier. Probably because of the rights of the Montpellier Society to publish in the *Histoire et mémoires* of the Paris Academy, some academicians may have hesitated over an identical connection to the Cercle des Philadelphes, and the business went back to committee for clarification. A supplement to the committee report suggested the compromise that the Academy might treat the resident associates of the Cercle in the same manner as its individual *correspondants,* allowing them to attend the private sessions of the Academy while visiting in Paris for up to one year. In other words, resident associates of the Cercle des Philadelphes in Saint Domingue would become *correspondants* en masse of the Academy of Sci-

ences in Paris. The status of *correspondant* of the Paris Academy represented a signal honor in science in the eighteenth century, and the Academy's act to confer it on a body collectively was unprecedented.[80]

The Academy approved these propositions for an "association" between the Academy of Sciences in Paris and the "Royal Society of Cap
François" at the meeting on April 1, 1789.[81] The oldest and the newest of
France's science academies thus forged a remarkable new association, one
full of promise, and the Cercle des Philadelphes thereby came to occupy a
unique and extraordinary position within the orbit of the world's leading
institutions of science. In making the scientific connection to Paris, the
Cercle des Philadelphes achieved a great honor for itself that com-
pounded to abundance the honor already forthcoming from the highest
levels of government. The newly chartered royal scientific society in
faraway Cap François could not have enjoyed any more prestigious backing at its new beginning. Is it not telling for the history of science and
colonialism in the eighteenth century that the Royal Society of Sciences
and Arts in colonial Saint Domingue became a formal extension of the
Royal Academy of Sciences in Paris?

Barré de Saint-Venant's mission to France on behalf of the Cercle des
Philadelphes thus produced two stunning triumphs in 1789: royal letters
patent and affiliation with the Paris Academy. Awaiting word of developments in France, the society in Cap François held a public meeting
on February 21, 1789. The new governor-general of the colony, Du
Chilleau, and the intendant, Barbé de Marbois, both in Cap François, attended together as the Cercle's honored guests. Once again both the
colony's administrators graced a reunion of the Cercle des Philadelphes,
along with the rest of official Cap François, as was customary by 1789.[82]
After an undistinguished series of papers, Arthaud closed the session
with the peroration:

> The approbation of the administrators, the flattering encourage-
> ment they have given the Cercle, and the applause of the most dis-
> tinguished and enlightened citizens are the most honorable fruits
> that the Cercle can harvest from its works and the most satisfying
> recompense it can receive for its efforts to make itself useful to the
> colony.[83]

The passage reads as if Arthaud were taking a bow after the years of effort
to secure a place for the Cercle des Philadelphes.

Word of the association with the Academy of Sciences in Paris
reached Saint Domingue and the Cercle des Philadelphes in time to be
highlighted at the fifth anniversary meeting held in Cap François on Au-

gust 17, 1789.[84] J.-B. Auvray, president for 1789, began the anniversary meeting on August 17 by reading an extract from the minutes of the Paris Academy of Sciences, wherein the Academy established its tie with the Cercle, "a tie so much more satisfying for our society as the Academy accords such an honor only rarely." Arthaud read four necrological notices, including Mlle le Masson le Golft's *éloge* of the late Abbé Dicquemare. A dozen presentations by various associates followed, many with a medical slant. A committee report detailed the communications received by the Cercle concerning the phenomenon of freezing in the mountains of Saint Domingue, an inquiry initiated in response to the inquiries from the Paris Academy and the Bordeaux Academy. The meeting closed with a reading of the prize program for the year upcoming, 1789–90.

The prize program for 1790 makes plain that the Cercle des Philadelphes looked forward to continuing the institutional agenda begun in 1784. The Cercle reiterated its general call for observations pertaining to the colony. It solicited memorial *éloges* for Christopher Columbus and for the founders of the two poorhouses in Cap François. The Cercle posed two major questions: one, for 1790, asked about the use of agricultural machinery; the other, for 1792, again concerned slavery and inquired about "the diverse peoples of Africa, . . . the regime best suited to individuals transported to the colonies, and the type of work most fitting for them." The Cercle also posed a detailed question concerning indigo, and one about the use of fertilizers. Two medical questions closed the contests for 1792; one, to distinguish various tropical fevers, the other on the cantharis fly (Spanish fly), whose use had, apparently, become fashionable in certain medical circles.

The anniversary meeting of August 17, 1789, doubtless represented another grand social success for the Cercle des Philadelphes, but it was a shame that the royal letters patent sent from France missed arriving in time for the meeting by just a few days. The Cercle des Philadelphes knew indirectly that letters patent were "in the works," but, unfortunately, the Philadelphians could not publicly unveil their elevated status as a royal society along with the association with the Paris Academy of Sciences at the time of the meeting on August 17, 1789. Arthaud thanked La Luzerne for the letters patent in a letter dated August 22, so they arrived in Cap François at least by that date, five days after the public meeting.[85] The Cercle seems to have realized the opportunity it had missed, for having already printed the prize program and an account of the anniversary meeting with the titles, "Program of Prizes of the Cercle des Philadelphes" and "Public Meeting of the Cercle des Philadelphes," upon receipt of the letters patent the royal press reset the type to announce the

"Program of Prizes of the Royal Society of Sciences and Arts of Cap François" and the "Public Meeting of the Royal Society of Sciences and Arts of Cap François."[86] At the top of the revised publications, the Cercle announced the important facts as they should have been at the public meeting: "The king has granted to the Cercle des Philadelphes the title of Royal Society of Sciences and Arts, and with this title it has become affiliated with the Royal Academy of Sciences of Paris."

August 1789 marked the pinnacle of the Cercle des Philadelphes and of organized science and medicine in colonial Saint Domingue. The glad tidings coming that month of the official charter and the approbation of the Paris Academy represent the apogee of *both* the Cercle des Philadelphes *and* the Royal Society it had become. At this peak, old and new within the Cercle des Philadelphes stood poised, but perilously and for only an instance, for the French Revolution was simultaneously about to break upon Saint Domingue and the colony's new Royal Society. The Société Royale of Cap François would spiral downward for nearly three more years before disappearing altogether, but at the momentary apex of the institution in August and September of 1789 the dark days of the future lay ahead, even if immediately ahead. Moreau de Saint-Méry offered a prayer for the Cercle des Philadelphes at this, its climactic moment: "Happy association, may you last as long as the New World, and may my feeble pen preserve for your courageous and generous founders the debt of gratitude that is their due."[87]

·❧ FOURTEEN ·❧

Profile of an Institution

A S PIERRE PLUCHON has indicated, sources exist—notably annotated membership lists—that "give an exhaustive view of the sociology of the learned society of the Cap."[1] Several such lists are extant from the spring of 1785; the Cercle des Philadelphes published a separate *Tableau* of its members in 1787, and three other lists of Cercle members appeared in the *Almanach de Saint-Domingue* for 1789, 1790, and 1791.[2] Each of these lists gives members' names, occupations, and other descriptive information by membership category. With modest computer capabilities, it becomes possible to manipulate that information to reveal the sociology underlying science and the Cercle des Philadelphes in Old Regime Saint Domingue.[3]

One hundred sixty-two men and one woman belonged to the Cercle des Philadelphes/Société Royale in the period 1784–92.[4] These raw numbers reveal a Cercle des Philadelphes completely comparable with its sister academies in the French provinces, which averaged a total of 194 members. The Cercle was the same size overall as the academies in Besançon, Châlons, and Metz; but these institutions were founded in the 1750s, which makes the scale of the Cercle des Philadelphes, founded in 1784, more impressive.[5]

Table 4 presents a breakdown by membership category. Twenty honorary associates formed the smallest subgroup of members at 12 percent of the overall membership. Indicative of the high level of connections forged through the position of honorary member, half of the honoraries served in the military or as government administrators, and 60 percent were noblemen. Half of the Cercle's honoraries lived in or

TABLE 4. Cercle Associates

Category of Associate	Number of Members	Percentage of Total
Honorary	20	12
Resident	31	19
Colonial	70	43
National/foreign	42	26
Total	163	

around Cap François, and only three could be found outside the West Indies. These findings confirm what other evidence has shown, that the Cercle used the honorary position to build a local and regional base of contacts with government authority and other centers of power.

Not quite a fifth of the total membership, resident associates constituted the active core of the organization in Cap François. Resident associates overwhelmingly came from the third estate; only one, Fournier de la Varenne, was a nobleman, a chevalier. Thirteen had careers in medicine, five in the law. Eleven had some association with the military, but mostly as colonists in the militia or as military medical personnel. Nine of the Cercle's associates resident in Cap François were also labeled *habitants*.

Colonial associates constituted the largest group of the Cercle's members, approaching half. Colonial associates—that is, members in the colonies from outside the Cap François region—distributed themselves evenly around Saint Domingue (80 percent) and elsewhere in the French West Indies (20 percent).[6] Roughly a third of colonial associates (twenty-four of seventy) worked in one of the medical professions; another third (twenty-four) came from the ranks of colonists (*habitants*). Colonial surveyors and engineers (at 14 percent of colonial associates) found solid representation in this group. Colonial members were decidedly bourgeois (85 percent) but less so proportionally than the Cercle's resident associates.

The single category of national/foreign associate included Cercle members living in France or elsewhere overseas, and it accounted for just over one-quarter of the Cercle's total membership. The overwhelming majority of this group (83 percent) lived in France, with a surprisingly balanced distribution of half in Paris and half in the provinces. A remaining half-dozen foreign associates resided elsewhere: in the West Indies, in Louisiana, in Philadelphia, in Vienna, and possibly in Geneva. A third of national and foreign associates (fourteen of forty-two) associated themselves professionally with medicine. The sources identify nine national associates (21 percent) as academicians, figures indicating the Cercle's ties with institutionalized science and learning in the academies in France.

Eighteen individuals or 11 percent of the total membership held one or another of the officer positions within the Cercle during its existence.

Graph 3 displays the growth of the Cercle's membership over time.[7] It indicates an initial period of building a membership through 1787–88. The category of national associate, especially, does not become firmly established until after provisional royal recognition at the end of 1786. The graph shows steady growth through 1789, at which point the membership peaks at its operational level of 110 to 115 members. The graph likewise reveals that, at best, the Cercle's membership stagnated after 1789. Documentation exists for only three new members after that date, and the sources probably do not give an accurate view of the effective decline in the active membership of the Cercle after 1789.

The three states of the Old Regime (the clergy, nobility, and third estate) each found representation in the Cercle des Philadelphes, but the institution was overwhelmingly bourgeois with 77 percent of its members (126 individuals) drawn from the third estate. This figure contrasts sharply with the membership of the Paris Academy of Sciences, where the third estate supplied 54 percent of the members, and the Academy of Sciences was decidedly the most "common" of the Parisian academies. Similarly, its large third-estate element puts the Cercle des Philadelphes

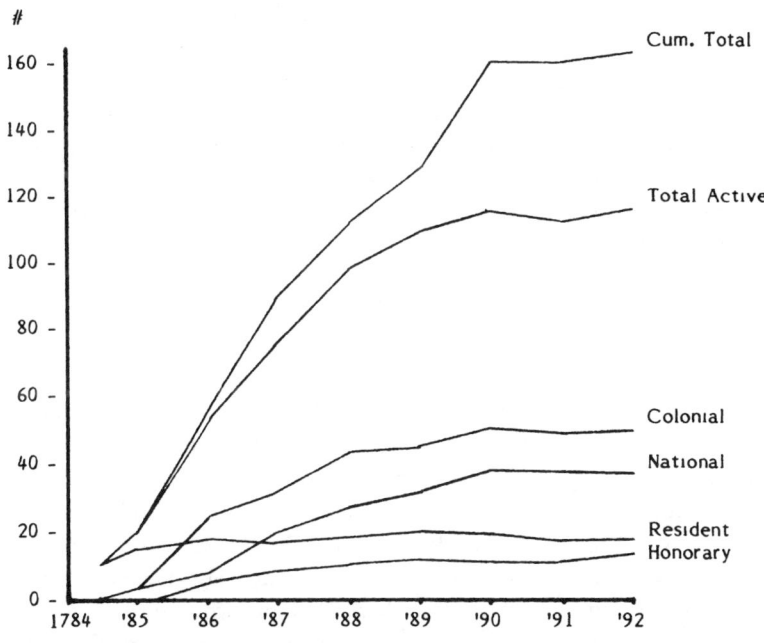

GRAPH 3. Membership of the Cercle des Philadelphes

among the most bourgeois of French provincial academies, comparable with those in Montpellier and Rouen.[8]

The nobility constituted 17 percent of the Cercle's members (twenty-eight men). Eleven of these twenty-eight nobles, or almost 7 percent of the overall membership, were from the high nobility (baron, count, marquis). The remainder of this group (10 percent of the total population) held lesser titles of nobility (chevalier, écuyer, conseiller). Noble associates of the Cercle predominated in the class of honoraries, and they likewise evidenced strong connections to the military.

Nine members of the regular and secular clergy belonged to the Cercle des Philadelphes, constituting not quite 6 percent of its total membership. Clerical members all came from the lower clergy and divided evenly into men of the church in the colony and in France.

Contemporary records shed valuable light on the occupations of members of the Cercle and types of economic employment that sustained them. Looking at the occupations of Cercle members allows one to be fairly precise about the constituent sociology of the institution. The ability to specify and compare occupations quantitatively is especially useful, since a small dispute has arisen over the sociology of the Cercle des Philadelphes. In particular, Pierre Pluchon takes issue with an interpretation put forth by Blanche Maurel. Mlle Maurel saw the Cercle's membership as composed primarily of physicians, talented amateurs, lawyers, planters, merchants, and high government functionaries. About the group as a whole she concluded, "Without exception, the members of the Cercle des Philadelphes belonged to the class of *grands blancs,* planters, royal officials, judges, and members of the liberal professions. Many were rich."[9]

What Maurel means to say is that the members of the Cercle came from the white elite of the colony, but Pluchon rightly criticizes her for seeming to incorporate so much into the category of *grands blancs,* a term more ordinarily limited to the planter class alone and even to sugar plantation blue bloods. But the issue is not simply semantic, for, while Pluchon recognizes that some of the more important members of the Cercle were indeed *grands blancs,* he argues the contrary point that overall the Cercle never succeeded in capturing the "sugar aristocracy" or in aligning itself with business or trading interests.[10] For Pluchon, the Cercle encompassed a more restricted group of urban professionals and intellectuals, having little to do with colonial planters or merchants.

How much the Cercle des Philadelphes was or was not connected to various groups in colonial Saint Domingue clearly affects any interpretation of the institution and its social base. The breakdown of occupations of members presented in Graph 4 allows one to weigh the alternatives presented by Maurel and Pluchon.[11]

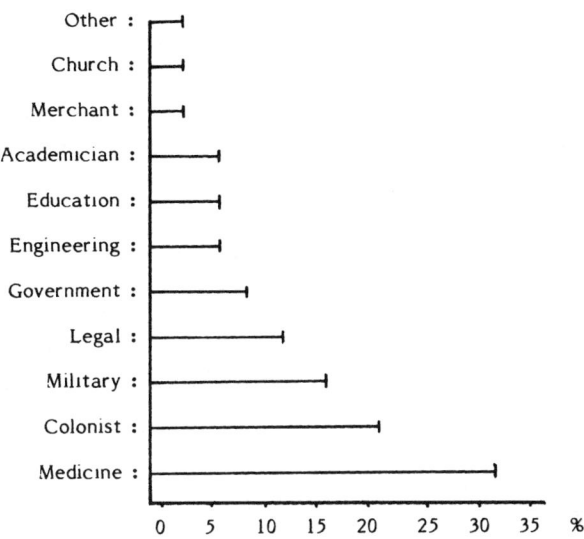

GRAPH 4. Occupations of Cercle Associates

Maurel and Pluchon do not disagree that physicians and those in the medical professions formed a large and important segment of the Cercle's membership, a segment confirmed in Graph 4 to be the largest and most important.[12] Not quite a third of the Cercle's members (31 percent) were medical professionals of one sort or another. Twenty-five physicians and seventeen licensed surgeons predominated in this group of fifty-one, which also included veterinarians, botanists, and professors of medicine. That nearly half of this medical subgroup of members (45 percent) held royal appointment and served in the medical bureaucracy further indicates the close ties of the Cercle des Philadelphes with official medicine in the colony and in France. To this extent, again, the Cercle des Philadelphes functioned as an official academy of medicine.

The category labeled "colonist" (*habitant*) stands out as the second largest component of the Cercle's membership. At first glance, so many *habitants* among the Cercle's members (thirty-six individuals representing 22 percent of the total membership) would seem to support Maurel and a link with the *grands blancs* of the colony. But Jean Tarrade distinguishes sharply between merchants and colonists as two opposed social groups with decidedly different interests and outlooks.[13] Finding so few members of the Cercle styled as merchants (*négociants,* only five individuals, a mere 3 percent) would seem more to support Pluchon's view of the Cercle as insulated from the commerce of the colony.

On balance, the weight of the evidence supports Pluchon over

Maurel. For example, only twelve of the thirty-six associates categorized as *habitant* can be shown to be plantation owners or rich proprietors, the rest may well have been more middle-class colonists, small-property owners or individuals employed to superintend and otherwise manage plantations they did not own, a finding that lessens any degree of linkage with the *grands blancs*. The majority of *grand blanc* planters in Saint Domingue no doubt went about raising cane and shipping sugar without much thought of the Cercle des Philadelphes. One should not overlook the several rich plantation owners who were among the leading members of the Cercle des Philadelphes (e.g., Auvray, Barré de Saint-Venant, Ingrandes, and Lefebvre-Deshayes), but the connection to merchant classes does seem very weak. Pluchon's notion of the Cercle des Philadelphes as an organization of urban professionals and intellectuals is very much on the mark. However, as the Cercle's agricultural survey and its other works indicate, the Cercle clearly associated itself with the social and economic interests of the *grands blancs* if less with the *grands blancs* themselves. Therefore, Pluchon may go too far in suggesting that the Cercle was entirely centrifugal, "fleeing the economic and social elite of the colony, its planters and merchants."[14]

Be that as it may, the general category of "military" constitutes the third major area of occupational activity pursued by members of the Cercle des Philadelphes. Twenty-nine men or 17 percent of the overall membership had some association with the military: thirteen as career officers (several at a high level), six ranking in the militia, and six as medical personnel serving in the military. Forty percent of members with military associations came from the nobility, indicating military connections at a high level. By the same token, a substantial portion (one-third) of the Cercle's resident members also had military associations, indicating the Cercle's local connections to the military in Cap François, which were reinforced through ties to the colonial militia.

Lawyers, judges, and others employed in the court system in Saint Domingue formed an important component of the membership of the Cercle des Philadelphes. Given what is otherwise known of the Cercle's robe element, it comes as a surprise to find that the Cercle drew only 13 percent of its members from robe circles.[15]

The Cercle's carefully crafted ties to institutionalized power are further seen in the fact that government administrators and bureaucrats account for fifteen of its associates (9 percent of the total membership). Ten of this group had military connections, and ten were nobles. The Cercle elected seven colonial governors-general and intendants.

Sources list ten "engineers" as members of the Cercle, including four royal engineers (*ingénieur breveté du roi*) and three royal surveyors (*arpenteur du roi*), all as colonial associates. The Cercle elected ten professors and others involved in teaching, almost entirely as national associates in France. Similarly as national associates, the Cercle elected another ten members identified as academicians.

Finally in this connection, ties between the Cercle des Philadelphes and the Catholic church would seem fairly weak; for, although the clergy provided nine of the Cercle's associates, the church gave employment to only four, and only three officiated as regular priests. The Cercle's notable colonial associate, the abbé de la Haye, somewhat belies the point, and one wonders about the Cercle's ties to Father Thimotée at Port-de-Paix, Father Balthazar at Port-Margot, and Father Bayard at Jérémie.

An important finding, not apparent in Graph 4, is that thirty-four men or 21 percent of the total population held a royal appointment of one sort or another, thus earning their livings as paid employees of the state. Most were royal physicians or surgeons, but the record notes several royal engineers and surveyors, a royal astronomer, a royal librarian, and a royal notary. This sizable group of technical specialists in royal employ reveals again just how much the Cercle des Philadelphes formed part of royal government and how much it was linked to institutions of colonial administration. If one adds military men and government administrators to this group of persons in royal employ, the Cercle's connections to government and the state become even more apparent.

Election to the Cercle des Philadelphes became a sought-after honor.[16] Jean Trembley, a long-time colonist, irrigation expert, and plantation manager from the Artibonite, made this point clear in one letter he wrote to his cousin in Switzerland, the famous naturalist Charles Bonnet.[17] Trembley dated the letter from the Artibonite, February 26, 1789, and it deserves to be quoted at length.

> I would ask you for a copy of Calendrini's discourse on comets and the sunspots, my good friend, because I want to use it in my relations with a learned society to which I have had the honor of being admitted. It formed in 1784 in Cap François, one of the two main towns in the colony, with royal approbation and the modest title of *Cercle des Philadelphes*. These Philadelphes were inhabitants of the town who with an enlightened taste for the useful sciences assembled to discuss such matters and to perform experiments. From meeting to meeting they saw that with their government approval they could form an official society like other academical bodies.

Since then, the society has been busy with several projects of the greatest utility to the colony. Although instituted only four or five years ago, it is already the equal of many European learned societies. Like them, it holds public meetings where it gives account of its work and announces prizes on questions of great utility. Private sessions are held regularly once a week, and the society has published good papers on natural history subjects of public utility. The class of resident associates, especially, honored itself by establishing at its own expense (through dues) a semipublic library, a cabinet of *physique* and chemistry, and a botanical garden. Two classes are not subject to dues, colonial associates and national or foreign associates. I had the honor of being elected to the former class, at the request of one of the principal founders [Arthaud?] and after sending a copy of my memoirs on hydraulics, per his request. This has made me a colleague of the celebrated Franklin and a few other distinguished men of learning.

As much as this is an honor for me, so much would it be for this learned society to count you among its members, and I could boast of you as a confrere. But you have so many distinguished literary credentials that this one would add nothing to your reputation. That's why I don't dare propose the small measure required by the rules of this company. If your zeal for the progress of science leads you to send the society your fine works, which would enrich its library, it would certainly receive them with plenty of gratitude. But I am persuaded that [even] without that, I have only to propose your admission, if you would accept it, for the society to send you a lovely parchment certificate like the one it sent me. I have already had occasion to mention you in one of my letters to the secretary.[18]

Baron Palisot de Beauvois, the *correspondant* of the Paris Academy of Sciences who ended up in Saint Domingue in 1788, provides another contemporary view of election to the Cercle des Philadelphes. In December 1788 Palisot wrote to his friend and confidant in Paris, the academician and botanist A.-L. de Jussieu.

Not only did I address myself to the Cercle des Philadelphes, a kind of academy established at Cap François, but I called upon several of its members and especially Monsieur Arthaud. I don't know if they were pleased with me or my zeal, but I have every reason to hope so. Monsieur Arthaud lately wrote to me that the Cercle has decided to make me an associate. You will guess that I accepted [immediately]. This offer was so respectable and flattering that one could not hesitate over it.[19]

Jean Trembley and Palisot de Beauvois give voice to the honor election as a Philadelphe had become, and they likewise make clear that the
Cercle welcomed all with any real interest in its activities.[20]

The Cercle des Philadelphes/Société Royale developed relations with
other scientific and learned academies, and these contacts linked the society in Saint Domingue to the contemporary mainstream of organized
and institutionalized science.

The special relation with the Paris Academy of Sciences, of course,
crowned its set of institutional associations. Given the catastrophic spiral
of events from 1789, the great promise of affiliating the scientific societies
in Paris and in Saint Domingue proved barren, and nothing substantial
came of the colonial connection to Paris. The Paris Academy did receive
descriptions and illustrations of colonial plants from de la Haye, a paper
from Gauché on "indigenous quinine," and an article from Dutrône la
Couture on the chemistry of sugar cane.[21] After their affiliation, the Paris
Academy contacted the Royal Society in Cap François through the abbé
Tessier about reports of freezing in the mountains of Saint Domingue,
and in early 1790 the Paris Academy heard a favorable report on a chain-
link suspension bridge designed by Barré de Saint-Venant.[22] The new
West Indian Royal Society elected academicians Tessier and Jérôme
Lalande in 1789. In September 1789 the Paris Academy received the vol-
ume of the Cercle's *Mémoires,* but after that the record is silent.[23]

Ranking after the tie to the Paris Academy of Sciences, important if
informal links forged through common membership and bureaucratic
channels connected the Cercle des Philadelphes with the Royal Society of
Medicine in Paris.[24] With Moreau de Saint-Méry effecting the previously
mentioned connection with the Paris Musée, the Cercle des Philadelphes
likewise made contact with the musées (free schools) in Bordeaux and
Toulouse and with the Royal Society of Agriculture in Paris. The Musée
in Bordeaux sent an official "diploma of association" to the Cercle des
Philadelphes in 1787, perhaps through Moreau's offices.[25] The Cercle des
Philadelphes likewise had dealings with the royal academies of science
and belles letters in Bordeaux and Toulouse, and several exchanges took
place with the Royal Academy of Sciences, Arts and Belles Letters in Dijon through its secretary, Hughes Maret. The Royal Academy at Rouen,
with the abbé Dicquemare as an associate, undertook experiments to test
dyes at the request of the Saint Domingue group, and the Royal Society
of Sciences and Arts in Metz asked the Cercle to publicize its essay contest on how to make French Jews more useful.[26] The Cercle similarly connected to the correspondence network centered on the Royal Academy in
Arras, as previously seen; and, at least through common members, the

Cercle des Philadelphes established relations with the French provincial royal academies in Caen, La Rochelle, Montpellier, Orléans, and Poitiers.[27] The active international connection to Franklin and the American Philosophical Society in Philadelphia has been detailed already, a connection strengthened by the elections of Arthaud and Moreau de Saint-Méry to the Philadelphia society in 1789. Informal ties through common members linked the Cercle to at least twenty-one other learned and scientific societies.[28]

At one point Arthaud mentions a campaign undertaken by the Cercle des Philadelphes to contact European academies and learned societies, so no one knows exactly how far the Cercle's institutional relations extended.[29] Regardless of their true extent, the associations enjoyed by the Cercle des Philadelphes/Société Royale show that the institution did not exist isolated at the edge of civilization but, rather, that the scientific society in Saint Domingue became well integrated into an international network and system of similar institutions spreading outward from Europe.

Having detailed in previous chapters the record of work undertaken by the Cercle des Philadelphes, that record needs only to be summarized to assess the different areas of institutional activity pursued by the Cercle. Judging from the Cercle's publications, papers submitted to the organization, and the general tenor of its record, one can say that the Cercle des Philadelphes functioned first and foremost as a colonial academy of medicine and center of expert medical knowledge. Agriculture and rural economy make up the Cercle's second-ranking set of concerns, and in this respect the Cercle des Philadelphes operated in a fashion more akin to a colonial agricultural society. Work in botany and natural history followed (much of it with a practical slant), after which came work in the physical sciences (*physique*). The Cercle proved very weak in mathematics and astronomy. The practical Philadelphes in Saint Domingue busied themselves very little with theoretical aspects of the sciences. Everything about the institution reveals it to have been closely adapted to its colonial circumstances and completely committed to public utility.

The character of the institution shows itself in two minor aspects of the Cercle's work: ethnography and mineralogy. The first concerned efforts to study native Amerindian culture extinct on Saint Domingue. Accounts of the native Arawaks and their destruction entered the historical record, of course, but in the eighteenth century little was known otherwise about the peoples who preceded the Spanish, the pirates, and the French in colonizing Saint Domingue. Charlevoix apparently promulgated a number of outlandish ideas about the Arawaks earlier in the eighteenth century, and in 1776 Father Nicolson characterized the whole subject as very obscure and called for collective ethnographical and archaeological research into "this unfortunate people." [30]

In 1786 the ever-active Charles Arthaud led the way in this research, publishing a small pamphlet entitled, "On the Native Inhabitants, Their Arts, Industry, and Means of Subsistence." Suffice it to say that, while dismissing several myths and presenting a substantially accurate view of the neolithic economy of the aboriginal Arawaks, Arthaud took pains to depict them as Noble Savages at peace in their natural surroundings: "These kindly and peaceable men needed neither our arts nor our sciences to be happy." The Arawaks were sober, virtuous, and animated by principles of equality and justice. "The destruction of the people of Hayti," Arthaud concluded, "is an epoch of misfortune for humanity." [31]

Arthaud returned to write two further articles about aspects of Indian culture. In the first, he rejected reports of head binding among the Indians, a stance that drew a critical response from a medical colleague in Guadeloupe. [32] His other paper discussed Arawak ritual phalluses and provided an occasion to meditate on Arawak natural religion. In these articles as well Arthaud sang the praises of the happy natural savages, and he asked, "Have we more virtues? Are we happier?" [33]

The Cercle des Philadelphes took up work in ethnography and anthropology in part through Arthaud, who styled himself prominently in these publications as the secretary of the Cercle des Philadelphes and who singled out the labors of other Philadelphes working in this area, notably Auvray. [34] But the Cercle itself strove to stimulate "research on the arts and industry of the original inhabitants," and it appealed to colonists directly for curiosities and antiquities. As a result, the Cercle's cabinet became a minor repository of Indian artifacts and a collection that marked "the beginnings of an Indian museum." [35] Auvray donated an Indian head recovered from a grave, Baudry contributed some stone axes, and the Cercle's cabinet also came to possess miscellaneous pieces of Indian pottery, jewelry of some sort, a stoneware mortar, other tools and utensils, several stone fetishes, and doubtless other artifacts as well. A rough kind of archaeological digging may have taken place under the aegis of the Cercle des Philadelphes, for Arthaud mentions that colonists found (and presumably disturbed) Indian tombs "many times." [36]

The cabinet of the Cercle des Philadelphes is of note, too, for its mineralogical collection. The Cercle early on sought to compile a catalog of the mineral resources of the colony, and it continued actively to solicit geological and mineral samples for its cabinet from colonists and other interested parties. The Cercle did assemble a modest mineralogical collection, and a separate catalog may have been prepared. To recall, the Cercle highlighted its mineralogical collection and its project to survey mineralogical resources at a special meeting in 1786 with the governor-general and intendant in attendance.[37]

Several articles touching on the mineralogy and geology of Saint Domingue appeared in the *Mémoires* of the Cercle des Philadelphes published in 1788. From these and other sources the rationale emerges for undertaking this research. Naturally enough, utility was the order of the day. In his memoir on a copper mine, for example, Gauché noted that "since France already pays a handsome tribute to Sweden for copper and since the government has decided to sheathe the ships of the royal navy with the metal, a good citizen can render a service to his country in showing it another means of saving money."[38] Regarding colonial coal deposits, Genton spoke of the utility of "this precious fossil" for local smithies, as a resource against wood shortages, and to power the extraordinary steam engine, whenever it arrived in the colony.[39] Arthaud indicated local sources for making good mortars and plasters, and he pointed to the need for local building stone that would resist decomposing like imported stone.[40] In a word, although the Cercle's work regarding colonial mineralogy did not constitute a major project, it represents yet another area of institutional activity, and it reveals the same practical spirit toward utilitarian ends that animated everything else the Cercle des Philadelphes pursued.

Several scientific papers and reports by Philadelphes from Saint Domingue, including Genton's on colonial mineralogy, appeared in the famous scientific journal published in Paris, the *Observations sur la physique, sur l'histoire naturelle et sur les arts,* otherwise known as Rozier's Journal.[41] At the time Rozier's Journal was a major scientific publication, the source of the most current news and substantive reports from the world of high science. Widely distributed and widely read, Rozier's Journal provided a major means for the rest of the world of science to learn about the little community in Saint Domingue and its work. In this regard one wonders what academicians in faraway Berlin, Stockholm, or Saint Petersburg thought when they read of the Cap hospital or of Arthaud's experiments with the crab-spider of the Antilles.

However much organized science achieved a foothold in the Americas through the institutionalization of the Cercle des Philadelphes, the colony in Saint Domingue remained a long way from the major scientific centers in Europe. The Cercle des Philadelphes amounted to a frontier fort for science, a post on the periphery. It had its place, and it served its purposes, as the Cercle's affiliation with the Paris Academy would indicate, but the institution in Saint Domingue and the men there who would be scientists stood cut off from the mainstream of contemporary science back in Europe.

Men like Arthaud and Nectoux held down full-time scientific posts in the colony, but Saint Domingue produced no indigenous or Creole scientists. At least two scientific amateurs in the colony, keenly aware of the limitations of their situation, lamented the career opportunities thus lost to them, for they would have done more with science had they been able. One, named Badoulé, concluded his 1788 report (on magnetic variations during earthquakes published in the *Affiches Américaines*) rather naively as follows:

> How many similar facts lie in oblivion that could lead to useful truths if persons trained to observe them and develop them were brought to the scene! My lack of wealth and my little knowledge do not allow me to take on such a task, but I would love to be able to sacrifice myself to that end.[42]

In his article on the sea anemone published not long before in Rozier's Journal, the second colonist, the more scientifically serious Étienne Lefebvre-Deshayes, voiced similar sentiments.

> That one cannot know everything or examine everything thoroughly is a true means of encouragement for those engaged in the immense undertaking [*carrière*] of natural history. But so as not to take wasted steps, or at least not to put out old knowledge as new, one must read all that has been written on the subject one is pursuing. Otherwise, how can one be assured of not simply repeating what has been said before? How can one know if one is adding anything to what our predecessors have established?
>
> There are nevertheless cases where it is not possible to follow this rule that reason prescribes. Such is the case, for example, of an inhabitant of a distant country, especially when he finds himself destitute of all scholarly resources. Should this man refuse to give in to his penchant for observation . . . ?
>
> It is our misfortune not to have any book that can instruct us in this very interesting part of natural history, and we don't even know if someone has already spoken about the strange animal that is the subject of this notice. . . . Unfortunately, there are no libraries in our area or even from far about. What a multitude of impediments and obstacles one encounters here at each step in the pursuit of science [*dans la carrière des sciences*].[43]

A successful plantation owner and colonist in Saint Domingue, Lefebvre-Deshayes was also a more than serious amateur scientist. Until his death in 1786 he numbered among the Cercle's most vital colonial associates. He became Buffon's official correspondent through the cabinet du roi. Deshayes also maintained contact with Lalande, Father Cotte, and the scientific community in France. He pursued his science and published a piece or two. He played his part as a Philadelphe, and otherwise fostered the development of science in the colony. Deshayes would have made even more professional contributions had the opportunity presented itself. In his *éloge* of Lefebvre-Deshayes the following year, Arthaud could not keep from mentioning Deshayes's "melancholic tendencies."[44] Clearly the poor colonial was depressed, at least in part because of barriers to a career in science in Saint Domingue.

One's inclination is to judge the history of the Cercle des Philadelphes positively. The creation of the institution, the royal recognition and financial backing it received, the energy and competence of its associates, the distinguished ties it formed with the Academy of Sciences in Paris and with Franklin and the American Philosophical Society, not to mention the Cercle's publication of four volumes of scientific memoirs (in addition to lesser pamphlets), the Cercle's prize contests, its active agricultural census, its botanical garden, and teaching efforts—all in an effective span of five years—argue strongly for thinking of the Cercle des Philadelphes as a vigorous and successful institution with plenty of promise left in 1789.

For Pierre Pluchon, however, the Cercle des Philadelphes failed.[45] Pluchon bases his judgment on the short history of the institution, the Cercle's lack of significant impact on practical knowledge or technique, what he judges the few publications of the Cercle and their minor interest, and the fact that the Cercle was not an educational institution. Pluchon might have considered more the background of organized science in France and the place of the Cercle des Philadelphes in the constellation of scientific societies internationally, but he is right about many of the particulars he mentions. The question is akin to asking whether the glass is half full or half empty. The present story gives a much more positive account of science in the Cercle des Philadelphes of old Saint Domingue.

FIFTEEN

The Fall of the Société
Royale of Cap François

M ANY SHIPS carried news from France across the Atlantic to Saint
Domingue in the summer of 1789. Two deserve particular notice.
The first sailed from France sometime in June and bore the letters patent
of the Royal Society of Sciences and Arts of Cap François. The second
flew on the wind with word of July 14 and the storming of the Bastille.
For a while in late July and early August 1789 the two ships plied the
Atlantic crossing together, the one with its momentous message bobbing
a month or so behind the other with its more parochial tidings.

As seen in a previous chapter, the first ship—the one with letters patent—just missed arriving in time for the anniversary meeting of the
Cercle des Philadelphes on August 17, 1789. When the first ship did arrive in Cap François, it brought its happy news for the associates of the
new Royal Society, and for the brief period through mid-September
1789, while the second ship was still at sea, the new Royal Society and the
colony as a whole enjoyed a last moment of normalcy (good or bad) before the tidal wave of revolution crashed down upon them.

The French Revolution did not come unheralded to Saint Domingue.
The colony had followed events in France closely, and planter interests
had sent their own deputies to the Estates General. But when word of the
fall of the Bastille reached Saint Domingue in late September 1789, and
even more so after the news of August 1789 and the Declaration of the
Rights of Man arrived in mid-October, an independent revolution began
in the colony, one that would not end for nearly fifteen years.[1] That process started with the spread of the French Revolution to Saint Domingue
in 1789, and it ended with the victory of the Haitian Revolution in 1804.

The extraordinary story of the fall of Saint Domingue and the winning of Haitian independence cannot be told in detail here.[2] Suffice it to say that circumstances were quite complex and involved several different groups, variously allied and antagonistic over the years: the French government (royal, then national), its bureaucrats, and its army; the white colonial elite (the *grands blancs*), the *petits blancs,* the free people of color, and, of course, the slaves. To this complicated mix one has to add invading British forces and remnants of Spanish power on Hispaniola, all acting and reacting against the flux of events in France, in Europe, and in Saint Domingue itself over the period from the fall of the Bastille to Napoleon's coronation. The process of revolution in Saint Domingue developed in several tortuous stages and contained several different revolutions within itself: a struggle among whites over colonial autonomy and independence; mulatto rebellions for representation (the first in 1790); the great slave uprising and revolt beginning in 1791; race war and the collapse of the white elite in 1793; invasion and partial occupation by British forces from 1793 to 1798; the beginnings of a war of national liberation; invasion by Napoleon's armies; and finally, Haitian independence.

The immediate impact of the revolution in Saint Domingue consolidated power on the political right, that is, in the hands of the *grands blancs*. Provincial assemblies of the white colonial elite formed in all three departments of the colony, and any notions of political rights for free people of color, suppressing the slave trade, or, perish the thought, freeing the slaves remained totally out of the question.

The coming of revolution to Saint Domingue brought a radically altered political context in which the Royal Society of Cap François suddenly found it had to operate. Within the Société Royale/Cercle des Philadelphes the impact of the revolution was immediate and explosive. Wild rumors circulated that Arthaud and Moreau favored emancipating the slaves and opposed seating the colony's representatives in France. A counterrevolutionary purge of the Cap scientific society occurred on September 26, 1789, at the first word of the outbreak of revolution in France. On the initiative of J.-B. Auvray, a pillar of the *grand blanc* establishment and sitting president of the Cercle des Philadelphes, Arthaud, Moreau, Mozard, and de Coupigny were expelled from the Royal Society as "traitors."[3] The Couseil Supérieur of Saint Domingue followed by stripping Moreau of his post of *conseiller,* and on Sunday, October 18, 1789, things got worse for Arthaud. Several reports exist of the incident, including one by Cercle associate Palisot de Beauvois.

> The revolutions here are almost equal to those in France. Our poor
> secretary, Monsieur Arthaud, suffered a miserable public humilia-

tion [*anerie*] here. Convinced that he wrote in favor of slaves, the populace took him from his rooms in his nightshirt and drawers, put him on an ass and promenaded him about town. If that were not enough, they took him to the cemetery where he had to apologize to all the dead he had sent to the grave in his capacity as royal physician.

In the meantime a Provincial Assembly has formed, and Monsieur Arthaud presented himself to learn of any complaints against him. The Assembly responded that he could continue in his functions, as it had nothing to reproach him for. Nevertheless, further consequences are to be feared for Arthaud.

I don't know what the Royal Society of Sciences of Saint Domingue is going to do with regard to its secretary, but I know what to think of him. If he doesn't step down, I am handing in my resignation, not wishing to be associated with an organization having a member who so *merited* public disgrace.[4]

Palisot's strong opinion reveals the deep tensions that rent the colony's little scientific society. Arthaud did petition the newly elected Provincial Assembly of the North with a bitter tale of seventeen years of self-sacrifice in the colony, asking at one point apropos of the Cercle des Philadelphes, "Did I betray the colony to have sacrificed my time and all my interests to form an establishment that proved to all Europe that the colonists of Saint Domingue are a polished people who unite a taste for the arts and sciences with a desire for wealth?"[5] Auvray and his party reelected themselves in the Royal Society on October 21, 1789, but the Provincial Assembly seems likewise to have restored Arthaud to his functions as royal physician and secretary of the Royal Society.[6] But thereafter, regardless of whether or to what extent the Royal Society of Cap François patched up its internal relations, the organization was never the same again, for the times were never the same again.

However, it is not true that after October 1789 the Royal Society simply "quit the colonial scene without anyone paying any attention."[7] Rather, the Cap Royal Society became another minor player caught up in the swirl of revolution, and for two and a half more years it did what it could to preserve itself in a wildly fluctuating situation. The record is less substantial for this part of the history of the Cercle des Philadelphes, but the available historical "snapshots" speak volumes not only about the fate of one institution but more generally about the history of science in revolutionary circumstances.

For example, in November 1789, while feuds were still fresh, the Royal Society of Sciences and Arts with Auvray at its head appeared as a delegation before the Provincial Assembly of the North, sitting in Cap François. On the part of the Royal Society Auvray paid homage to the Provincial Assembly, and he congratulated the body on bringing calm to the area.[8] After requesting support for the work of the Society, Auvray went on to pronounce against "these supposed *philosophes* who, closing their eyes to the evil all around them, declaim grandiloquently against the evil that only exists in their lofty imaginations." He then asked Bussan, president of the Provincial Assembly, to accept election as an honorary associate of the Society. This Bussan did, enlisting the Society in turn to "unmask these incendiary fanatics who abuse the sacred name of humanity and who come in that name to ravage the colony."[9] Blanche Maurel, who discusses this matter, makes the key point that the Cap Royal Society paid homage to the new local power of the *grands blancs* in the same manner it formerly reserved for the officers of the king.[10] Not for the first or last time, the protean political character of the institution reveals itself.

Against the background of an increasingly tense political situation, the Société Royale/Cercle des Philadelphes probably kept up at least the semblance of regular meetings through 1789 and into 1790, but the next indication that the organization still operated dates only from July 1790.[11] At a meeting sometime that month the Society awarded a prize, its last, to Moreau de Saint-Méry for his *éloges* of the founders of the poorhouses in Cap François, Turc de Castelveyve and Dolioules. Arthaud may also have read his paper on Indian heads at this meeting.[12]

Some otherwise unknown local disturbance prevented the Society from holding the anniversary meeting scheduled for August 17, 1790. Extraordinarily, on August 30 at the Society's next meeting, the membership voted that the secretary should publish a program of what was to have happened at the disrupted anniversary meeting, and the result—an imaginary set of minutes, really—sheds an eerie, nether light on the Royal Society in the midst of the political turmoil unfolding in the colony.[13]

The notice says only that the Society regretted that circumstances proved unfavorable for the anniversary meeting of 1790 and that it published the notice "to let the world know that if the adversity of the times disturbs its works and lessens the interest one can bring to them in more leisurely moments, the Society is still carrying them forward with zeal and with the same spirit of patriotism that motivated it initially." Had it taken place, the scheduled meeting would have been very similar to other public meetings held over the years. The *éloge* of de Foulquier, the late intendant of Martinique and honorary associate, was to have opened the meeting along with a tribute to Benjamin Franklin, who had died earlier in 1790, an honorary associate since 1786. The intention was to announce the July award to Moreau de Saint-Méry and to remand the sought-after *éloge* of Christopher Columbus, but, notably, the Society declared no new prize contests at the time. Scheduled for presentation: Barré de Saint-Venant's plans for an iron bridge, approved by the Paris Academy of Sciences and "offered to the [royal] Administrators of the Colony." A series of communications were to have followed, including one from the Royal Society of Agriculture in Paris regarding the Cercle's inquiries into the preservation of flour and a reprise of the inquiries from the Bordeaux Academy and the Paris Academy regarding freezing in the mountains. Arthaud stood ready with an ethnographic paper, and De Morancy had a report from Rozier's Journal regarding a scientific exchange between the abbé Spallanzani and Charles Bonnet. The latter may have been part of the effort to get Bonnet elected to the Royal Society of Saint Domingue, for at the same meeting Trembley was to have communicated Bonnet's request for information regarding plant lice in the tropics.[14] The Society made Bonnet's request its own and solicited colonists regarding the life cycle of such insects.

In hindsight it is easy to question the relevance of anniversary meetings (real or phantom) or investigating plant lice, while revolutionary violence developed further in the colony. In August 1790, *grand blanc* and *petit blanc* forces narrowly averted civil war, and in October 1790 united white forces brutally suppressed the first mulatto uprising, the Ogé revolt.[15] Yet, in this threatening context the Cap Royal Society angled to pursue every avenue of patronage in order to sustain itself. On May 2, 1791, the Society entertained the recently installed governor-general, Blanchelande, and in this instance at least Arthaud and the Royal Society had something more concrete to offer: medical reform.[16] Arthaud's proposal forms part of related debates in revolutionary France over reorganizing medicine, restructuring educ on, and reconsidering the place of the learned academies, and, indeed, later in 1791 Arthaud addressed his expanded remarks to the National Assembly in France.[17] In this respect Arthaud modeled himself after the permanent secretary of the Royal Society of Medicine in Paris, Vicq d'Azyr, who also wrote on medical legislation and who similarly argued for a strong and overarching medical academy.

In his presentation to Blanchelande, Arthaud asked for a chemist to improve colonial manufactures and a botanist to superintend a new botanical garden for his institution. But the heart of the proposal lay in the idea of having the Royal Society take control over the public health service and organized medicine in the colony.[18] The proposal left existing medical legislation intact, but physicians and surgeons of the public health corps would be required to be members of the Royal Society and to report to it annually, as would physicians and surgeons in training.

The same would apply to the staffs of the colony's poorhouses, the inspector of mineral waters, and veterinarians. Those who did not maintain a satisfactory correspondence with the Royal Society would be stripped of their offices. Appointment to medical positions in the colony would be reserved to those approved by the Royal Society. The ways in which the Cercle des Philadelphes/Société Royale served as a colonial medical academy have been noted at several junctures. Arthaud's proposal of 1791 was a further, formalizing step in that direction, one wholly in line with like developments in France.

For all of these grandiose plans for a revitalized role for the Royal Society of Sciences and Arts, Arthaud ended his presentation before Governor-General Blanchelande with a rather pathetic appeal for funds. The Society's letters patent, not quite two years old, granted the Society 10,000 livres on the "Liberties" account in Saint Domingue, but, said Arthaud, the account was depleted and for more than a year the Society again had to rely on dues to support its activities. Many members had left for France, and the expense of maintaining the institution had become a burden to the Society's remaining resident associates. "If it were deprived of the funding accorded by the king, one can doubt whether the Society could survive, despite the zeal of its members."[19] Arthaud went on to point out what was only temporarily true, that the academies were being preserved in France, and he appealed to the governor-general for support of the colony's academy of science. Unfortunately for the Cap group, Blanchelande's power was waning fast, and soon to face further fighting in the colony, the French governor could do nothing to help them.

The Provincial Assembly of the North did toss a public-spirited bone to the Cap Society of Sciences and Arts in June 1791, when it called on literate citizens of the colony to submit to the Society inscriptions honoring Castelveyve and Dolioule, the neglected founders of the two official poorhouses in Cap François. The Provincial Assembly offered a cash award, and it proposed to mount commemorative plaques with the winning inscriptions within the poorhouses in question. Submissions to the Society were to be in French, and they needed to be received in time to be judged before the Society's public meeting scheduled for August 16, 1791.[20]

New possibilities in economic botany arose for the Royal Society in 1791. The details emerge in a letter from André Thouin, chief gardener at the Jardin du Roi in Paris, to Hippolyte Nectoux, the government botanist in Saint Domingue and the director of the royal botanical garden in Port-au-Prince. In his letter, dated August 15, 1791, Thouin described a plan for a set of commercial gardens in Saint Domingue to raise spices and exotic plants.[21] The Paris Royal Society of Agriculture and the botanical establishment in France supported the plan. Dutrône la Couture pushed the plan, expected to be heard in the National Assembly shortly. Thouin did not explicitly mention any role for the Royal Society in Cap François, but he did underscore Dutrône's and Nectoux's status as members of "the Society of Sciences of the Cap" and that their voices would carry great weight in deciding on new commercial gardens for the colony. Had the plan had any chance of success, one can be sure the Royal Society would have become involved.

But time was running out for the Royal Society and the whole of the French colony as it had existed in Saint Domingue for over a century, because by August 1791 the slaves of the northern department stood poised to rise up and begin their own revolution. The formidably great mass of the colony's slaves could not be kept ignorant of the principles of liberty, equality, and fraternity that so filled the air after 1789. Neither was the slave community blind to the violent conflict that had already broken out between whites and mulattoes. Besides, slaves had their own traditions and reasons enough of their own for taking up arms. At last, the time had come for their historic rebellion, and in a dramatic ceremony on the stormy Sunday night of August 14, 1791, in a wood not far from Cap François, Boukman, the maroon slave and vodun priest, sealed a revolutionary conspiracy of slaves with the blood and body of a black pig. They would begin their revolt in one week, on August 22.[22]

The *Affiches Américaines* announced a *séance publique* of the Royal Society of Sciences and Arts of Cap François for Tuesday, August 16, 1791 at 3:00 P.M.[23] Unlike the previous year, nothing untoward prevented the scheduled anniversary meeting, and so, incredibly, while the slave revolt spread clandestinely on the northern plain behind Cap François, the savants of Saint Domingue once again assembled before the public in their rooms on the rue Vaudreuil.[24] Representatives of the Provincial Assembly of the North attended, as no doubt, yet again, did the cream of Cap François society, whatever that amounted to in August 1791.

By and large, the Society's program that day appears undistinguished and gives no indication of anything other than business as usual within its confines. Guyot, master surgeon in the Cap militia and vice-president for 1791, presided over the meeting. The unspectacular but necessary series of presentations followed: on tampons, on a snail, on an extrauterine pregnancy, on a new aerometer, on Saint Domingue fruits, on a soldier who ate money, on irrigation sluices, and on elephant teeth from "Kentucky in Virginia." A committee of the Royal Society reported on experiments with a new alkali for use in producing sugar, and Arthaud read the preface to the collection of papers about coffee sent to the Society, a collection the Society presumably had in mind publishing.

Arthaud also read a paper on the "physical and moral character of the mulattoes of Saint Domingue to refute some assertions of M. Clavière in his address from the Société des Amis des Noirs to the National Assembly." This sort of overtly political paper would not have been heard at any meeting of the Society before 1789. Four days later, on Saturday, August 20, 1791, with the slave revolt still impending, the Provincial Assembly of the North officially requested a copy of Arthaud's speech "to serve as an instructional guide when it comes a question of the *gens de couleur.*"[25] Arthaud and the Royal Society doubtless hoped for more of this kind of solicitation of their expertise in the future.

The anniversary meeting of August 16, 1791, closed with the presentation of a new prize program, the Society's first in the two years since 1789.[26] A substantial part of the prize program of 1791 merely repeated the scientific research program that the institution first put forward in 1784. The Society asked for various meteorological observations; it wanted information about "illnesses of whites and slaves in the mountains, on the plains, and in different industries, and on the illnesses of animals," and, as usual, "The Society will gratefully receive all works, memoirs and observations on agriculture, natural history, manufactures, and works of rural economy." Newer questions, stimulated by Bonnet and Lalande in Europe, asked about plant lice and about tides in and around Saint Domingue. On behalf of M. Bessaignet of Petit Goave, the Society also inquired about the apparent decline in sources of fresh water.

But if the prize program of 1791 was unexceptional in most respects, in at least one other respect it was extraordinary, for the colony's Royal Society of Sciences proposed to take up the political sciences and to study the revolution then underway in Saint Domingue.

> As an institution, the Society shares the concerns of every citizen that the best political and civil order be established in the colony, and its right to speak out like every citizen is measured by its stake in public affairs. The Society did not believe it could more usefully serve the country to which it has dedicated its existence than by drawing the attention of colonists to the means to prepare and assure their prosperity.
>
> The Society wants an analysis of the colonies, considered by themselves and in their relation to France. What are the benefits they procure for metropolitan France, and what political conduct must France maintain toward the colonies in order to assure their prosperity? What were the circumstances in the colony at the time of the Revolution, and what led people to change the government of the colony?

One seeks a succinct and faithful account of the facts and the succession of events in the colony down to today. What were the works of the assemblies? Their views? Their governing principles? What could they provisionally or definitively institute or legislate regarding government, justice, finance, police or internal affairs, and regarding their relations with the National Assembly?

What are the commercial links between France and its colonies, and within the colonies what are the connections, collectively or separately, between merchants and traders on the one hand and growers on the other?[27]

These questions are remarkable for their political frankness and penetration. The clear spirit they manifest for colonial independence is likewise impressive. They also evidence a certain naiveté that the political equation of the revolution in Saint Domingue could be solved through rational, scientific inquiry, and in this respect something of Condorcet and his science of politics seems to be reflected in this new area of institutional activity. However, and the reservation is everything, except as covered by the code words "police or internal affairs," the Royal Society wholly neglected to consider the largest factor in the political equation before it: the thousands and tens of thousands of slaves passing word of the impending revolt among the plantations of the northern plain, while the white people speechified and published pamphlets in Cap François.

The great uprising of the slaves began as planned on August 22, 1791. Overnight 50,000 slaves armed with whatever weapons they could find or fashion rose in righteous revolt.[28] Another 50,000 slaves immediately left their plantations to make their way in the chaos sweeping across the northern plain. Fire, death, and destruction raged everywhere. Uprising slaves destroyed several hundred sugar plantations and two thousand coffee plantations almost immediately. They massacred a thousand whites, including women and children, in scenes of horrifying brutality that matched the brutality of slavery in Saint Domingue.

In ten days the slaves were masters of the northern plain, and they drove thousands of white inhabitants and other refugees into Cap François. At this point, regular army units and loyal mulatto detachments defended the city and pushed the relatively disorganized slave forces back into the bush and the Spanish part of Saint Domingue. But they could not contain the revolt of the slaves to the north. Revolt spread to the other departments of the colony and soon counted 100,000 active insurgents. The slave and great leader of the Haitian revolution, Toussaint Louverture, joined the revolt in November 1791. The white colony remained a long way from its complete collapse in 1793, but a new and bloody stage began in August 1791, as revolution and ongoing revolts in turn gave way to race war and a war of national liberation. The stability and social relations that made possible the Royal Society of Sciences were quickly disintegrating.

Still, the Royal Society of Cap François and the structures of French colonialism in Saint Domingue clung to what remained. A new, all white Colonial Assembly began meeting in Léogane in August 1791, just as the slave revolt broke out, but it soon transferred its seat to Cap François. Having previously made its overtures to the defunct Provincial Assembly of the North, the Royal Society of Cap François proved no less anxious to present itself to this new Colonial Assembly. But the Colonial Assembly, like the colony as a whole, faced deep and mounting turmoil as 1791 progressed and was otherwise intent on colonial independence. It was also not as sympathetic as the Provincial Assembly of the North earlier had been to a royal institution like the Royal Society of Sciences and Arts. Nevertheless, sometime in the fall of 1791, with obvious reservations Arthaud addressed the Colonial Assembly on behalf of his failing institution:

> Mister President and Sirs: The Royal Society of Sciences and Arts of Cap François has the honor to offer you its respectful compliments. This Society is already seven years old. Protected and encouraged by the government from its birth, it was constituted in letters patent given at Versailles on May 17, 1789. When it formed, the Society could not imagine ever to see such an imposing assembly as yours in the colony. It never considered it would have the satisfaction of reporting to you and saying to you, "Here is the plan for public utility that I followed, here are the means I employed, the sacrifices I made; here's the tribute I have paid to the fatherland." I am pleased to acknowledge this in an assembly of chosen men who, uniting Enlightenment and Wisdom, will be occupied only in restoring order and peace to the colony and doing so in good faith. . . .
>
> The limitation of its means and the little time it has existed have not yet permitted the Society to produce very important works. [Still,] in the sciences it takes only a moment to mature and to make valuable discoveries. But talent must be applied, genius must get involved to fan the animating fire, and to seize a fleeting truth, the Society must continue in the slow path of observation.
>
> True to its commitments, the Society has progressed steadily toward the goals it has set for itself. Already it has produced several works that have been judged useful. The Society has the honor to present them to you, asking that you look at them.
>
> It also entreats you, Sirs, to consider its public utility, to settle

its existence, and to oversee its organization. It cannot doubt that you wish to conserve an institution that through its works can serve the colony and produce for it men distinguished by their talents and virtues.

Too eager to take in the principles of a Revolution that would prove fatal to us, we have taken the intoxicating drink of novelty without seeing that its poison must rend our guts. Sirs, you are going to measure with wisdom what the colony must adopt from the French constitution and you will adapt it to the colonial constitution you are going to frame. . . . Finally, in the horrible danger that threatens us, your courage and your vigilance assure us resources, and you are forging the route that each citizen must follow. . . . The colony in peril can still hope."[29]

Desperate for support, the Cap Royal Society would align itself with whomever white came to power, and thus the fact of this overture to the Colonial Assembly is more important than any response to the Royal Society, which was bound to be token at best.

The almost reflexive tendency of an increasingly powerless and stagnating institution to reach out to every source of support is seen again in one of the last vignettes available of the scientific society of Saint Domingue: a meeting in Cap François on January 30, 1792, in the presence of Mirbeck and Roume, two of the three civil commissioners sent from France the previous fall to restore peace in Saint Domingue.

One of the last acts of the Constituent Assembly in France in September 1791 had been to renege on its commitments to mulattoes and to send the "national civil commissioners delegated by the king" (the First Civil Commission) backed up with a promise of eighteen thousand soldiers to Saint Domingue to reestablish tranquillity and to reassure white colonists of France's desire to do nothing to upset the prosperity of Saint Domingue.[30] The commission arrived in Saint Domingue in late November 1791 with six thousand men. Two months later, having made some progress in temporarily damping down the fires of war and revolution in the colony, the two commissioners found themselves entertained by Arthaud and the Cap Royal Society, as had been so many of their various, honored predecessors.

The Royal Society of Sciences and Arts elected Mirbeck and Roume as national associates at its meeting on January 30, 1792.[31] The election of the French commissioners, doubtless the last associates to join the Society in Saint Domingue, was an appropriate step for what was still a royal institution to take toward men who were technically royal representatives. It was also completely characteristic of the Society's long-

standing habit of granting government officials a place in its membership. The gesture of associating itself with men whose mission was to fall
apart over the spring of 1792 likewise seems fitting for an organization
soon to dissolve into the mists.

A bloody and complicated three-way civil war and race war overran
the colony as the spring of 1792 progressed. Everywhere white power
and white institutions found themselves under assault and in retreat. The
National Assembly in France prepared a second and more powerful Civil
Commission to reestablish authority. The British considered overtures
received from Saint Domingue's planters to invade the colony. On a
lesser scale, but like many others, Charles Arthaud abandoned Saint
Domingue at this time. Forsaking his posts and the Royal Society,
Arthaud returned to France.[32]

The last glimpse available of the Royal Society of Cap François dates
from June 30, 1792, and comes in a letter to Arthaud in Paris from
Geanty in Cap François.[33] Geanty, a lawyer, royal notary, and resident
associate of the Royal Society, may have taken over from Arthaud as secretary of the Royal Society. He began his letter to Arthaud, whom he
addressed as "my dear doctor," with a report of various revolts and
recent military movements, particularly those of Governor-General
Blanchelande fighting the slaves in the south. Then, Geanty turned to a
rather substantial request of Arthaud to secure scientific instruments for
the Royal Society in Saint Domingue. As if they had discussed matters
previously, Geanty wrote to Arthaud:

> My friend, we forgot a few things for the Society, necessary above
> all else. Please send two thermometers and two barometers, and you
> probably had better add two or three hygrometers and as many at-
> mospheric electroscopes and an assortment of aerometers. As we
> will want instruments that can be compared very exactly, it is ad-
> visable to take some care in obtaining them. I will pay for a quarter
> or a third of everything the Society will not want for itself. Send us
> some lamps for our sessions, too.

The image that Geanty evokes of the Royal Society as an active institution, meeting at night and seemingly at last about to undertake meteorological research, comes as a surprise. But when Geanty asks
Arthaud to add a shipment of pistols, a different and more ominous picture emerges. Geanty requested two guns for himself "regardless of
price, and if the Society wanted to keep one, I would gladly give it over."
This note of violence more accurately anticipated the *Götterdämmerung*
ahead. Geanty's melancholy closing to Arthaud is also the last word from
the Royal Society of Cap François: "Farewell, my dear doctor. Remem-

ber me to Monsieur and Madame Moreau [de Saint-Méry] and Baudry [des Lozières]. Send me your news and a report on the state of things in France quickly. Farewell, farewell. All to you."

The Royal Society of Cap François may well have formally greeted L.-F. Sonthronax and the Second Civil Commission from France after its arrival in Cap François in late September 1792.[34] That would have been entirely in keeping with the Society's style, but the Society may also have faced a hostile reception from the Jacobin Sonthronax. After the fall of the monarchy in August 1792 and the creation of the French Republic, the status of a royal institution like the Royal Society of Sciences and Arts would have been increasingly dubious. But all that remains a matter of speculation, for no one knows when the Royal Society of Cap François last met or when someone last acted on its behalf.

The Royal Society certainly did not survive the sack of Cap François that took place in June 1793. That disaster began on June 20, 1793, when two thousand men loyal to the *grand blanc* cause and to General Galbaud, a recently arrived governor, attacked Sonthronax and his mulatto forces in Cap François.[35] Driven from the town, Sonthronax, the legal authority in the colony, promised freedom and pardon to the bands of insurgent slaves around Cap François for their help in regaining the town and suppressing the illegal revolt. Wholesale massacre and the destruction of the town resulted. Over ten thousand people, most of them whites, died in the carnage that followed. Another ten thousand escaped with the fleet bound for Baltimore. After a few days only a quarter of Cap François remained standing. It was the end. The white colony in Saint Domingue had more or less collapsed.

Technically, the Royal Society of Sciences and Arts of Cap François, with its treasured letters patent, survived as a legal entity. But in Paris, just days after the sack of Cap François, the organization that started out as the Cercle des Philadelphes in 1784 died a second death, for it clearly came within the purview of the famous decree of August 8, 1793, passed by the Jacobin Convention in France suppressing "all academies and literary societies patented or endowed by the nation."[36]

Incredibly, the history of science in colonial Saint Domingue does not quite end with the fall of the Cercle des Philadelphes. In fact, against the background of continuing war and revolution in the colony after 1793, individuals put forward at least three proposals to recreate a scientific institution in Saint Domingue akin to the former Cercle des Philadelphes. Even in a revolution, the ingrained notion of the utility of organized science was not easily abandoned.

A small step may have been taken toward realizing the first of these plans for a new scientific society for Saint Domingue. The idea for a colonial "Société Libre des Arts et des Sciences" originated in 1796 with Alexandre Giroud and Julien Raimond, members of the Third Civil Commission sent to Saint Domingue by the Directory to reinvigorate loyal French forces of all hues.[37] Both Giroud and Raimond brought solid credentials as founders of the proposed new organization: Raimond as Saint Domingue's distinguished mulatto leader and a man who had long lobbied in Saint Domingue and in France for mulatto rights and interests; Giroud as a geologist, mineralogist, and a nonresident associate of the newly reformulated Académie des Sciences of the Institut de France.[38] With slavery (temporarily) abolished in Saint Domingue in 1793–94, Raimond's presence as a civil commissioner and as founder of this proposed Free Society of Arts and Sciences signals that, technically at least, race no longer figured as part of French or colonial politics or of the further scientific development envisioned for postrevolutionary Saint Domingue. Equality and fraternity were the order of the day among the French and their allied forces. Out of the ashes of the Old Regime in Saint Domingue, the phoenix of science could emerge again to embrace all races in the further progress of mankind. Such would seem to be the thrust of Giroud's remarks sent to the Institut de France, remarks wherein Giroud seems to say that concrete steps had been taken to establish this new society.

> You will receive . . . a letter from the Free Society of Arts and Sciences that has just formed here at Cap François through the efforts of your colleague, Raimond, and my own. You will see from the nominal list of members of this Society that it is composed of citizens from each of the three colors that nuance the human skin in this colony. A fairly large number of black and light-skinned citizens here are knowledgeable and very worthy of figuring in the Republic of Sciences and Letters. Announce this important truth to Europe, misled on the point by the infamous cupidity of a few colonists and merchants whom avarice has transformed into cannibals.[39]

Meanwhile, a complex and shifting set of political and military circumstances continued to unfold across French Saint Domingue and the whole of Hispaniola through the middle and later 1790s.[40] Toussaint Louverture led insurgent slave forces, first for Spain and then for the French. After 1793 the Spanish invaded from the east, and the British invaded from the south. Independently, mulatto insurgency flared up around the colony. In light of these circumstances the new colonial academy put forward by Giraud and Raimond was bound to prove stillborn. The French were simply too besieged.

Still, the idea was not to be suppressed, and Michel-Étienne Descourtilz, a young physician, sometime during his occasionally perilous stay as a naturalist in the colony between 1799 and 1803, established the Lycée Colonial of Saint Domingue. Descourtilz styles himself as founder of such an establishment in his book, *Voyages d'un naturaliste,* published in France in 1809, and both Ulrich Duvivier and Rulx Léon, who mention Descourtilz's Lycée Colonial, emphasize its character as a learned society. For Léon, the Lycée was "a kind of learned circle like the Cercle des Philadelphes that had done such good scientific work in the colony a few years earlier."[41] But fate likewise doomed the Lycée Colonial.

None other than Barré de Saint-Venant, the former president of the Cercle des Philadelphes who had done so much to win letters patent for the Société Royale du Cap François and to forge relations with the Paris Academy in 1789, made the third proposal for a new scientific society in Saint Domingue in 1802. Writing from Paris, Barré put forward an elaborate program for revitalizing Saint Domingue, a program that included an Institut Colonial analogous to the Institut de France in the motherland. Barré's Institut was to unite "geographers, *physiciens,* chemists, mechanists, naturalists, botanists, engineers, physicians and surgeons."[42] Barré thought his Institut might naturalize the opuntia cactus for cochineal production, and in general he pledged practical support for colonial agriculture and manufactures. In other words, he wanted to bring back the Cercle des Philadelphes.

None of these proposals for a colonial institution of science stood any even medium-term chance of success during the violent period in Saint Domingue from 1793 to 1804, but the fact that the idea of a learned organization cropped up again and again, even in the midst of further war and revolution, reveals the hold of science in certain circles and the tenacity of the promise of applied learning in the service of governance and economic enrichment.

Barré's proposals coincided with the Leclerc expedition to Saint Domingue, a last push by Napoleon to reassert the national authority of France and to recover the colonial jewel fast being lost. Thirty-five thousand or so seasoned French troops arrived in Saint Domingue in stages through 1801 and 1802.[43] The French came to reestablish their colonial system, and, conversely, with Napoleon's restoration of slavery in 1802, the roused and dedicated Haitian masses became determined to drive the French into the sea. Both sides fought the ensuing battles fiercely, and intense and widespread destruction resulted. Cap François, for example, was totally destroyed once again on February 4, 1802.[44] Leclerc and his generals had more than their share of successes, including the dishonorable capture and deportation of Toussaint. But the heroic men and women who sacrificed themselves for the Haitian Revolution were aided by a powerful ally: yellow fever. French troops died by the thousands virtually as they disembarked, some twenty-five thousand or more deaths.[45] Leclerc himself succumbed to the disease, leaving General Rochambeau ultimately to withdraw the French from Saint Domingue. The Haitians prevailed but, ironically, they were aided in their cause by disease and the forces of the natural world that the French had fought so long and so hard to overcome in colonizing Saint Domingue in the first place. Victory secured, Haitian independence was declared on January 1, 1804.[46] Thus ended a century and a half of French colonialism in Saint Domingue.

Science and Colonial Development

O N THE SIMPLEST level this study has shown that eighteenth-century French science and medicine formed integral elements of the colonizing process in Saint Domingue. From the first missionary naturalists in the colony to the last proposals for a restored scientific society after the revolution, every instance reveals that science and medicine did not arrive in Saint Domingue as cultural afterthoughts, but came part and parcel with the other agencies and instrumentalities of eighteenth-century French colonialism. The powers that be in the Old Regime made every effort to exploit science and medicine in the promotion of economic growth and colonial development.

One corollary follows immediately. At its fiery birth in 1804 the Republic of Haiti ceased to be an active seat of European colonialism and simultaneously found itself largely beyond the pale of Western science. The fact that Haiti in the nineteenth century was not the scientific or colonial center Saint Domingue had been in the eighteenth makes clear that in this case, at least, institutes of science marched at the vanguard, not the rear guard of colonialism.

A summary is not required to make the point that utility and the promise of utility constituted the reasons why science and medicine became enlisted in the service of colonial development in eighteenth-century Saint Domingue. One slightly less obvious point informing this study is that the French state—and, more particularly, the central government—was the entity primarily responsible for employing science in the colonial process. Virtually every agency of French colonial and national government has appeared in this account: the ministry of the navy and the colonies in Versailles, local colonial administrators in Saint Domingue, various branches of the French navy, the French army, the local Conseils Supérieur in Cap François and Port-au-Prince, and Louis XVI himself.

In addition, given that the state organized and institutionalized science and medicine, French scientific institutions need to be counted as additional government entities that had a hand in colonial development. In Saint Domingue itself the official public health corps, the Jardin du Roi and related gardens, and, of course, the Cercle des Philadelphes emerged as government scientific and medical institutions charged specifically to apply useful knowledge to colonialist ends. The number of technical specialists possessing royal appointments of one kind or another (21 percent of the members of the Cercle des Philadelphes) only underscores the close connections between science and the state in the history of Saint Domingue. The existence of technical, scientific experts working as paid employees of the colonial bureaucracy supports the thesis that colonialism was backed up by cadres of such experts who brought rational, scientific judgments to bear on problems affecting colonial development.[1]

Government scientific institutions in metropolitan France also played a part: the Royal Academy of Sciences in Paris, the Marine Academy at Brest, the cartographical bureau in the navy ministry, the medical branch of the navy, the Observatoire, the Jardin du Roi, the Alfort veterinary school, and the Royal Society of Medicine in Paris.[2] In this vein one is tempted to add the several provincial academies of science that had contact with Saint Domingue. In a word, science and society in colonial Saint Domingue cannot be considered apart from contemporary science and scientific institutions in France: the empire of French science expanded along with the rest of the French colonial empire in the eighteenth century, and in the end, institutional ties linking Saint Domingue and metropolitan France indicate that colonial science in Saint Domingue, while on the periphery, was anything but peripheral.

In tapping science to serve colonial development in Saint Domingue, the government of France had in mind a simple set of immediate goals: to promote the growth of the colony and to expand its economic base and output. National and colonial governments were naturally interested in advancing the prosperity of individual colonists and individual colonial enterprises, but still and all, the underlying rationale for government colonial policy was more mercantilistic and self-interested: to bolster the political and economic strength of France and the royal government, and not especially of Saint Domingue. As A.-P. Blerald emphasized and as Jean Tarrade's magisterial study of French colonial trade previously demonstrated, the principles and practice of mercantilism rather than capi-

talist or industrial expansion per se governed the government's vision toward its American colonies virtually down to 1789.[3] From our point of
view it becomes clear that the central government of France enlisted the
intellectual and technical services of contemporary science and medicine
in support of its larger mercantilist policies.

The extraordinary success of Saint Domingue no doubt reinforced
government attitudes about the correctness of its mercantilist approaches
toward the colony. Yet, the very success of the colony created new economic and political conditions much at odds with the policy of mercantilism and the priorities of the old order. A historical contradiction developed. As Mintz, Pluchon, and others have stressed, plantations and the
plantation economy that developed so remarkably in Saint Domingue
constituted large-scale agroindustries more akin to the later factories of
the Industrial Revolution than to the trading and mercantilist contexts of
an earlier age.[4] Mercantilist colonial policies forged new capitalist conditions that ultimately collided with Old Regime economic thinking.[5] The
manifestation of a movement for colonial economic and political independence in Saint Domingue after 1789, for example, supports this analysis.

The idea that feudalism was somehow an element in the colonial development of Saint Domingue is a red herring that should not put us off
the trail. True, certain feudal trappings of the Old Regime found their
place in Saint Domingue: honorary benches in churches, the public procession, preferential taxes on certain "noble" land.[6] But the essential nature of the society and economy of Old Regime Saint Domingue lay elsewhere. With regard to science, for example, the Cercle des Philadelphes
strove not to preserve ancient privileges but to promote colonial prosperity. Similarly, botanical gardens arose in Saint Domingue, not so that
the nobility might promenade through them but to provide nurseries for
useful plants and profit. The authorities created the Eaux de Boynes spa
not for the pleasure of the rich and idle, but to economize on medical bills
for the army. In a word, Saint Domingue was in every way a modern
colony on the cutting edge of colonial development in the eighteenth
century.

A more significant antagonism did exist, nonetheless, between the
mercantilist policies of government and the inherent economic development of capitalism and the plantation economy. As far as this study of
science in colonial Saint Domingue is concerned, it is crucial to recognize
that organized knowledge was not strongly aligned with planters, merchants, or capitalists themselves. Although government colonial policy
worked to promote productivity and profit in the plantation system,
how much colonists, plantation owners and operators, or merchants directly concerned themselves with the applications of science and learning

is another question. Several progressive plantation owners did actively involve themselves in the rational and scientific study of aspects of the plantation economy—one thinks again of Auvray or Lefebvre-Deshayes. Similarly, the examples of smallpox inoculation and the readiness of slaveholders to apply mesmerism to their captive property indicate that plantation owners were more than willing to employ science and medicine when benefits could be demonstrated to them. Brulley and his cochineal farm represent the best of private planter initiative. On the whole, however, planters left it to the state to invest in the necessary research and institutions, and on balance this study shows that most of Saint Domingue's plantations churned out their loaves of sugar, sacks of coffee, and bales of cotton with little regard for science, medicine, or public or private efforts to apply these profitably to colonial manufactures. In an equally telling fashion, essentially no evidence suggests that merchant or trading interests more narrowly conceived ever looked to science or medicine to improve their endeavors. The point emerges again that the interventionist French state was the primary agent that saw in science a force that could be harnessed for its own interests. Thus, in this case at least, state science of the eighteenth century did not serve wholly as a progressive force to advance capitalism directly, but furthered the retrograde system of mercantilism, a system essentially at odds with the victorious industrial capitalism that triumphed in the nineteenth century.

Colonial science and medicine clearly served to support the institution of slavery in Saint Domingue. In so many instances great and small, colonists and colonial powers deployed science to bolster the colony's slave system. Physicians especially directed a great deal of colonial medical work—concerning smallpox or tetanus, for example—toward preserving the health of slaves. Research on poisons stemmed from a perceived problem of significant proportions derivative of slavery. In many of its prize questions and especially in its agricultural survey of 1787, the Cercle des Philadelphes looked to exploit knowledge to enhance the regime of slavery. Indeed, the credo of the Cercle des Philadelphes to improve the lot of slaves "without harming the interests of colonists" epitomizes the attitude of that institution and French colonial science generally toward efforts to apply useful knowledge to the colony's system of slavery. Slavery did not constitute the foremost area of activity or state support for colonial science—cartography and economic botany loom larger—but slavery nevertheless represents an important and revealing aspect of French colonial science in the eighteenth century. As the nineteenth century approached, chattel slavery was doomed to be supplanted by other forms of human and productive organization, and advanced thinkers could see that future.[7] The sad but not surprising moral of the present tale is that French colonial science and medicine did not serve to advance the cause of human freedom so much as to reinforce the chains of human bondage.

The history of science and medicine in colonial Saint Domingue shows that the French government of the Old Regime used expert knowledge virtually as a "productive force" in building its colonies, and recruiting science and medicine in the cause of colonial development no doubt seems progressive and a key element in the astounding success of the colony. Yet, when one considers the effect of science and medicine in underpinning the ultimately retrogressive policies of mercantilism and slavery in Saint Domingue, one at least has to temper the Enlightenment's and our assumption that science and medicine inevitably have been the agents of progress in modern history.

That subject—historical progress and its ambiguities—lies beyond the bounds of the present study, but a more restricted question can be asked about "the Enlightenment" as a potentially useful concept for interpreting colonial science in the eighteenth century. David Wade Chambers posed the problematic of the Enlightenment for the history of colonial science in eighteenth-century Mexico and Latin America.[8] The history of Old Regime Saint Domingue shows that, while elite colonial society had access to the full range of eighteenth-century French intellectual and social currents, the colony was not especially liberal or much affected by the contemporary Enlightenment. One finds little or nothing in Saint Domingue of the Enlightenment's anticlericalism, for example, or Rousseau's naturalistic enthusiasms characteristic of the period in France. Socially, economically, and politically the Enlightenment and Rousseau were simply irrelevant in Saint Domingue. By the same token, the beliefs and institutions of French science and medicine became well rooted in the structures of society and culture in eighteenth-century Saint Domingue. The point emerges that "science" more than "Enlightenment" seems the appropriate criterion by which to analyze French colonialism in the eighteenth century.[9] The commitment of colonial powers to science in the eighteenth century is again seen to be essentially conservative.

That theoretical science essentially does not figure in this account reveals something of the dynamics of patronage not only for science in a colonial context but for science generally in the eighteenth century. Not only are the higher reaches of the more difficult mathematical sciences scarcely to be found, but the theoretical dimensions of any of the sciences—exact or otherwise—are not a part of the story of science in colonial Saint Domingue. From the problem of longitude, through the natural history of the cochineal insect, to perfecting lightning rods or insect-proof paper, the "science" uncovered here is almost entirely practical and applied in orientation. Put another way, scientific work in the colony never rose to disinterested speculation (by state-supported academicians, say, or university professors), but such work always derived from immediate practical problems and always looked to have a socially beneficial impact. Without public support, an established tradition, or a sufficient institutional base, pure science did not take hold in colonial Saint Domingue.

To a very small degree, it is true, natural philosophy tagged along with the generally useful. This aspect of the story is most evident with regard to contemporary astronomy. To be sure, astronomy was seen to be a useful science, and early astronomical observations made in the colony in conjunction with finding the longitude of Saint Domingue and the later calendars and astronomical ephemerides produced by Pingré indicate the ways in which scientific astronomy could be turned to colonial utility. Still, contemporary astronomers managed to undertake some important "pure" research. In this regard one thinks of Des Hayes's voyage in 1699 to check the slowing of the pendulum, the French expedition to Peru to measure the shape of the earth, or the observations of the transit of Venus in 1769 to determine the astronomical unit. However, the center of expertise in these more technical and mathematical domains remained in France, the Academy of Sciences first and foremost.

These astronomical examples are the exceptions that prove the rule: the French state supported science in Saint Domingue for the practical benefits it saw as forthcoming and not for any lofty notions of natural philosophy or advancing theoretical knowledge. This finding is so clear and unencumbered in colonial Saint Domingue that it would seem to illuminate the general question of state support for science in Europe as well. Historians have already noted an effective exchange—a quid pro quo—wrought between science and government in general in the eighteenth century, an exchange clearly evident in Saint Domingue.[10] On the one side, government supported, legitimated, and empowered science, scientists, and scientific institutions. In return, science and organized knowledge served the state and colonial government by providing expert advice and service. The ink tests performed for the Conseil Supérieur by the Cercle des Philadelphes in 1785, for example, or the medical testimony that decided a legal case in Saint-Marc in 1786 are minor but telling examples of the authority and expertise of science at work in the service of state authority.[11] Because state support for science and useful knowledge in the colonies came without the additional costs of natural philosophy or disinterested theoretical speculation, the case of Saint Domingue would further suggest that support for "pure science" and "pure scientists" at scientific centers in Europe—the disinterested pursuit of new knowledge by someone like Euler or Lagrange, for example—was an additional toll extracted from European governments by scientists as the price for practical benefits. True, the French especially made much of the cultural *gloire* accruing to the state for its subvention of useless intellectuals, but the government would have dispensed with them if it could otherwise have secured the practical benefits of applied knowledge.[12]

The overwhelming emphasis on the utilitarian, the practical, and the applied in Saint Domingue has a notable corollary regarding problem choice in colonial science. Clearly, colonial researchers took up problems not on account of any internal scientific tradition but because of the exigencies of the local context in Saint Domingue. The Cercle des Philadelphes and other colonial medical investigators undertook extensive studies of tetanus, for example, first and foremost because the disease posed a serious threat in Saint Domingue. The research done in the colony on poisons and their effects is likewise to be ascribed to colonial anxieties over poisonings by slaves. Similarly, progressive colonists studied coffee and the indigo plant not simply as students of botany but because coffee and indigo constituted important commodities in the economy of the colony. Induction is not required to make the point that commitments to utility and practical application drove problem choice in colonial science.

The present study shows that colonial science per se grew out of colonial medicine and the prior institutionalization of medicine in Saint Domingue. The success of science in taking root in a colony and ultimately achieving a level of autonomy would seem to require the existence and success of other colonial social institutions, some of which, like the medical, at early stages may have been functionally more important than science and may have provided science with an initial means to weave its own way into the fabric of the colony in question. One envisions here a variant of the Merton thesis for colonial science, where the scientific enterprise rides "piggyback," not on the tenets of Puritan religion, but on the structures of colonial medicine. The case of Saint Domingue would suggest as much.

This study demonstrates that colonial Saint Domingue needs to be seen as "part" of eighteenth-century France and French history. Native-born French constituted only a small minority of the colonial population (5 percent), and the social order that arose in Saint Domingue was unlike anything in France. Yet overwhelmingly the colony was geared toward France. The dominant language and culture of the colony was French. The structure of government and administration, again, was typical of any French *généralité,* and Louis XVI was king. The economy of Saint

Domingue, although it fostered the interests of local planters and centrifugal political tendencies, centered on producing colonial commodities for shipment to France in French bottoms in return for more slaves and French staples and luxuries. Specific economic transactions occurred at more than one point along these paths, and Africa, to be sure, was an essential part of the cycle, but the orientation of the colony's production and trade was overwhelmingly toward France. One needs to remember Saint Domingue when thinking about eighteenth-century France; one cannot forget eighteenth-century France in thinking about Saint Domingue.

Models for how science evolves in a colonial context have attracted considerable attention in the recent literature concerning science and colonialism. The "simplified" three-stage model offered by George Basalla in 1967 has been the focus of debate.[13] According to Basalla, the first stage of colonial scientific development is characterized by initial geographical, natural historical, and like encounters with a new land or people by a distant mother science. Basalla's second stage, "colonial science," is typified by a scientific colonial outpost dependent on the traditions and institutions of the motherland. Finally, in the third stage the colony establishes an independent scientific tradition.

To a considerable degree the story of science in colonial Saint Domingue fits Basalla's model. Saint Domingue was first an unknown land and a scientific frontier whence came interesting new reports, particularly in botany, and these initial scientific investigations by the distant mother science correspond to the first stage of Basalla's model. Even as Saint Domingue matured, the colony continued to serve as a scientific outpost where scientists from France stopped over and "did" science in one fashion or another.

The case of Saint Domingue also meshes with what Basalla has in mind for the second stage of his model: science weakly rooted in a colonial setting, unable to survive without the institutional, intellectual, and moral resources supplied by organized science in the colonizing power. By the time of the French and Haitian revolutions, organized science in Saint Domingue remained far from achieving an independent existence and scientific autonomy, hallmarks of Basalla's third stage in histories of colonial science. Nevertheless, by the end of the eighteenth century science in colonial Saint Domingue had passed several noteworthy landmarks along the road to Basalla's third stage. Institutionalized medicine provided an important base for colonial science at this stage, as has been mentioned.[14] The creation of government botanical gardens, the full-time employment of scientists, and their semipermanent stationing in Saint

Domingue enhanced the state of science in the colony. (The career of Hippolyte Nectoux is again paradigmatic.) Most significantly, the creation of a royal colonial scientific society marked a solid step in permanently implanting science in Saint Domingue.

Despite its usefulness, Basalla's model of colonial scientific development has been seen as flawed, and historians have proposed several more sophisticated alternative models: models of pure intellectual diffusion, sociological and geographical models exploring center-periphery relations, models based on political history, models of technological determinism, and more complicated models built on a more nuanced sense of the intricacies of colonial history.[15] Indeed, while colonial science in French Saint Domingue roughly fits Basalla's model, the complexities of the case are in other ways more revealing of a uniquely Old Regime style of colonization and scientific development.

By and large, however, published critiques of Basalla, while advancing understanding of the complexities of colonial science, still share with Basalla's model a primary emphasis on the impact of colonialism upon science and the scientific enterprise, rather than the other way around. That is to say, historians of science writing on this topic have generally been concerned with how the experience of colonialism affected science; they have not especially focused on the role and function of science in the development of colonies.[16]

Too often in the literature to date the emphasis has been on science as some sort of transcendent entity whose connection to colonialism is almost accidental. But the present survey of science in the French colonial experience in Saint Domingue indicates that the dynamic in the history of colonial science stems primarily from the role that science plays in colonialism and not the other way around. In other words, the historical motive force for change in colonial science derives first and foremost from the various strategies of colonialist and imperialist powers involving science. Can anyone doubt that the history of science and medicine in colonial Saint Domingue depended chiefly on the demands of French colonialism and not on some logical or institutional necessity of French science? To understand the history of science in a colonial context like Saint Domingue, one needs first to understand the practical and policy functions science and organized knowledge served for a country like France. One hopes that future studies of science and colonialism will continue to rectify this historiographical imbalance and that they will uncover more of the ways and reasons colonial powers enlisted science in the service of their empires.

ABBREVIATIONS

AA	*Affiches Américaines.*
	PauP Port-au-Prince edition.
	PauP Supplement Supplement to the Port-au-Prince edition.
	Cap Cap François edition.
	Cap Supplement Supplement to the Cap François edition.
APS	American Philosophical Society, Philadelphia.
AN	Archives Nationales, Paris.
	SdOM Section d'Outre-Mer.
	BMSM "Bibliothèque de Moreau de Saint-Méry."
BN	Bibliothèque Nationale, Paris.
	DE Département des Estampes
	Mss. Département des Manuscrits.
	n.a.f. Nouvelles acquisitions françaises.
CA-HdC	Charles Arthaud, Ms. "Histoire du Cercle des Philadelphes du Cap Français" (1788). Cited is fair copy: AN, Colonies F³ 81, fols. 148–84. Original is AN, Colonies F³ 152, fols. 249–89.
CCP	"Correspondance du Cercle des Philadelphes depuis son Établissement," AN, Colonies F³ 81, fols. 184–95.

DPF   Moreau de Saint-Méry, *Description de la Partie Française de l'Isle de Saint-Domingue (1797–98)*, ed. Maurel and Taillemite, 1958/1984.

HMARS   *Histoire et mémoires de l'Académie Royale des Sciences* (Paris).

NYPL   New York Public Library
        SC   Schomburg Center for Research for Research in Black Culture.

PA   Archives of the Académie des Sciences, Institut de France, Paris.

PA-PV   Procès-verbaux de l'Académie Royale des Sciences, Archives of the Académie des Sciences, Institut de France, Paris.

NOTES

Published sources are cited by author and year of publication as listed in the bibliography. Citations of folio pages indicate the recto unless otherwise noted. All translations are the author's. Where given, eighteenth-century French is rendered with original spelling, punctuation, and accents.

INTRODUCTION: The Case of Saint Domingue

1. Lewis, p. 123; Cauna (1987), p. 12; Genovese, p. 85; Devèze, p. 267. These points are developed further in chapter 4.

2. Abeille, p. 3; J. Fouchard in Cauna (1987), p. 9; Barré de Saint-Venant, p. 471; see also James (1963), p. 49; *DPF*, p. 313; Geggus (1989), p. 1291.

3. Maurel (1935), p. 1; Lokke, p. 29.

4. Moreau de Saint-Méry puts the population in French Saint Domingue at six and one-half times that in the Spanish part of Hispaniola; see *DPF*, p. 28.

5. Klein's recent figures (p. 295) put the slave population of Saint Domingue about equal that of slaves in the United States. The whole of the American south was 43 percent slave on average, and the United States in toto did not surpass 18 percent slave at the end of the eighteenth century; see Montague, pp. 5–6; Hughes, pp. 23, 39, 112; Finley, p. 182.

6. Hoetink, p. 55.

7. This figure includes transients; see *DPF*, p. 478; Julien (1977), p. 64; Girod (1972), p. 77; Mercier, p. 107. *Cap François* was and is more commonly spelled with an *o*, but sometimes the name appears as *Cap Français*. It is pronounced as *Cap Français*, however.

8. Hughes, p. 33; Cobban, p. 48; Roche (1978), 2:358.

9. The population of these cities in 1790: approximately 44,000, 42,000, 33,000, and 16,000, respectively. See previous note and Fouchard (1955b), p. 6; Devèze, pp. 329–31; Butel, p. 28; Knight, p. 60. Not everyone would rate Cap François quite so highly; see Geggus (1982), p. 7.

10. One should not overestimate the vitality of these noteworthy North American institutions. All things considered, the Philadelphia and Boston societies remained unexceptional provincial centers in the world of eighteenth-

century science, not manifestly superior to their academical cousin in the Caribbean, the Cercle des Philadelphes; compare Hindle (1956), pp. 142–45; and McClellan (1985), pp. 35, 140–44, 265, 276–77.

11. See J. G. Taylor, pp. 85–90; McClellan (1985), pp. 124–25, 263.

12. The contrast between the more than century-old colony in Saint Domingue and the famous French expedition to Egypt (1798–1801) is revealing. The Napoleonic expedition to Egypt, the "scientific" investigation of the country by a team of French savants, the creation of the Institut d'Égypt, and the subsequent publication of the results of the expedition represent a mighty demonstration of French colonial and scientific power. But the remarkable French adventure in Egypt hardly compares to the lesser-known but much more substantial and long-term investment of public resources in science and its institutions evident throughout the eighteenth century in Saint Domingue. On the French in Egypt, see Gillispie (1987); Lokke, pp. 187–230; Basalla, pp. 613, 618.

13. See MacLeod, pp. 223–29, 246; Brockway (1979). Brockway's bibliography does not evidence any significant precursors to her work.

14. Johnson (p. 414) characterizes Pyenson's work (1985) as "pioneering," and he calls for further scholarly contributions in this area.

15. Billed as a "cross-cultural comparison," the volume from the Australian conference deals mostly with science within the circle of British colonialism and within the past century or so; see the editors' Introduction, pp. vii–ix; see also MacLeod, p. 217, for reference to the increased pace of research over the past decade into the topic of colonial science.

16. On this point see esp. Chambers, who suggests (p. 316) that further progress must await "a multitude of case studies." MacLeod, p. 217, laments that research on colonial science to date "has revealed enormous gaps in our knowledge." See also Johnson, p. 415.

17. MacLeod, p. 217, and Jarrell, p. 348, emphasize the better-developed if still-nascent state of studies of science and colonialism in the Anglo-American world. Osborne, the recent exception proving the rule, also notes the poverty of work regarding French colonial science (pp. xii, 391).

18. Pyenson (1982), p. 1.

19. See Chambers, pp. 299, 306–10; and the Conclusion to the present work.

20. The precise location lies several miles to the east of Cap François/Cap Haitian. See DPF, pp. 172–73, 196, 213–15; Butel, p. 19.

21. In the period of concern here Spanish and Portuguese colonies constituted empires on the decline. Also, early in the eighteenth century (if not slightly sooner) Holland lost its trading preeminence first to England and then to France; see Liss, pp. 48–50, 58–59; Hyma, pp. 3–4, 37, Webster et al., pp. 866, 869.

22. On French colonization in the Old Regime, see primarily Priestley and Saintoyant.

23. Contemporaries did not view the French loss of Canada (in exchange for the return of Guadeloupe) as so great when compared with the value of retaining the West Indian sugar island. In this connection one recalls Voltaire's famous repudiation of Canada as a "few acres of snow"; see Lokke, p. 40; Stein (1988), p. x; W. B. Cohen, p. 162; Williams, pp. 114–15.

24. Brockway, pp. 20–27; Williams, p. 37.

25. Devèze, pp. 255, 273; Williams, p. 123; Knight, p. 87; Dunn, p. 35.

26. Only 25 percent of British imports come from the West Indies, compared with 35 percent of French imports; only 9 percent of British exports went there, compared with 17 percent for the French. See Brockway, p. 31; Williams, pp. 54–55, 58; Devèze, p. 275.

27. Stein (1988), p. x; see also Priestley, p. 291.

28. M. I. Finley provides the most explicit and analytical typology of colonies. Finley defines *colonies* strictly as transplanted settlements of people dependent on and subservient to a mother country. Saint Domingue might be considered the colony-type of the second of Finley's three-stage model (2:3), early-modern overseas colonies importing slave labor. See other remarks on colonial typology by Curtin, pp. 22–23; Jarrell, pp. 327–28; Bordier, pp. 30–32; Knight, chap. 3.

29. On colonial agroindustries and their importance, see Mintz (1974), p. 60; Lewis, pp. 2, 7, 13–14; Tarrade, pp. 27–29.

30. On the plantation system, see esp. Dunn; also Mintz (1985b), pp. 128–29; Devèze, chaps. 9–12.

31. Marx, pp. 270–72; Fox-Genovese and Genovese, chap. 1.

32. On trading companies, see esp. Chailley-Bert, passim and pp. 21–25, where he identifies seventy-eight companies chartered by the French crown in the period 1599–1787; see also Priestley, pp. 127–28, 131; Boxer, pp. 23–24, 43–53; Hyma, p. 58; Marx, pp. 41–50; Saintoyant, 2:448–54.

33. The authority on French mercantilism is Cole; see Cole (1939) and (1971). On eighteenth-century French mercantilism, see also Tarrade, pp. 83–95; Priestley, pp. 128–34; Chailley-Bert, pp. 7–11; Mathieu, pp. 16–21; Lokke, pp. 27–30; W. B. Cohen, pp. 156–57.

34. Fox-Genovese and Genovese, pp. 392–95; Bernal, 2:560–67; Coquery-Vidrovitch, pp. 350–53; Brockway, pp. 20–24; Headrick.

35. Pluchon, ed. (1980), p. 32.

36. On contraband trade, see Tarrade, pp. 95–101, and my chapter 4.

37. Mercantilism and the essentially conservative aims of the reforms of 1784 are the principal themes of Tarrade. Blerald, pp. 12–18, likewise makes the "mercantile colonial system" central to his analysis of the Caribbean economy in the eighteenth century.

38. Two factors played a role in this collapse: a long period of intense and bloody warfare and various land policies instituted in the new Haiti that broke down large colonial plantations into subsistence holdings; see Dorsainvil, pp. 147, 167; Cauna (1987), pp. 247–50; Cornevin, p. 57.

39. This point emerged in conversation with M. Gérand Fombrun, but the barman I encountered in Port-au-Prince, who recited speeches by Toussaint Louverture and other early Haitian patriots, vividly instantiated the point. On literacy in Haiti, see Stone, p. 27.

40. On this point with all of its disheartening historiographical implications, see Mintz (1974), p. 51; Mintz and Price, p. 6.

41. The point about modern French impressions of Saint Domingue derives partly from a popular and semipopular literature about the former colony, as in,

for example, works by Hugo, Rebell, and Girod (1972). I owe M. Jacques Cauna for the latter point about French confusions of Saint Domingue with Santo Domingo and Tahiti; see also Cauna (1987), pp. 26, 193.

42. In the eighteenth century one located Saint Domingue in the Greater Antilles and "under the wind" (*sous le vent*) as opposed to the more easterly *isles du vent*, or the Windward Islands or Lesser Antilles; see Butel, pp. 294–95.

43. Cabon (1916b), pp. 65–68; Rodman, pp. 1–2. Lokke, p. 21n, recognizes this problem of nomenclature and the necessity of the term *Saint Domingue* to refer to the former French colony. To call the colony Haiti before 1804 is anachronistic, and Williams, Geggus, and other modern historians consistently use the correct, if less familiar, Saint Domingue; see Spencer in Moreau de Saint-Méry (1985), p. iv, in this regard. French usage hyphenates Saint-Domingue, but here we follow English usage and leave the hyphen out.

44. The Société de l'Histoire des Colonies Françaises (from 1959 the Société Française de l'Histoire d'Outre-Mer) began in France in 1913. The Société Haïtienne d'Histoire de Géographie et de Géologie commenced in 1925. Both societies have regularly published important series of variously titled *Revues*. The French Académie des Sciences d'Outre-Mer dates from 1922, and several of its publications have concerned Saint Domingue and Haiti. Finally in this connection, the Franco-Haitian literary and intellectual journal, *Conjonction,* published in Haiti by the Institut Français d'Haïti since 1945, needs to be mentioned as a notable resource.

45. For more complete lists and discussions of the Debien oeuvre, consult Shannon, pp. 51–55; Maurel in *DPF,* p. xlvi; Anglade (1973b), p. 140.

46. See esp. the fine work of Georges Anglade and Jean Saint-Vil. The substantial annotated bibliographies by Ulrich Duvivier (1941), Max Bissainthe (1951), the Michel Laguerre (1982) are key starting points for any study of Haiti or Saint Domingue. Darondel discusses other aspects of the historiography of Haiti and Saint Domingue.

47. See Maurel (1938) and (1961); Pluchon (1985a). The remark made in 1955 by the eminent Jean Fouchard (1955b, p. 57) that the scientific, artistic and literary accomplishments of Saint Domingue have not been sufficiently appreciated must still be considered just.

48. Another indication of this sad state of affairs is the (1985) publication of the first, partial translation in English of the major source for eighteenth-century Saint Domingue, Moreau de Saint-Méry's *Description de la Partie Française de l'Isle de Saint-Domingue* (1797–98), about which more momentarily. Ivor D. Spencer's translation is self-styled "a vastly abridged edition" of only 275 typed pages; the complete, modern edition in French runs to over 1,500 pages of very small print; see Spencer in Moreau (1985), p. iii.

49. Mintz focuses on the British case because of the key role that nation played in the Industrial Revolution, and Mintz hopes particularly thereby (1985a, p. xxix) "to show the special significance of a colonial product like sugar in the growth of world capitalism." But not to consider the role of sugar in other contexts, particularly in such a nontrivial case as Saint Domingue, stacks the deck in favor of a historically less nuanced thesis linking sugar and the rise of capitalism.

To note (as Mintz does, pp. 189–90) that sugar did not and does not figure prominently in French cuisine misses the point: the issues for French sugar in the eighteenth-century concern not French consumption, but French production and export, their significance for the Old Regime economy, and their impact on patterns of consumption in the rest of Europe. Happily, Stein's study of the French sugar business appeared in 1988 to remedy this situation.

50. See Liss, passim and p. xiii. Although the problems are hardly as serious, one might criticize Parry (1971) along these same lines. Williams's pioneering volume likewise calls for comparative material from the French case. Stein (1979), p. xvi, remarks on a similar imbalance in how historians have treated the French and British slave trades, and Hoetink, pp. 56–57, likewise complains about a "bias" and "narrowness of focus" among Caribbeanists slanted toward the English-speaking colonies.

51. See Canny and Pagden; the essay by Elliott defines the concept of colonial identities (pp. 4–5). See also the review by Anderson.

52. Canny and Pagden, pp. 272, 275–76. That is not to say that Canny, Pagden, or their contributors are insensitive to racial issues or the fact of slavery. The article by Jack P. Greene on Barbados belies any such criticism. The point is more that the dramatic and relevant cases of Haiti and Saint Domingue simply do not enter the picture as they should.

53. The abbé Raynal's critical and highly popular *Histoire philosophique et politique des Établissements et du Commerce des Européens dans les deux Indes* (the first of many editions published in 1770) especially provoked a storm of debate. Raynal's later *Essai sur l'Administration de Saint Domingue* (1785) brought the focus of his criticisms even more directly to bear on Saint Domingue; see Lokke, pp. 45–49, and responses to Reynal by Hilliard d'Auberteuil, Dubuisson, and vols. by Barbé de Marbois.

54. Moreau de Saint-Méry will appear throughout this account; see biography by Elicona; and Taillemite in *DPF,* pp. vii–xxxvi. While a law student in Paris Moreau supposedly studied astronomy with Lalande and Messier.

55. Moreau's atlas contains detailed maps of the colony and a dozen of its towns along with priceless illustrated scenes of contemporary colonial life. Much of the iconography used here finds its source in Moreau's *Recueil de Vues;* see further Duvivier, pp. 114–15; Maurel in *DPF,* pp. xlv–xlvi.

56. Taillemite calls these materials a mine for historians, and Maurel notes that many items are absolutely unique. Two deposits are in question. The French Archives Nationales proper house the so-called Collection de Moreau de Saint-Méry (series F^3, 287 volumes). The French Ministry of Overseas Affairs (Ministère de la France d'Outre-Mer) superintends the other, the so-called Bibliothèque de Moreau de Saint-Méry; see Taillemite and Maurel in *DPF,* pp. xxxii, xxxix, xli–xlii; May; and Moreau himself, *DPF,* p. 354.

57. Not to slight the Mangonès research collection on colonial and national history in the Bibliothèque Saint Louis de Gonzague in Port-au-Prince. The now-vanished library of the Petit Séminaire Collège Saint Martial in Port-au-Prince might someday reappear, holding scholarly surprises.

58. See Maurel in *DPF,* pp. xliii and following. Administrators normally sent

dispatches to and from Saint Domingue in multiple copies, and in 1776 the French established an entire set of duplicate colonial records at Versailles; see DPF, p. 377; Pluchon (1985a), p. 167.

59. Chapter 5 treats the *Affiches Américaines* in detail; here one can note that the *Affiches* dwarfs the major contemporary Mexican journal that included scientific and technical notices, the *Gazeta de Literatura;* see Chambers, p. 303.

60. Moreau de Saint-Méry claimed that his *Description . . . de Saint Domingue* was the most complete and detailed account of any region ever made, and modern opinion supports his view as more or less correct. See DPF, p. 9, and Maurel in DPF, p. xlvi. (On equivalent descriptions of towns in Old Regime France, see Darnton (1985), chap. 3 and pp. 273 n.1, 274 n.7.) On the one hand, it is astounding to consider that Moreau's *Description* depicts eighteenth-century Saint Domingue better than so many more well-known locales; on the other hand, one can hardly imagine the bounty of an equivalent 1,500 pages given over to a systematic description of, say, New Jersey in 1789. The availability of significant source materials in addition to Moreau's *Description* only compounds this embarrassment of scholarly riches.

CHAPTER 1: Material Factors

1. Regarding distance as a factor and the definitive imprint of the sea on colonial development in Saint Domingue, see Mathieu, p. 132; Gordon, pp. 1, 21; DPF, p. 597. For a suggestive geographical and climatological comparison of the Caribbean and the Mediterranean, see Lewis, pp. 16–20.

2. Moreau de Saint-Méry calls the great tall ships "the most astonishing machines created by the genius of man"; DPF, p. 465.

3. The record reports various sailing times; see, e.g., Cornevin, p. 44; Cauna (1987), p. 18. See also DPF, pp. 294, 1409–22. Passage to French outposts in the Indian Ocean took six months; Lokke, p. 103.

4. Consider that it took from two days to two weeks to turn the Môle peninsula en route to Port-au-Prince; to reach the southern part of the colony one had to sail completely around Hispaniola; see DPF, pp. 761, 1166.

5. See DPF, p. 462; Neufchateau, p. ii; Desfeuilles, p. 367.

6. For scientific descriptions of the territory occupied by Saint Domingue and Haiti, see DPF, pp. 26–29; Anglade (1982); Tippenhauer; Watts, chap. 1.

7. Regarding the mountains of Saint Domingue, see Anglade (1982), p. 41; Centre d'Études de Géographie Tropicale, pl. 4; Barré de Saint-Venant, p. 173; DPF, pp. 581, 713, 716, 962, 1163, 1222; Cornevin, p. 7.

8. On the coastal plains of Saint Domingue, see Gordon, p. 4; Centre d'Études de Géographie Tropicale, pl. 2; Anglade (1982), p. 17; Cauna (1987), pp. 34–38; DPF, pp. 117, 811, 938, 1273ff.

9. DPF, pp. 27, 1409.

10. On Saint Domingue's climate, see Duvivier, 1:1–3; Bordier, pp. 488–89; Cornevin, pp. 10–11; DPF, pp. 499, 962–64; Watts, pp. 20, 24; Anglade (1982), pp. 54–55. The ecology of western Hispaniola has changed considerably in more modern times; not all of these climatic or ecological conditions still apply.

11. Barré de Saint-Venant, p. 183. The Artibonite was not the only desert area in Saint Domingue; see DPF, pp. 722, 801, 807.

12. *DPF,* p. 812, and Taillemite and Maurel in *DPF,* pp. 1427, 1430–31.

13. *DPF,* pp. 226, 683.

14. Duvivier, 1:386–93, lists the major storms recorded for the period 1495–1821. Tarrade, pp. 25–26, and Watts, pp. 531–32, emphasize the influence of natural disasters on colonial development.

15. See *DPF,* pp. 719, 965, 1061, 1238, 1262, 1320.

16. *DPF,* pp. 512, 777; Cauna (1987), pp. 137–38.

17. Duvivier, 1:110–15; Tippenhauer, pp. 170–73; Watts, passim, per p. 593. Dunn, pp. 42–44, describes the famous earthquake of 1692 that destroyed Port Royal in Jamaica.

18. See *DPF,* pp. 720–21, 979, 1039, 1062–67, 1116, for these details; see also Corvington, pp. 23, 44, 69; Saint-Vil, pp. 17–18.

19. See *DPF,* pp. 341, 472, 873, 744, 907, 1004, 1262, for these and related details; see also Julien (1977), p. 126; Cauna (1987), p. 33.

20. Only five days earlier the royal frigate *Inconstante* burned in Saint-Louis with over eighty casualties; see *DPF,* pp. 318, 470, 1263.

21. See *DPF,* pp. 472, 889, 1039; *AA,* Cap Supplement, 1789, no. 6 (February 7).

22. CA-HdC, fol. 164, Lokke, p. 82. Exacerbating problems, traders stayed away when drought or calamity struck, knowing no profitable business could be found.

23. On these points, see *DPF,* pp. 291, 514–23, 1181, 1367; Cauna (1987), pp. 121–30; Chaume in Lind, p. xix.

24. Crosby; Gentilini and Nozais, passim and pp. 45, 57; on disease patterns elsewhere in the Caribbean, see Dunn, pp. 302–6.

25. DPF, p. 427; *AA,* PauP, 1788, no. 17 (February 28).

26. Cauna (1987), pp. 123–27; *DPF,* p. 557; Caillet. Among the dermatological remedies advertised in the *Affiches Américaines,* the list for 1789 alone includes Ailhaud's famous powder, Bruckman's cosmetic vinegar, the Swedish elixir, "Russian water," common green unguent, the powder of Mlle Aubin Genier of Marseille, and others.

27. See Léon (1931), pp. 86–87; *DPF,* pp. 61, 520, 672–73, 1272; Cauna (1987), pp. 101–5.

28. See *DPF,* pp. 521–22, 1068; also Gentilini and Nozais, pp. 53–56; Cabon [1928–33], 2:438; Gros, p. 345.

29. See Gentilini and Nozais, pp. 55, 60–61, 63–64; Carron, pp. 3–12; Cauna (1987), p. 123; Pluchon (1985b), p. 106; Léon (1931), pp. 84–85.

30. *DPF,* pp. 291, 952, 1272.

31. *DPF,* p. 515; Dubuisson, pp. 8–26.

32. On these ailments, see Cauna (1984), p. 35; Pouppée-Desportes, 1:25; *DPF,* pp. 478, 522–23, 619, 1035.

33. See *DPF,* p. 29; Butel, p. 21; Lewis, pp. 4–5.

34. Moreau repeatedly refers to pests; see *DPF,* pp. 265, 509, 957, 1070, 1198, 1229, 1262, 1271, 1277, 1297. Stein (1988), p. 63, and Cauna (1987), pp. 58, 141, also recognize the problem of insect and other infestations and provide additional details.

35. *DPF,* pp. 1404–5; see also Hicks and his marvelous photographs of a simi-

lar natural phenomenon on Christmas Island in the Indian Ocean.

36. See *DPF,* pp. 465, 689, 1239, 1371, 1404–6; Fouchard (1955a); Watts, p. 33. In addition to red tide, crabs posed a real danger of poison because they ate and transmitted the poison from the fruit of the manchineel tree, which was vigorously exterminated; offshore whaling took place around Saint Domingue, but that was apparently left to the Americans.

37. See *DPF,* pp. 805, 896–97, 968, 1070, 1117, 1253, 1271, 1403.

38. *DPF,* pp. 209, 264, 301, 688, 782, 1359, 1403; Girod (1972), p. 227; Saint-Vil, p. 16; Knight, p. 27. Exotic birds were imported from Île de France, Senegal, Guiana, and Louisiana; the existence of domestic cats is inferred from Moreau's mention of wild cats.

39. Lacroix makes this point, 3:138–41; see also Crosby.

40. On these points see Moreau's many remarks in *DPF,* pp, 633, 635, 684, 697, 707, 806, 856, 886, 962, 1224, 1263, 1270, 1330, 1393, 1406; Charles. With so little wood in Haiti today, charcoal is made from cactus.

41. *DPF,* pp. 910, 1091; Spencer in Moreau (1985), pp. 167–74; Stein (1979), p. 42; Anglade (1982), p. 55. On the superior fertility of Saint Domingue compared with plantations in the British Caribbean colonies, see Williams, pp. 113–14.

42. See Mintz (1985b), p. 136; Knight, p.25; Stehlé (1966), pp. 34–36; and references in *DPF,* pp. 185, 221, 224, 686, 890, 897, 1272, 1400, 1406.

43. See *DPF,* pp. 111, 155–56, 243, 265, 806, 1404; Cabon [1928–33], 2:443.

44. Moreau tells the story in, *DPF,* pp. 933–34.

CHAPTER 2: Historical Development

1. On French colonization in Canada, see Salone, esp. pp. 15–53.

2. See esp. Crouse (1977), chap. 2. The French settlement on Saint Christopher was disbanded in 1690, with a thousand or so emigrants coming to Saint Domingue. On these latter points, see Crouse (1966), pp. 165–66, 176, 309; Knight, pp. 36–38; Dunn, chap. 4.

3. Crouse (1977), chaps. 3–4, 9–10; Crouse (1966), pp. 2–3.

4. Crouse (1966), pp. 128–32; Crouse (1977), pp. 85–88; *DPF,* p. 17.

5. Dorsainvil, p. 9, from Bartolomé de las Casas; others put the aboriginal population closer to one million; see *DPF,* pp. 29, 1370; Gordon, p. 11; Butel, p. 21. Knight, pp. 5–6, 14, gives the figure of half a million. See also Devèze, pp. 13–34; Ewen, pp. 41–47; Centre d'Études de Géographie Tropicale, pl. 2a. The Arawaks seems to have been a pacific people who had achieved a (temporary) harmony with their environment. They possessed an essentially neolithic economy: garden/hoe agriculture, village living, and, in this case, "gathering" the natural abundance around them.

6. By 1514 only thirty thousand Arawaks remained, 1 percent of the original population. By 1533, the number had fallen to six hundred; by 1570, only one hundred remained. Late in the eighteenth century, in 1773, the sad report appears of the last pure-blooded Arawak. See Houdaille, p. 860; Montague, p. 4n; Devèze, chap. 2; Dorsainvil, chaps. 2–3, and *DPF,* pp. 1085–86, 1130; *Entrennes Américaines [a.k.a. Almanach Historique et Chronologique de Saint-Domingue],* 1773, pp. 70–71.

7. Crouse (1977) makes the point about the Caribs; see listing, p. 284.

8. Devèze, p. 144; see also Hoetink, p. 60; Ewen, p. 42; Knight, p. 31.

9. On seventeenth-century pirates, see the pioneering study by Burg and comments by C. Hill. See also Butel; Doucet; Lewis, pp. 78–83. For Moreau on pirates, see listing, *DPF*, p. 1435.

10. Crouse (1977), pp. 80–98; Crouse (1966), pp. 122–28; Butel, pp. 226–27; *DPF*, pp. 644, 669; Mathieu, pp. 13–14; Cauna (1987), pp. 22–23.

11. The settlements of Jérémie and Jacmel followed in 1673 and 1680; see *DPF*, pp. 297–98, 577, 787; also Crouse (1966), p. 129.

12. Butel, pp. 121, 226–27, 278; *DPF*, p. 671; Lokke, pp. 17, 60.

13. See *DPF*, pp. 865–73. Annual salt production from Saint Domingue reached 200,000 barrels, three-quarters of which went to salt cod in North America.

14. Authorities finally chased out the Jérémie pirates in 1778 when they refused to serve in the American War; see *DPF*, pp. 905, 1357, 1394; Crouse (1966), p. 309; Cabon [1928–33], 2:453.

15. See Stein (1979), pp. 4, 11; Knight, pp. 79–80; Crouse (1966), pp. 128–29, 145; Crouse (1977), chap. 13 and p. 98.

16. On seventeenth-century Saint Domingue, see Crouse (1966), chap. 5; Devèze, pp. 136–43; Butel, pp. 159–93; Doucet, pp. 77–99; *DPF*, pp. 669–71, 1531.

17. Butel, p. 167; *DPF*, p. 45; Duvèze, p. 140; Stein (1979), p. 9; Crouse (1966), p. 133; Williams, p. 9. The three-year term of service contrasts with the seven years required of the indentured in British North America.

18. Convoys of "ladies" (*filles*), often conscripted off the street in France, continued to arrive in Saint Domingue at least through 1720; see Julien (1976), p. 162; Crouse (1966), p. 132; James (1963), p. 30.

19. See Moreau de Saint-Méry's list of governors-general and their formal titles in *DPF*, pp. 17–18; see also Tarrade, pp. 65–72.

20. *DPF*, pp. 671, 708, 1088; Frostin (1975), chap. 1; Crouse (1966), p. 142. Tortue reverted to the wild and became famous for its wild pigs, killed by the thousands by hunting parties later in the eighteenth century.

21. See Chailley-Bert; Doucet, pp. 78–81, 87–89.

22. See *DPF*, pp. 1143, 1165, 1232, 1241–44; Raynal (1781), vol. 7, p. 149; Julien (1976), p. 167; Frostin (1975), chap. 3.

23. See Crouse (1966); Dunn, pp. 22–23.

24. *DPF*, p. 31; Stein (1979), p. 23; Ott, p. 6.

25. *DPF*, p. 112, and listing, pp. 1562–65; parishes divided unofficially into 550 cantons in 1789; see listing, *DPF*, pp. 1422–33. Parishes grouped into the three administrative departments of the colony.

26. On the churches in Cap François and Port-au-Prince, see detailed descriptions in *DPF*, pp. 331–40, 995–96; on churches in outlying settlements, see, e.g., *DPF*, pp. 724, 743. See also Corvington, pp. 47, 69; Saint-Vil, p. 17; Castonnet des Fosses (1884), p. 21.

27. Colonial religious orders have yet to receive systematic study, but Jesuits, Dominicans, and the Brothers of Saint John numbered among the orders established in Saint Domingue; see *DPF*, pp. 419–22, 994–95; Debien (1949),

pp. 557–75; Cabon [1928–33], 1:160–61; Dorsainvil, pp. 331–35.

28. A religious calendar appeared annually from 1773 in the *Almanach de Saint-Domingue;* see also Cauna (1987), p. 117.

29. See *DPF*, pp. 335, 512, 877, 996, 1007; see also Darnton (1985), chapt. 3.

30. See *DPF*, pp. 338, 995; Debien (1974), p. 120; James (1963), p. 32.

31. See *DPF*, pp. 18–19, 482–83, 715, 1382; also Chastonnet-Desterre, pp. 51–53; Cobban, pp. 12–14, 41; Tarrade, pp. 68–72.

32. See esp. the accounts of the greatest intendant of Saint Domingue, Barbé de Marbois, who served from 1785 to 1789; see also Corvington, pp. 145–72.

33. Cabon [1928–33], 1:272; James (1963), pp. 34–35.

34. Corvington, p. 23; *DPF*, pp. 481–82, 676.

35. On the military in Saint Domingue, see Corvington, p. 98; Bellegarde (1923), p. 105. See also Moreau's constant attention to military defense, *DPF*, pp. 589–608, 737, 747, 1078.

36. *DPF*, pp. 601, 1006–7; Frostin (1973), p. 332; Mathieu, p. 34.

37. Located throughout the colony, the number of *chasseurs royaux* could easily have been one thousand; they served with distinction around Savannah in the American War of Independence; see *DPF*, pp. 85, 103, 747, 758, 1008.

38. See *DPF*, pp. 417, 608, 990, 1029, 1243.

39. See *DPF*, pp. 31, 116, 483; see also Frostin (1973), pp. 317–43; Pluchon (1985b), p. 125; Castonnet des Fosses (1884), p. 12; Cauna (1987), p. 12.

40. See *DPF*, pp. 484–87, 899, 913. Moreau gives figures for the *milices* in each parish; the number, twenty thousand, is based on the conservative estimate of four hundred per parish. The institution of a standing militia in 1769 precipitated yet another round of unrest in the colony; see Frostin (1975), chap. 6.

41. On the Maréchaussée in Saint Domingue, see *DPF*, pp. 440–43, 691, 805, 855.

42. On police units in the colony, see *DPF*, pp. 473–78, 886, 1035–39; Pluchon (1985b), p. 127. Despite an official inspector of weights and measures, Cap François was nevertheless notorious for false weights.

43. On the conseils, see *DPF*, pp. 373–84, 999–1005; Tramond [1928], pp. 18–19.

44. The Conseils Supérieurs sat for one hundred days a year with Paris law prevailing. Moreau presents the century's worth of cases handled by the Conseils Supérieurs of Saint Domingue in his six-volume *Lois et Constitutions des Colonies François de l'Amérique Sous-le-Vent.*

45. The Conseils of Saint Domingue evidenced a streak of parlementary independence similar to their counterparts in France; in 1769, for example, the august Prince de Rohan, serving as governor-general, banished the Port-au-Prince conseil back to France. See *DPF*, p. 1000.

46. *Almanach de Saint-Domingue*, 1773, pp. 79–80 (and subsequent years). *Conseillers* had to be university trained and at least twenty-seven years old. Their pay varied from 9,000 to 13,500 livres; see *DPF*, pp. 380, 1001; Neufchateau.

47. On the Sénéchaussée courts, see *DPF*, pp. 378–79, 477–78, 670, 1015, 1168.

48. See *DPF*, pp. 379–80, 877, 1017, 1248, 1307. The jurisdictions of the

colony's various courts overlapped in a haphazard way that caused great confusion; see *DPF*, pp. 805, 1144–45.

49. See *DPF*, pp. 889, 915, 942, 1016, 1028–30.

50. On colonial Chambers of Agriculture and Commerce, see Tarrade, pp. 74–81; Nicolas; Frostin (1975), pp. 355–58; *DPF*, pp. 490–91. On agricultural societies in France, see McClellan (1985), pp. 38, 136, 326.

51. The Cap Chamber of Agriculture produced one hundred or so articles over a twenty-five-year period, but its main attraction seemed to be the fees it brought; one Bourgeois supposedly received 6,200 livres as secretary of the Cap chamber; see *DPF*, pp. 293, 317, 490; Arthaud (1785), pp. 10–11, 16.

52. *DPF*, pp. 316, 491–92; Chambre d'Agriculture.

53. *DPF*, pp. 450–51; cf. Darnton (1985), chap. 3.

54. See *DPF*, p. 969; Cauna (1987), pp. 12, 35. The existence of colonial receivers of *droits domaniaux* further evidences the point; see *DPF*, pp. 145, 891, 1030. Tarrade, p. 146, however, minimizes the extent of feudal privileges in the colony; see also Crouse (1977), p. 168; Vaissière, chap. 2; and the Conclusion in the present work.

55. Cauna (1987), pp. 170–71.

56. On this trend, see Cauna (1987), pp. 26, 45–50, 197.

57. Moreau tells this story with disgust; see *DPF*, pp. 697–98, 1158, 1325–26.

CHAPTER 3: Population and Sociology

1. In a famous discussion (*DPF*, pp. 86–100), Moreau de Saint-Méry identifies 128 racial combinations and 13 different racial types. In practice, he admits (p. 29) to 3 constituent groups: whites, blacks, and mulattoes. Williams, p. 7, and Dunn, p. 225, distinguish underlying economic causes to slavery and racial segregation.

2. See *DPF*, pp. 102, 110.

3. See Madiou, 1:29; Thompson, 2:25, from contemporary Spanish sources; see also Spencer in Moreau de Saint-Méry (1985), p. 41.

4. Montague, p. 5.

5. Graphs 1A and 1B derive from sources listed in Table 1, note a.

6. Fox-Genovese and Genovese, p. 43.

7. Cf. Blerald, p. 28; Pouzet, p. 28; Doucet, p. 151; Martinière, p. 289; Houdaille, p. 871; Stein (1979), p. 120; see also Le Bihan (1974), p. 48.

8. Montague, pp. 5–6; Darby, p. 268; and comparable figures in Frostin (1962), p. 299; Devèze, pp. 282, 365; Pouzet, p. 23. The same racial proportions would seem to hold throughout the Spanish Caribbean; see Klein, pp. 295–96.

9. *DPF*, p. 44; Humphreys, p. 418; James (1963), p. 56.

10. The major authority on the French slave trade is now Stein (1979). See also Klein, pp. 139–61; Pluchon (1987), appendix 5; Lewis, citing Philip Curtin, p. 5; Geggus (1989), pp. 1294–96; Knight, pp. 86–87; Dunn, pp. 229–38; Stein (1988), pp. 20–21, 33; Bohannan and Curtin, pp. 25, 27.

11. These conservative figures are based on Chastonnet-Desterre, pp. 41–46. Stein (1988), p. 22, and Stein (1979), p. 211, puts the number of slaves imported annually at closer to forty thousand; Tarrade, table xi and p. 783, reports the fig-

ure of thirty thousand. See also Pluchon (1987), appendix 5.

12. *DPF*, pp. 901, 1296; Barré de Saint-Venant, p. 441.

13. The best study of an eighteenth-century plantation is now Cauna (1987); see also Girod (1970); Debien (1956); Cauna (1984), pp. 55–78; Knight, pp. 95–103.

14. On these points, see Stein (1979), p. 113; *DPF*, pp. 221, 739, 745, 795, 843, 1055, 1385; Bellegarde (1950), p. 11.

15. Dazille (1792), p. 2; see also *DPF*, p. 1271.

16. On the *journée de nègre*, see *DPF*, pp. 1206, 1267; Cauna (1987), p. 170.

17. See Pouzet, 33; Cauna (1987), pp. 101, 257; Castonnet des Fosses (1893), p. 12; *DPF*, pp. 61, 1272; Butel, pp. 234, 246; compare Bush.

18. Cf. Father Nicolson, p. 54, and Cauna (1987), pp. 119–21; see also Cauna (1984), pp. 18–55; Girod (1970), pp. 112–17.

19. *DPF*, p. 1295; Girod (1970), pp. 105–6, 121–32; Cauna (1987), pp. 91–92, 106.

20. An extensive literature documents sadism and the torture of slaves in Saint Domingue. A moderate example: Girod (1970), p. 118, reports the case of a slave who received fifteen hundred lashes; he survived, only to die later chained in his master's private prison. Pluchon (1984), pp. 164–88, discusses these matters at length; Vasty, pp. 40–63, provides other horrible details. See also Spencer in Moreau de Saint-Méry (1985), pp. 263–72; the anonymous *Manuel théorique*, and Rebell, passim. One wonders if the Marquis de Sade himself ever thought of refuge in Saint Domingue.

21. See Butel, p. 239, James (1963), pp. 22–24; *DPF*, p. 1103; Spencer in Moreau de Saint-Méry (1985), pp. 269–71. The much less severe treatment of slaves in Spanish Saint Domingue deserves emphasis; the distinction is sometimes drawn between "patrician" slavery of the Spanish and "plantation" slavery of the French; cf. Moreau de Saint-Méry (1796a), 1:58–60, 119; 2:39; Cauna (1987), pp. 134–35.

22. On the Code Noir and slave regulations, see Devèze, pp. 283–85; Lokke, pp. 32–34; Cauna (1987), pp. 118–21, 132; *DPF*, pp. 548, 1038. Doucet, appendix 1, prints in abridged version of the Code Noir; cf. Dunn, pp. 238–46.

23. Devèze, p. 281; L. Mercier, p. 108; Cauna (1987), p. 105; *DPF*, p. 59; and surveys of slave prices in *AA*, November 8, 1787, and in BN, Mss., n.a.f. 22085, fol. 145v. Stein (1979), pp. 84, 141, reports the price of slaves in Africa at upward of 750 livres tournois.

24. *DPF*, pp. 1166, 1295, 1297, 1319; also Greggus (1989), p. 1291n.

25. Moreau (*DPF*, pp. 64–70) provides an early and vivid description of vodun in Saint Domingue; Pluchon (1987) presents the most current and extensive study; see also the controversial anthropological investigation by Davis. As Davis notes, pp. 11–12, the English term *voodoo* has developed derogatory connotations and has been replaced by *vodun* (also *vodoun*). On vodun, see also Cornevin, pp. 42–43; Lewis, pp. 188–97.

26. On slave society and culture, see portraits by Moreau in *DPF*, pp. 44–83, 544, and Nicolson, pp. 49–59. See also Pluchon (1987), esp. pp. 58–60, 64, 75–78; Devèze, pp. 281–91; for views of Caribbean slave cultures generally, see

Klein, pp. 163–87; Lewis, pp. 184–205; Knight, chap. 5. Bush provides an insightful study of female slaves.

27. Debien (1967); *DPF*, pp. 1055, 1237, 1272, 1316, 1352.

28. *DPF*, pp. 80–83, 320; Lewis, p. 12; Préval. On the different significations of the term *creole,* see Mintz and Price, p. 6.

29. Pluchon, ed. (1980), p. 77; see also W. B. Cohen, pp. 56–58, 128, 291–93.

30. Chastonnet-Desterre, p. 300, and Castonnet des Fosses (1893), p. 79. For the larger background to New World slave revolts, see Genovese; Klein, pp. 189–215.

31. The recent volume by Pluchon (1987) supersedes all previous studies and accounts of poison and poisoning in Saint Domingue. See also Davis, pp. 198–99; Lewis, pp. 177–78; James (1963), pp. 15–17; W. B. Cohen, pp. 103–4.

32. Moreau recounts the Macandal conspiracy, *DPF*, pp. 629–31; see fuller account and analysis in Pluchon (1987), chap. 7–8 and appendix 7.

33. The major source on maroon groups is Price, ed. (1979); on the meanings of the term *maroon,* see p. 1n; for the French case, see pp. 105–48. For Moreau's accounts of maroon groups, see *DPF*, pp. 157, 183, 207, 210, 697, 956, 1271, 1276, 1353, 1395. See also Fouchard (1972); Lewis, pp. 175–77, 230–32; Cauna (1987), pp. 130–35.

34. The colonial constabulary and mulatto military units captured 48,000 runaway slaves between 1764 and 1792, an average of 1,700 a year; see Thésée, p. 179; see also Davis, p. 194.

35. Moreau, who uses the term *forest people* (*créols de bois*), discusses this group in detail, in *DPF*, pp. 1131–36, translated and reprinted in Price, ed. (1979), pp. 135–42. See also *DPF*, p. 972; Debbasch; Spencer in Moreau de Saint-Méry (1985), pp. 247–53.

36. The campaign of 1776–77, for example, involved 180 men, lasted three months, and cost 80,000 livres; *DPF*, pp. 1131–33; also Davis, p. 193.

37. Even this estimate may be high; see Cornevin, quoting Girod, p. 32. On whites and their social makeup, see *DPF*, pp. 29–44; Devèze, pp. 179–89; Geggus (1982), pt. 1; Girod (1972), chaps. 2–3.

38. *DPF*, pp. 259, 400, 475, 478, 480, 1318–19; Stein (1979), p. 9.

39. Houdaille, p. 864; also, *DPF*, p. 32. Another contemporary source says that "almost all" whites were born in France; see Debien (1974), p. 120.

40. Houdaille, p. 863; Saint-Vil, p. 22. Moreau, *DPF*, pp. 119, 722, puts the ratio at 3:2, male to female.

41. See Houdaille, p. 864; *DPF*, pp. 523; Cauna (1987), p. 257.

42. *DPF*, pp. 480, 519. Saint-Vil reports (p. 22) that in 1777 only eleven white children and five slave children under twelve lived in all of Cap François.

43. *DPF*, p. 1358; see also pp. 641, 1300, reporting ladies who married five and seven times. Bellegarde puts the number of white marriages at two thousand; see Bellegarde (1950), pp. 9, 13, and (1923), pp. 104, 107.

44. Fouchard (1955a), pp. 62–63, and (1955b), pp. 26, 36, 125; Vignols, p. 364; *DPF*, pp. 518, 722, 890, 1223, 1238, 1403; Mintz (1985a), pp. 136–37.

45. Dermigny, pp. 58–60; Dunn, pp. 9–10; Girod (1970), pp. 12–16; *DPF*, p. 1282; Robert Darnton, private communication of December 2, 1985.

46. Charles Arthaud, mss. "Discours . . . 3 X 1786," AN, Colonies F³ 152, fols. 234–238v. On cupidity as the driving force in the colony, see also Raynal (1781), 7:192; Neufchateau, pp. 54–56; Castonnet des Fosses (1884), p. 3.

47. On the *affranchis*, see *DPF*, pp. 100–110; Klein, pp. 217–41. See also James (1963); Knight, pp. 105–9; W. B. Cohen, pp. 52–56, 101–8.

48. *DPF*, p. 102. Girod (1972), chaps. 4–5; compare W. B. Cohen, p. 108.

49. Lewis, pp. 8–10; also Cauna (1987), p. 54. Overwhelmingly, mulattoes were the offspring of white men and black women. Such contact was permitted, legally and socially. Moreau notwithstanding to the contrary, evidence suggests some sexual contact between white women and black and mulatto men. For white women to have sexual relations with nonwhite men was a crime, however, punishable by death for the man; see Pluchon (1984), pp. 214–25; *DPF*, p. 95; Dunn, pp. 228, 252.

50. See Table 1; Houdaille, p. 865; *DPF*, pp. 119, 722. Klein's figures, p. 296, show only 13,000 mulattoes in the British West Indies and only 32,000 in the United States late in the eighteenth-century, compared with our figure of 28,000 for Saint Domingue. Klein further confirms that the major concentration of mulattoes in the Caribbean (80,000) was in Spanish Saint Domingue.

51. See *DPF*, pp. 103–4, 222, 676, 1197, 1237, 1389, 1400; Spencer in Moreau de Saint-Méry (1985), pp. 243–47; Chastonnet-Desterre, p. 51; Castonnet des Fosses (1893), p. 11; note the several hamlets called Fond-des-Blancs and Fond-des-Nègres.

52. See extended discussion in *DPF*, pp. 104–10. Moreau's widely reported portrait is no less interesting and relevant for its obviously sexist bias.

53. Bellegrade (1950), pp. 9, 13, and (1923), p. 104; see also Hilliard d'Auberteuil; Pluchon (1984), p. 215; James (1963), p. 32.

54. *DPF*, p. 105; see also *DPF*, pp. 95, 104; Pluchon (1984), p. 215.

55. See *DPF*, pp. 108–9, 516. Sumptuary laws were widely ignored.

56. *AA*, PauP Supplement, 1786, no. 1 (January 7); also, Corvington, pp. 106–7, 110.

57. Regarding Jews in the colony, see Pluchon (1984); Loker (1976a) and (1976b); see also Dunn, pp. 106–8, 183; Crouse (1977), p. 179.

58. See Pluchon (1984), pp. 91, 106; Loker (1976a) and (1976b); Launay, p. 33; *DPF*, pp. 1196, 1236, 1251–52, 1339, 1518–19.

59. *DPF*, p. 95; Stein (1979), p. 9.

60. See *DPF*, pp. 734–37; Lokke pp. 18–21; Chaia, p. 139; Broc, pp. 276–77. The expedition to Kourou cost 30 million livres and ten thousand lives.

61. *DPF*, pp. 232–33, 267, 735, 752–61, 912; Castonnet des Fosses (1893), p. 30.

CHAPTER 4: Industry and Economy

1. Most commentators agree on this point; see Lewis, p. 123; Pluchon, ed. (1980), p. 26; Doucet, p. 129; Stein (1979), pp. 23, 109; Cauna (1987), p. 11; James (1963), pp. 51, 55; Fox-Genovese and Genovese, p. 43; Williams, p. 122. Those who do not award Saint Domingue quite the highest laurels do place it among the world's top colonial possessions at the time; for views of this sort, see Geggus (1982), pp. 1, 6; Humphreys, p. 417; Julien (1977), p. 66.

2. All agree on this point; see Tarrade, chap. 19; Geggus (1989), p. 1291;

Mathieu, p. 21; Priestley, p. 315; Saintoyant, 2:471–72; *DPF*, pp. 25, 694. See also the revealing, "Mémoire du roy pour servir d'Instruction au Comte de la Luzerne [1786]," AN, C^{9B} 36 (1).

3. Jean Tarade represents the ur-source for eighteenth-century French colonial trade; see esp. his table 3, p. 739; on the present point, see also Geggus (1989), p. 1291; James (1974), p. 551; Knight, p. 150; Gordon, p. 20.

4. On employment in colonial trade, see Abeille, p. 3; Fouchard in Cauna (1987), p. 9; James (1963), p. 49; Barré de Saint-Venant, p. 471. Geggus (1989), p. 1291, puts the number employed in colonial trade at one million. For other trade statistics, see *DPF*, pp. 4, 597–98; Chastonnet-Desterre, p. 40; Lokke, pp. 72, 80; James (1963), p. 49; Stein (1979), p. 36; Pluchon, ed. (1980), pp. 30–31.

5. On these points see Tarrade, table vi, p. 752; Stein (1979), pp. 118, 197; Stein (1988), pp. ix–x, 110; Pluchon (1985a), p. 157.

6. Citing contemporary sources, Lokke, p. 29, puts tax revenues from colonial goods at over 15 million livres; see also Chastonnet-Desterre, p. 40; Stein (1988), pp. 158–61; Laurent, pp. 4, 9–22; Cauna (1987), p. 195.

7. Geggus (1989), p. 1290n; Bellegarde (1923), p. 108. James (1963), p. 49, values the French colonies at 3 billion. The contrast with French Guiana is revealing: Guiana exported 440,000 livres worth of goods annually, less than 1 percent of Saint Domingue's productive capacity, but imported over 500,000 livres worth of goods, making it a net drain on France; see Stehlé (1966), p. 31. According to Lokke, p. 32, Cayenne, St. Lucia, and the Indian Ocean colonies also lost money.

8. Lewis, pp. 4–5, 13–14, echoes this point for the whole of the Caribbean.

9. On tobacco and the tobacco boom in the seventeenth century, see Butel, pp. 167–69; Julien (1976), p. 52; (1977), p. 96; Fournier, p. 69.

10. Gordon, p. 20; Thompson, 2:25; Tarrade, pp. 773–83. See also remarkable graphical representations of the colonial economic output of Saint Domingue in Anglade (1982), p. 17; and Centre d'Études de Géographie Tropicale, pl. 2.

11. *DPF*, p. 111. Several sources provide these or like figures for Table 2: Raynal (1785), p. 22; Barré de Saint-Venant, p. 459; Bellegarde (1950), p. 2; James (1963), p. 45; Pluchon, ed. (1980), p. 32; Stein (1988), p. 42.

12. Knight, passim and here p. 63.

13. Mintz (1985a), pp. 51–52. On the plantation system, see also Mintz (1985b), pp. 128–29; Klein, pp. 50–66; Dunn, esp. chap. 6; Lewis, pp. 5, 105–9.

14. Although dating from 1680, sugar production did not begin seriously in Saint Domingue until 1699, and by 1710 only thirty-five sugar plantations existed in the whole of the colony. From that point, however, the number and output of sugar plantations increased steadily and dramatically. On the history of sugar and sugar production, see Hagelberg, p. 88; Stehlé (1966), pp. 29–31; Stein (1979), pp. 7, 22, 25; Stein (1988), p. 41; Cauna (1987), pp. 13, 15; Dunn, pp. 60–61; Devèze, pp. 201, 256; *DPF*, pp. 120, 240, 962, 1091.

15. Hagelberg, p. 91.

16. Geggus (1989), p. 1291; Cauna (1987), p. 11; Tarrade, p. 36; Mintz (1985a), pp. 188–89; Stein (1988), pp. 99–100; Hagelberg, p. 96; Gordon, pp. 20–21.

17. Stein (1988), pp. 10, 67, 75; Devèze, pp. 256, 267. Hagelberg, p. 96, cites

an equivalent 80,000 metric tons of sugar from Saint Domingue in 1791, "a colossal amount for the production methods of that time."

18. Doucet, p. 163; Mintz (1985a), p. 36.

19. Williams, pp. 113-14, 122-23, 145.

20. Goubert (1952), p. 330; Lewis, pp. 2, 7, 14; Knight, pp. 41-49.

21. Cauna (1987), pp. 141-55; Klein, p. 62.

22. On the sugar production process, see Stein (1988), chap. 7; Cauna (1987), chap. 6 and figs. 13-29; Mintz (1985a), pp. 46-50 and figs. following p. 78; Vilaire et al., pp. 78-79, 84-85; Williams, pp. 73-78; Dunn, pp. 189-201; James (1963), pp. 48-49; Moreau de Saint-Méry (1791), pls. 30-31.

23. See Barré de Saint-Venant, pp. 433, 435; Dubuisson, pp. 119-20; *DPF*, pp. 1091, 1220, 1260, 1275, 1293, 1309, 1386. A working water wheel has been preserved in Haiti by M. Gérand Fombrum at Moulin-sur-Mer on the former Augier plantation near Montrouis. Figure 9 is taken from Dutertre's *Histoire générale des Antilles* of 1667. Most colonial sugar mills were of this type, but advanced eighteenth-century mills became even more factorylike and capital intensive; see illustrations in Moreau de Saint-Méry (1791), pl. 30, 31; Vilaire et al., pp. 71-99.

24. Brockway, pp. 51-52; Stehlé (1966), p. 33; Lacroix, 3:25-29; *DPF*, p. 173. Coffee likewise passed from the Amsterdam Botanic Garden to Dutch Guiana and thence to French Guiana and Brazil.

25. See Devèze, p. 268; *DPF*, pp. 264, 1117, 1211. Stein (1979), p. 25, notes a tenfold increase in coffee production between 1740 and 1770.

26. *DPF*, pp. 683, 959; Geggus (1989), p. 1291; Devèze, p. 267.

27. See Girod (1970), pp. 59-70; Brockway, pp. 55-56, Nicolson, p. 31; and Devèze, p. 268, who puts Saint Domingue's export total of indigo in 1753 at approximately 2 million pounds. See also *DPF*, pp. 620, 817, 1359.

28. Cornevin, p. 75; Brockway, pp. 54-55; *DPF*, pp. 794, 817-18; James (1963), pp. 45-46; Devèze, pp. 257, 269.

29. *DPF*, p. 1301; also Williams, p. 72.

30. Fournier, p. 69; *DPF*, pp. 111, 816, 1199, 1371, 1377; Raynal (1781), 7:148; Nicolson, p. 22; see also Brockway, pp. 53-54.

31. Dutrône la Couture (1790) and subsequent editions. Dutrône became a notable associate of the Cercle des Philadelphes.

32. See papers of Baudry des Lozières, BN, Mss., n.a.f. 22085.

33. BN, Mss., n.a.f. 22085, fol. 199.

34. "Essais Théoriques sur l'agriculture des Colonies," BN, Mss., n.a.f. 22085, fols. 258r-258v.

35. *DPF*, pp. 905, 911, 1260.

36. On these points, see *DPF*, pp. 111, 795, 902, 917, 1054, 1238, 1239.

37. Arthaud, CA-HdC, fol. 175; Stein (1979), pp. 56, 136-37.

38. Chastonnet-Desterre, pp. 38-40; Durant, p. 935; Mathieu, pp. 22, 24. Hector and Moise, p. 137, and James (1963), p. 50, give a higher figure of 1,587 sailings; Tarrade, p. 731, and Cauna (1987), p. 12, give slightly lower figures. On colonial shipping, see also Stein (1979), pp. 36, 135, 170, 233 n. 22; Doucet, p. 267; Basket, p. 60; Humphreys, p. 417; Lokke, p. 28.

39. Saintoyant, 2:394; also, Hector and Moise, p. 137.

40. *DPF*, pp. 1116, 1148, 1313. One plantation owner had his own ship that sailed directly between the south coast and France; *DPF*, p. 1354.

41. On smuggling and illicit trade, see Tarrade, pp. 95–101, 782; Laurent, p. 4; Mathieu, pp. 16–21; James (1963), p. 45; Knight, pp. 82–83; Stein (1979), p. 6; Stein (1988), p. 76; *DPF*, pp. 739, 1300.

42. Montague, pp. 29–32; Williams, pp. 80–81, 148; *DPF*, pp. 589, 1329.

43. Tarrade, passim and p. 782; also, Laurent, p. 4; *DPF*, p. 739; Lokke, pp. 72–76; Blerald, pp. 12–18.

44. Raynal (1781), 7:1934; *DPF*, pp. 1166, 1339, 1363.

45. See Tarrade, pp. 666–77; Richard; Lacombe, pp. 25–28; *DPF*, pp. 1078, 1313; Stein (1979), pp. 6, 42, 147; Stein (1988), p. 85; Cauna (1987), p. 196.

46. *DPF*, pp. 111, 803, 960, 1216; Nicholson, pp. 46–47.

47. Moreau comments emphatically to this effect, *DPF*, pp. 717, 794–95, 929, 932; see also remarks by Barré de Saint-Venant, p. 183.

48. *DPF*, p. 949.

49. See *DPF*, pp. 1083, 1091–92, 1237; Gordon, p. 6.

50. *DPF*, p. 1206; see also *DPF*, pp. 1294–95.

51. E.g., *DPF*, pp. 941, 1281; see also Cauna (1987), p. 90.

52. *DPF*, pp. 940–53; Cauna (1987), pp. 37–38; Aubin, pp. 9–11, Cauna (1981), p. 65; Hector and Moise, p. 117.

53. *DPF*, pp. 921–26, 1279–95.

54. *DPF*, pp. 1294, 1557; Cabon [1928–33], 2:442.

55. Moreau discusses the Artibonite case at length, *DPF*, pp. 819–44; see also Cauna (1987), fig. 8.

56. *DPF*, pp. 826, 861–62, 901. The Artibonite River was navigable and engineers constructed levees for slaves to haul boats up the river.

57. Debien (1955); Fouchard (1955b), p. 58. Trembley's manuscript, "Essais hydrauliques pour la plaine de l'Artibonite," is preserved, AN, Colonies F^3 187, fols. 161–259.

58. Boyden; Cabon [1928–33], 2:448–49; Maurel (1961), p. 248; AN, Colonies F^3 87, fol. 279: "Notes de M. Hesse relatives . . . à la machine à feu établie à l'Artibonite." Genton, p. 176, mentions coal deposits in Saint Domingue and their applicability to steam power. Williams, p. 103, notes that in 1785 Boulton and Watt discussed applying steam power to run sugar mills in Jamaica.

CHAPTER 5: The Urban Context

1. Saint-Vil, pp. 18–19.

2. *DPF*, pp. 480, 1053, 1316. Ott, p. 7, puts the 1791 population of Cap François at a staggering fifty thousand. On towns in Saint Domingue, see also Saint-Vil, pp. 6–11, 19, 31; Girod (1972), p. 78; Julien (1977), p. 64; Mercier, p. 107; Corvington, p. 173; James (1963), pp. 31–32.

3. Based on *DPF*, pp. 478, 1053, 1316; Saint-Vil, pp. 18–20; see also Table 1. On urban populations, see also James (1963), p. 86; Ott, p. 7.

4. See Map 2; Saint-Vil, pp. 6–11.

5. See *DPF*, pp. 1140, 1214, 1234. The names of some of the lesser hamlets

NOTES TO PAGES 78–83

convey something of the raw spirit of their settings, e.g., Camp [con] de Louise, Pensez-y-bien, Trou-d'Enfer, Étron-de-porc, Source à Zomby; see *DPF*, pp. 909, 1219, 1335, 1369.

6. Based on thirty-two parish villages and twenty-two hamlets estimated at three hundred and one hundred inhabitants, respectively; see Map 2 and Table 1.

7. Debien (1974), p. 120; *DPF*, p. 1259.

8. See *DPF*, pp. 217, 245, 616, 627, 651, 692, 745, 1212, 1313.

9. *DPF*, pp. 760, 917, 1150.

10. *DPF*, pp. 969, 1219. The carriage route linking Saint-Marc and Arcahaye began operations in 1725. The southern route west out of Port-au-Prince was also "une belle route de voiture"; see *DPF*, pp. 717, 929.

11. Saint-Vil, p. 17; *DPF*, pp. 493, 747, 785, 911. The English set Brabant down on the Saint Domingue coast in 1759; he survived.

12. *DPF*, pp. 1205, 1379. About the road outside Jérémie, Moreau comments (p. 1389) "the prudent traveler will prefer to admire on foot, rather than on horseback, several spots along this masterpiece of a route."

13. *DPF*, pp. 786, 908, 916, 1111, 1265, 1387; Cabon [1928–33], 2:444–47. The shortcut in the north across the Spanish part of Saint Domingue, e.g., entailed a half-day of real danger.

14. *DPF*, pp. 638–41, 789, 803–4, 1074, 1266; Tramond [1928], p. 4.

15. *DPF*, p. 1072.

16. *DPF*, pp. 1120, 1332.

17. *DPF*, pp. 149, 1076–77. About one inn, Moreau commented, "one only has to glance at this asylum to realize that comfort, much less luxury, has never been part of it. The thought of the next day's route is the only thing that counsels sleeping there." *DPF*, p. 803; see also pp. 895, 915, 1221.

18. Institutionalized racism dictated that whites paid twice the fare charged free people of color; see *DPF*, pp. 149, 640–41, 1077; Saint-Vil, p. 17.

19. *DPF*, pp. 444–53, 815, 845, 1387; Laurent, p. 20.

20. On bridges in the colony, see *DPF*, pp. 286, 449–51, 849, 876, 892, 894, 971, 1123, 1192; Hector and Moise, pp. 118–19.

21. On the colonial post, see Bellegarde (1950), p. 2; Maurel (1935); Desfeuilles; *DPF*, pp. 1077, 1120; the count of post offices derives from listings in the annual *Almanach de Saint-Domingue*.

22. *DPF*, pp. 645, 670, 676, 1088, 1143.

23. *DPF*, pp. 128, 298, 873, 1303.

24. *DPF*, pp. 733, 797, 937, 976–77, 1379; Ott, p. 78.

25. *DPF*, pp. 670–71; Fournier, p. 66.

26. Saint-Vil, pp. 23, 117–18; *DPF*, pp. 408, 739.

27. Compiled from data presented in *DPF*, pp. 478, 981, 1304; see also, Corvington, pp. 23, 44, 69, 179.

28. *DPF*, pp. 426, 678, 1305. Moreau, *DPF*, p. 1417, speaks with disgust about unplanned settlements outside Saint Domingue.

29. Moreau mentions the trees planted in the forest in Boynes that only further cut off the sun's light; *DPF*, p. 772. See also *DPF*, pp. 224, 724–26, 743.

30. Baudry des Lozières manuscripts, BN, Mss., n.a.f. 22086, fol. 151v.

31. APS, map collection: 725 [Bois St. Lys, G.J.J.] [1800] Sa 27cfid. Undated

and unsigned, this map resembles that of Cap François published by Moreau and René Phelipeau in Moreau's *Recueil de Vues* (1791); see also *DPF*, following p. 531. One Dupuis drew the map of Cap François for Moreau's *Recueil*, and he may have drawn this original. On these points, see *DPF*, pp. 851, 1323, 1536; Cabon [1928–33], 2:450; Tooley, p. 503; APS Archives, Donation Books, vol. 2, November 20, 1835.

32. *DPF*, pp. 464, 471. The *Affiches Américaines* gives the number of ships in port on the order of one hundred in 1788–89. On the port in Port-au-Prince, see *DPF*, pp. 471, 1052; Corvington, p. 187.

33. Not a problem in Cap François, silting did affect Port-au-Prince where an imported dredging machine proved unsuccessful; *DPF*, p. 1051.

34. *DPF*, pp. 331–40; see also Figure 2, with the Masonic emblem below the cross.

35. *DPF*, pp. 927, 996, 1120; Castonnet des Fosses (1884), p. 21. Master forgers cast bells locally; a minor passion for bells reigned in the colony.

36. *DPF*, p. 338; Debien (1974), p. 120; Saint-Vil, p. 30; Pluchon (1987), pp. 38, 40.

37. *DPF*, pp. 320, 327–28; Fouchard (1955b), p. 29.

38. *DPF*, pp. 299, 886, 1037, 1054; Girod (1971), p. 226 and (1972), p. 82; Saint-Vil, pp. 12–13. The names of streets were modern, mostly descriptive (e.g., rue de la Fontaine) or named for a prominent personage; an exception is the rue du Pet au Diable in Cap François.

39. *DPF*, pp. 539, 1052; Castonnet des Fosses (1884), p. 20.

40. *DPF*, pp. 299–300, 743; Saint-Vil, p. 13; Cabon [1928–33], 2:563; Fouchard (1955d), pp. 1–2.

41. *DPF*, p. 426. Segregated neighborhoods existed in larger colonial towns.

42. *DPF*, pp. 432–39; Cabon [1928–33], 1:311.

43. *DPF*, pp. 370–75. The Government compound is marked *A* on Map 4.

44. Pluchon (1987), pp. 43, 237–42, shows that Jesuit support of slaves provided a special colonial factor in their expulsion from Saint Domingue.

45. The clock atop le Gouvernement kept better time than the one at the church; it had an attendant who received 1,200 livres. Only one public clock existed in Port-au-Prince, installed in a private merchant's establishment in 1776 at a cost of 16,800 livres; see *DPF*, pp. 372, 387, 991.

46. *DPF*, p. 356.

47. *DPF*, pp. 310, 1381; see also Gordon, pp. 8, 74.

48. Moreau counts eight to nine hundred wells in Cap François; *DPF*, p. 505.

49. *DPF*, pp. 327, 417–18, 501–7, 538. Beyond the nine public fountains, the conduit system also brought water to le Gouvernement, the prison, the convent, and the barracks.

50. *DPF*, pp. 302–3, 311, 501–2; Cauna (1987), fig. 9.

51. See *DPF*, pp. 132, 677, 990, 1012, 1045; Corvington, pp. 44, 69, 87, 99, etc.; Fouchard (1955d), pp. 1–2.

52. *DPF*, pp. 336–37, 428–31; Debien (1967); Duvivier, 2:117. A first cemetery existed behind the church, but burials shifted to this site, known as La Fossette, "the dump"! Government authorities outlawed church burials in Saint Domingue in 1763, as they did in France.

53. *DPF*, pp. 408–9, 431. A second cemetery in Cap François, located behind the town on the west, became the last resting place for Protestants, Jews, nonbaptized slaves, paupers, executed criminals, and bodies washed up on shore. A Protestant burial cost 190 livres; for Jews, 180 livres. Animals sometimes dug bodies out of colonial cemeteries; *DPF*, pp. 408–9.

54. *DPF*, pp. 473–74, 541–45. Next to La Fossette stood the quarters where newly arrived slaves were held and sold. Duels were fought at La Fossette, and slaves danced there on Sunday nights.

55. *DPF*, pp. 416–47, 474, 533–34, 538, 1009, 1037; Corvington, p. 184.

56. See *DPF*, pp. 311–13, 418, 547, 585, 1038, 1047; Mercier, pp. 121–22; *AA*, Cap, 1787, no. 32 (August 11); and Figure 6A.

57. See *DPF*, pp. 443, 480, 548–49, 676, 1038, 1053, 1316.

58. See *DPF*, pp. 314, 480, 709, 1053, 1316; Zupko, pp. 46, 99.

59. *DPF*, pp. 417–19; Saint-Vil, p. 26; and Figure 3 above.

60. Mulatto troops were stationed across town on the place Royale; *DPF*, p. 439.

61. *DPF*, pp. 419–23, and Debien (1949), esp. pp. 558–59, 561.

62. On colonial prisons, see *DPF*, pp. 390–92, 580, 1014, 1035, 1247.

63. *DPF*, pp. 393–409; see also Moreau de Saint-Méry (1790); Cabon [1928–33], 1:161; 2:369–70.

64. Moreau, *DPF*, pp. 404–7, lists 128 benefactors of the Providence through 1788. Most were local judges, merchants, and planters in about even balance; fifteen were women, five widowed; one was identified as a Jew, two as free mulattoes (male and female), and two as free blacks (male).

65. *DPF*, pp. 388–89. The women's poorhouse had an infirmary of ten beds and housed twenty women and ten children in 1788. Not all were ill, but all were otherwise absolutely without resources.

66. The Providence for whites in Port-au-Prince began in 1768 and received letters patent in 1789. It housed both sexes and was particularly charged to take in immigrants who arrived impoverished from France. A Providence in Les Cayes sheltered 150 persons and operated for twenty years until the will establishing the hospice was overturned in 1765. Citizens also undertook charity efforts in Léogane, but no official hospice developed; see *DPF*, pp. 1031–35, 1105–6, 1317–18.

67. *DPF*, pp. 411–16. The colored poorhouse sheltered thirteen individuals; mothers sometimes abandoned babies there.

68. These "hospitals" primarily treated slaves who arrived ill in slave ships. Slaves were also sent to Durant's Maison on private account for 4 livres a day. His center treated 672 individuals in 1783 and suffered a mortality rate of 16 percent, comparable with that of white hospitals; see *DPF*, pp. 410, 1068; Debien (1981); Cauna (1984); Cabon [1928–33], 2:438.

69. On the Charité Hospital, see *DPF*, pp. 549–76; Arthaud (1792); Pluchon (1985b), pp. 115–26; Girod (1972), pp. 91–96, Frostin (1973), esp. pp. 334–36. The Charité Hospital can be seen on the left in Figure 15.

70. Pluchon (1985b), p. 120; Brau, pp. 70–71, 91. Tension existed between regular physicians and the Brothers of Charity over administering medical care.

71. Pluchon (1985b), p. 124, provides complete figures on patients treated in

the hospitals at Cap François, Port-au-Prince, and Les Cayes from 1774 to 1786.

72. See *DPF*, pp. 1023, 1102, 1310; Pluchon (1985b), pp. 116, 118, 124; Brau, pp. 92, 94–97. A listing of other hospitals appears in *DPF*, p. 1437.

73. The theater in Saint Domingue has been exhaustively studied by Jean Fouchard; see Fouchard (1955c), (1955d), and (1955b). Cale recapitulates and amends Fouchard in part. Moreau devotes many pages to Saint Domingue's various theaters; see esp. *DPF*, pp. 356–67, 879–84, 985–89, 1099–1101, 1308–9; see also Girod (1972), pp. 97–98; Mercier, p. 122; Corvington, p. 122; Vigoreux, pp. 18–19. As Jean Fouchard realized, the *Affiches Américaines* provides a rich source for the study of the colonial theater.

74. The social status of actors was questionable in Saint Domingue, as in Europe. In one instance clerical authorities refused an actress a church burial on account of her profession; see *DPF*, pp. 366, 1099.

75. Fouchard (1955c); Cale, p. 51 and appendix.

76. *DPF*, p. 361. Chevalier died the death of Molière, on stage, declaring, "The farce is played."

77. *DPF*, pp. 343–44, 356–59. Figure 11 depicts the place Montarcher.

78. Free blacks were admitted to the theater only in 1775, after a group of black mothers protested they could not attend with their mulatto daughters. The daughters, mostly prostitutes, objected to having their mothers sit with them, so the authorities instituted the rule, "ebony to the left, copper to the right"; *DPF*, pp. 358, 362.

79. *DPF*, pp. 366–67, 885, 1384–85; Girod (1972), p. 97; Vigoureux, p. 21.

80. Brockway, p. 32; Girod (1972); p. 104; Fouchard (1955b), pp. 125–30; *DPF*, pp. 456–57, 885, 1030, 1385; Vigoureux, p. 19.

81. See Vignols, passim and pp. 360, 364; Fouchard (1955b), pp. 26, 32; Mercier, p. 125; Girod (1972), p. 100; *DPF*, p. 475.

82. The proprietors of these shows also sold glass eyes. *AA*, Feuille du Cap, 1787, no. 49 (December 8); 1789, no. 24 (May 2); Fouchard (1955b), pp. 150–63; *DPF*, p. 366.

83. Corvington, p. 117; *DPF*, p. 478; *AA*, Cap, 1772, no. 32 (August 8).

84. Eighty fatal duels were fought in 1777–78; see Frostin (1973), p. 342; also Castonnet des Fosses (1884), p. 28; *DPF*, pp. 63, 231.

85. On *cabinets littéraires*, see Girod (1972), pp. 104–5; Fouchard (1954), p. 110; *DPF*, pp. 321, 344–45, 891, 1308.

86. For these details, see *DPF*, pp. 350–54; Fouchard (1954), pp. 106–7; Girod (1972), pp. 102–5; Ménier and Debien.

87. Separate editions of the *Almanach* appeared for Port-au-Prince and for Cap François; see *DPF*, pp. 495–96, and the *Almanachs* themselves, BN, Lc³² 18, 19.

88. *DPF*, pp. 354, 493–99; Fouchard (1954), pp. 109–10.

89. Fouchard (1954), p. 110n; Cabon [1928–33], 2:478; *DPF*, p. 352.

90. Ménier and Debien, p. 424 and passim; Bissainthe, pp. 755–58, 859; cf. Darnton and Roche.

91. Fouchard (1955b), p. 98; *DPF*, pp. 495–97; Bissainthe, p. 757.

92. The *Affiches Américaines* has been fairly well studied; see Ménier and Debien; Bissainthe, pp. 755–58; *DPF*, pp. 492–96; Girod (1972), pp. 102–5; Fouchard (1955b), pp. 113–14; Corvington, p. 197.

93. The BN houses a mint complete run of the *Affiches Américaines*, sent to France through official channels; the primary call number is Lc^{12} 20.

94. On the colorful and progressive Mozard, see Maurel and Taillemite in *DPF*, pp. 1528–29; Fouchard (1955d), pp. 265–71; Moreau in *DPF*, pp. 494–95.

95. Figure 14 presents a representative front page of the *Affiches*. Because the figure likewise illustrates the problem of insect damage, this number of the *Affiches* was chosen from a handful in the Kurt Fisher/Haitian History Collection at the Schomburg Center of the New York Public Library.

96. See *Affiches Américaines*, passim. The experience of following news of the American War, for example, or the unfolding of the French Revolution at a remove of weeks or months in the *Affiches Américaines* is a startling one that gives an unusual sense of the pace of contemporary life.

97. *DPF*, pp. 494–95.

98. Fouchard (1954); also *DPF*, p. 321; Debien (1974), p. 121.

99. Fouchard (1954), pp. 104, 107–9; and e.g., *AA*, PauP Supplement, 1784, no. 18 (May 1), which lists thirty-five books.

100. Compare Furet; Darnton (1982), chap. 6, esp. pp. 173–82.

101. See Fouchard (1954), pp. 108–9; *Gazette de Saint-Domingue*, August 3, 1791 (BN: 4° Lc^{12} 25); *AA*, PauP, 1780, no. 45 (October 31); *AA*, PauP, 1782, no. 31 (August 3); *AA*, PauP Supplement, 1784, no. 8 (February 21); *AA*, PauP, 1789, no. 15 (February 19).

102. See Fouchard (1954), pp. 106–7; *DPF*, p. 478; Corvington, pp. 125–26. For "medical" titles, see *AA*, PauP Supplement, 1784, no. 18 (May 1), where one finds *La Nymphomanie* and *De l'Homme & de la Femme considérés physiquement*.

103. *AA*, Feuille de Cap, 1791, no. 38 (May 11).

104. See Fouchard (1955b), pp. 62–67; Cale, p. 50; *AA*, Feuille du Cap, 1788, no. 15 (April 12); 1789, nos. 2 (January 10), 3 (January 17), 5 (January 31), 8 (February 23), 10 (March 7), 11 (March 14), 14 (March 28), 48 (July 25), 59 (September 2), 60 (September 5).

105. See *AA*, Feuille du Cap, 1789, nos. 2 (January 10), 3 (January 17), 60 (September 26), and passim. See also Caillet; *DPF*, p. 966.

106. *AA*, Feuille du Cap, 1789, no. 38 (June 20); 1788, no. 36 (September 6). Missing and rotten teeth were a commonplace; *DPF*, p. 518.

107. *AA*, PauP, 1788, no. 101 (December 18).

108. See *AA*, PauP, 1786, no. 10 (March 11); Feuille du Cap, 1789, nos. 1 (January 3), 2 (January 10), 9 (April 15), 57 (August 26); 1788, no. 18 (May 3), etc. See also Arthaud (1788), p. 178; Cauna (1987), p. 89.

109. See Pressoir, p. ix; *DPF*, pp. 529–31.

110. See *AA*, PauP, 1780, no. 18 (May 2); *AA*, PauP Supplement, 1786, no. 28 (July 15); *AA*, Feuille du Cap, 1787, nos. 2 (January 13), 18 (May 5); 1789, nos. 9 (February 28), 11 (March 14), 43 (July 8). Moreau relates many of these details, but he dates Dorseuil's establishment from 1784; see *DPF*, pp. 530–31. Private tutoring was also available in the colony; see *AA*, Feuille du Cap, 1789, no. 34 (June 6). On the lack of schools in Les Cayes, see *DPF*, p. 1300.

111. The *AA*, Feuille du Cap, 1789, no. 43 (July 8), suggests the curriculum for girls; an updated prospectus from Dorseuil details that for boys; *AA*, Feuille

du Cap, 1789, no. 9 (February 28); see also *AA*, Feuille du Cap, 1789, nos. 12 (March 21), 13 (March 25), 50 (August 1), 54 (August 15).

112. The literature surrounding eighteenth-century Freemasonry is vast. Le Bihan's encyclopedic survey of French Masonic lodges and his special study of colonial Freemasonry, including in Saint Domingue, provide the starting points for the present remarks; see also Jacob; Faÿ (1942).

113. Le Bihan (1974), pp. 41, 44; (1967), pp. 389–400; see also Cabon [1928–33], 2:343; *DPF*, p. 1202. Lodges existed, e.g., in Petit Trou and Fond-des-Nègres.

114. Le Bihan (1974), p. 46; Pluchon (1985a), p. 157.

115. Derived from Le Bihan (1967), pp. 389–400; see also Le Bihan (1974), pp. 45–46, and *DPF*, pp. 1309–10.

116. Le Bihan (1967), pp. 389–95, and (1974), p. 44; *DPF*, pp. 427, 539–40; [Tramond] (1927), p. 605; Girod (1972), p. 106.

117. *DPF*, pp. 540, 1055; Le Bihan (1967), pp. 399–400; [Tramond] (1927), p. 606.

118. See chapter 11 and the refutation of the argument that the Cercle des Philadelphes was itself a Masonic organization.

119. Fouchard (1955d), p. 6.

CHAPTER 6: Missionary Naturalists

1. Fournier, the major source, makes these points; see pp. 23, 35, 93–95. See also Lewis, p. 63; Lacroix, 3:xii, ix, 3, 136, 142. Cabon (1916a), pp. 155–69, and Crouse (1977), pp. 273–81, provide valuable background.

2. On Breton, see Fournier, pp. 35, 40; Stehlé (1970), p. 86; Lherisson, p. 8; Crouse (1977), p. 45.

3. On Dutertre, see Fournier, pp. 36–40; Duvivier, 1:175–76; Stehlé (1967), pp. 281–82, and (1970), pp. 86–87; Lewis, pp. 62–69; Broc, p. 79; Lherisson, p. 9; Lanier, p. 96; Crouse (1977), pp. 105, 160–63. Much of Crouse (1977) is based on Dutertre and gives a flavor of his narrations.

4. One César de Rochefort stole and twice published Dutertre's manuscript in Rotterdam (1658 and 1665) as the *Histoire naturelle et morale des Isles Antilles de l'Amérique*. Rochefort's book was reportedly read at one of the scientific salons in Paris in the 1650s, possibly Thévenot's; see Fournier, p. 40; Stehlé (1967), pp. 281–82; Pressoir et al., p. 52; Lewis, pp. 62–63.

5. On Labatt, see Fournier, pp. 63–68; Pressoir et al., pp. 65–71; Lewis, pp. 63–69; Negre (1965), pp. 4–6; Broc, pp. 80–83.

6. On Le Pers and Charlevoix, see Fournier, pp. 50–52; Pressoir et al., pp. 61–65, 71–73; Léon (1976), pp. 13–15; Lherisson, p. 9; Broc, p. 83; *DPF*, pp. 269–70. In the 1730s Father Margat, another Saint Domingue Jesuit, prepared a work like Le Pers'; see Duvivier, 1:188–89 and note.

7. On Fagon and the Jardin du Roi, see esp. Laissus, pp. 288–92 and appendix; Lacroix, 4:223; Lanier, p. 99; Osborne, pp. 3–5.

8. On Plumier, see Fournier, pp. 52–59; Duvivier, 1:134–36; Stehlé (1967), pp. 282–84, (1970), p. 88; Lherisson, p. 10; Chardon, p. 170; Léon (1976), pp. 10–13.

9. Leeuwenhoek confirmed Plumier's discovery in 1704; see Fournier, who seems to mistake the date, p. 57; Lanier, p. 100; Lherisson, p. 10.

10. See Feuillée; Fournier, pp. 60-63, 71-72; Lacroix, 3:15-20; C. Bourgeois, pp. 10-13. Other expeditions sent out by Fagon include those of Tournefort to the Levant in 1700-2; Lippi to Upper Egypt; and Laval (S.J.) to the Mississippi; see Laissus, pp. 291-92; Feuillée, 3:227.

11. Feuillée, 1:5, 243.

12. About these experiences Feuillée commented, "At sea one knows no one, and the safest route is to avoid everyone"; Feuillée, 3:368. See also Feuillée, 3:365-89; C. Bourgeois, p. 12.

13. On Nicolson, see Fournier, pp. 68-72; Léon (1976), pp. 29-35; Stehlé (1970), p. 97; Cabon (1917), pp. 118-19; Chardon, p. 172; Pressoir et al., pp. 85-86; Lherisson, p. 12.

14. Nicolson, p. 130; Duvivier, 1:136-37.

15. On de la Haye, see Maurel in DPF, p. 1473; and below chapter 13.

16. Chapter 9 details the system of French colonial botanical gardens.

17. Stehlé (1967), p. 284; Fournier, pp. 55-56, 93-94. Fournier notes that the new system of nomenclature and priority established by Linneaus obscured the accomplishments of seventeenth- and eighteenth-century French missionary naturalists.

18. Stearns, pp. 235-43; Dunn, pp. 309-10; Duvivier, 1:133; Chardon, p. 55; N. L. Bourgeois, p. 503; Armas, p. 13.

19. On Jacquin, see Chardon, pp. 66, 172; Duvivier, 1:136; Stehlé (1967), p. 285; Cabon (1917), pp. 118-19.

20. Chardon, p. 173; Duvivier, 1:137; Stehlé (1967), p. 288.

CHAPTER 7: Expeditions to Saint Domingue

1. On Richer's discovery, see Fontenelle, pp. 114-16; Martinière, pp. 361-405; Lacroix, 3:11-15; Devèze, p. 308; Chaia, p. 129; Hahn, p. 365.

2. Fontenelle in HMARS, 1701 (1704), Histoire section, pp. 111-12; Fleurieu, 1:800; Broc, p. 17. On a cartographic mission to Canada by Des Hayes in 1685, see Fontenelle, HMARS, 1699 (1702), Histoire section, p. 86.

3. Fontenelle noted that "all this enriches the Academy, but does not affect the public enough"; HMARS, 1701 (1704), Histoire section, p. 112. See also, HMARS, 1700 (1703), Histoire section, pp. 114-16. In Saint Domingue Des Hayes took readings at five spots along the coasts of the colony.

4. Book III, Prop. XX, Prob. IV, pp. 428-33; Newton discusses the full series of expeditions that went into confirming Richer.

5. Feuillée, 1:preface and p. 1; Fournier, pp. 60-61; Lacroix, 3:15; Lherisson, p. 10.

6. Feuillée, esp. 1:5; Fournier, pp. 60-61; Fleurieu, 1:449.

7. Feuillée published some sixty astronomical papers in the Mémoires of the Academy and served as first director of the royal observatory in Marseille; see C. Bourgeois, p. 11; Lacroix, 3:18; Fournier, pp. 61, 71-72.

8. See Fontenelle, HMARS, 1706 (1707), Histoire section, pp. 113-14; HMARS, 1707 (1708), Histoire section, pp. 82-83; La Hire; Fleurieu, 1:430-31. On Boutin, later curé at Cap François, see listing in DPF, p. 1457. A 15-foot-

high column later erected on a hill above Cap François established on official meridian; see *DPF,* p. 311.

9. For fuller accounts of these well-known expeditions, see R. Mercier; Chardon, pp. 59–64; Broc, pp. 37–42; Konvitz, pp. 4–12; McClellan (1985), pp. 200–202: Aiton, p. 106, disputes that the voyages proved immediately decisive, whereas Broc, pp. 41, 269, and Konvitz, p. 12, argue that they did.

10. *DPF,* pp. 1185–86. Apropos of the clockmaker, at this stage clocks were not reliable as chronometers and could be used only for local time.

11. See Bouguer, pp. 522, 528; La Condamine, p. 529; *DPF,* p. 1186.

12. Such would seem the gist of Godin's remark in a letter to Antoine de Jussieu dated Petit Goave, August 11, 1735: "M. Votre Frère se porte bien et a toujours fait de mesme, un peu d'humeur noire quelque fois, mais qui se passe sans que personne puisse s'en plaindre"; see letter reproduced in Lacroix, 4: pl. 53, following p. 160; see also Godin.

13. Godin letter to A. de Jussieu, cited in previous note.

14. Lacroix, 3:52; Institut de France, p. 78.

15. The "Voyage du Comte de ★★★★ à Saint-Domingue," in N. L. Bourgeois, pp. 85–170; here p. 144. The count escaped from France after killing his son.

16. Ulloa, 1:19. Don George returned to Cap François nine years later. His portrait of the colony stressed its economic abundance and its potential as a model for Spanish colonial development; he also took pendulum measurements. See Don George in Ulloa, 2:122–26 and 251–54.

17. For background, see E. G. R. Taylor; McClellan (1985), pp. 65, 214–15. Latitude is more easily determined, e.g., by the height of the pole star or the sun at noon.

18. Fleurieu, 1:ix–xii; McClellan (1985), p. 214.

19. Pingré (1773b) discusses all of these voyages; see also Pingré (1772), (1773a); Chabert, esp. pp. 49, 59; Broc, pp. 281–84. The highest levels of government and the Academy were involved in these tests. For example, the duke de Praslin, minister of the navy and academy honorary, coordinated the 1769 tests; Berthoud was royal clockmaker to the king and the naval ministry, and prize money from the Academy was at stake.

20. Woolf; see also McClellan (1985), pp. 202–6, 208–20; Broc, pp. 284–85; Konvitz, pp. 74–75.

21. Newton, e.g., set the astronomical unit at 70 million miles; with the Venus observations, eighteenth-century astronomers arrived at the modern value of 93 to 96 million miles; see Woolf, pp. 196–97; McClellan (1985), p. 220.

22. *DPF,* p. 1410; Cabon (1916b), p. 57, and [1928–33], 2:350, repeats the reference. The Mercury transit episode in Saint Domingue is not mentioned in Woolf.

23. The present account draws from Pingré (1772) and (1773a); Fleurieu, 1:124–37, 412–27; 2:121–69; *DPF,* p. 540; Woolf, p. 160.

24. Pingré (1772), pp. 516–17, and (1773a), p. 513; Fleurieu, 2:132; and insertion in the memoirs of the Paris Academy, *HMARS,* 1767 (1770), p. 643.

25. *DPF,* p. 540; *Almanach de Saint-Domingue;* for more on Pingré, see Armitage; Ronan; Lacroix, 3:177–79; Debien (1937), pp. 23–24.

26. See Cabon (1916b), pp. 51–59; Duvivier, 1:103; McClellan (1985), pp. 99,

184; Broc, pp. 22–25, 280–81; Konvitz, pp. 75–77. In granting permissions for cartographical works, e.g., the Brest Academy acted as the naval arm of the Paris Academy; see permissions heading Fleurieu and Puységur (1787b); Puységur (1787a), p. 18; Debien (1937), p. 5.

27. Bellin (1764); Bellin (1807), esp. pl. 75bis. Duvivier, 1:88, 98–104, presents an annotated listing of contemporary maps of Saint Domingue; see also Cabon (1916b), pp. 51–56; Cabon [1928–33], 2:353.

28. Fleurieu, 1:xlviii; see other remarks by Fleurieu critical of Bellin in Fleurieu, 1:xlviii–lxi, 433, 445, 451. The English tore down the navigation aids set up by the French on these cayes, and the French lighthouses depicted in the frontispiece are thus wholly imaginary. Moreau deplores the lack of agreement among the colonial powers regarding navigation in American waters; see DPF, pp. 1412, 1415, 1421.

29. On Puységur and his voyage, see Cabon (1916b), pp. 59–62; Maurel (1961), pp. 234–35; listing in DPF, p. 1540; and further chapter 10 in the present work.

30. The next year, one Lieudé de Sepmanville added an accurate map of the central island of La Gonave, the only area not covered by Puységur; see Puységur (1787a), passim, pp. 1–15, pl. 7, and (1787b); Duvivier, 1:89–91, 104; DPF, p. 1156.

31. Cabon (1916b), pp. 65–68.

32. DPF, pp. 216, 639.

33. DPF, p. 1422. For local maps, see Moreau (1791); Girod (1970), following p. 207; the Bois St. Lys collection of the American Philosophical Society.

34. AA, PauP, 1769, no. 23 (June 7); AA, PauP Supplement, 1786, no. 19 (May 13).

35. AA, PauP, 1769, nos. 35 (August 30), 40 (October 4); 1769, nos. 3 (January 18); 23 (June 7). Lacking a telescope, the editor stopped observing the comet, "even though persuaded it is still in our vortex [sic]."

36. AA, Cap, 1776, no. 3 (January 20).

37. AA, PauP, 1776, no. 31 (July 31); AA, PauP, 1785, nos. 23 (June 4), 26 (June 25).

38. AA, PauP, 1787, nos. 73 (September 13), 83 (October 18). The intendant supported the proposed magnetism survey. See Almanach de Saint-Domingue (1789), pp. 128–29, for list of official surveyors; and Figure 16.

39. AA, PauP, 1787, no. 73 (September 13); Cercle des Philadelphes (1789a and b), p. 4. On List, see also Maurel (1961), p. 265; DPF, pp. 1155, 1518.

40. DPF, p. 223. On colonial efforts to square the circle, see chapter 10.

CHAPTER 8: Medicine and Medical Administration

1. On diseases in Saint Domingue, see chapter 1: and Pluchon (1985b), p. 126.

2. On the system of colonial hospitals, see chapter 5.

3. Brau, chap. 1; Pouzet, pp. 37–39. On native American medicine in the Caribbean, see Joubert, pp. 15–20.

4. Léon (1976), pp. 9–10; Lewis, p. 79; Doucet, p. 55; Butel, p. 81; Burg, pp. 112–13, 129. Oexmelin's book was quickly translated.

5. Pluchon (1985b), pp. 98–100.

6. *DPF*, p. 488; Brau, pp. 59, 63; Pluchon (1985b), p. 100; Launay, p. 33; Dermigny, p. 66. A physician, Dautun acted informally as médecin du roi in Cap François from 1710.

7. Pluchon (1985b), pp. 90, 96. Pluchon's work in this area supersedes the previous literature.

8. Pluchon (1985b), p. 90, notes a dozen medical *entretenus*, but the *Almanach de Saint-Domingue*, (1789), pp. 121–28, suggests a somewhat higher total.

9. Compiled from Loker (1981).

10. Pluchon (1985b), p. 90.

11. See *DPF*, p. 488; Pluchon (1985b), p. 112; Brau, pp. 79–80, 82, 105; Broussole and Masson, p. 73.

12. Pluchon (1985b), p. 92. Léon (1953), p. 8, counts two hundred licensed medical personnel in the colony; Castonnet des Fosses (1884), p. 11, says over four hundred.

13. *Almanach de Saint-Domingue* (1789), pp. 121–28; *DPF*, pp. 488, 1030, 1102.

14. Pluchon (1985b), pp. 89–90; Joubert, pp. 27–32; Lacroix, 3:36–46.

15. Léon (1976), pp. 53–59; Pluchon (1985b), p. 96; Brau, pp. 71–73, 91.

16. Pluchon (1985b), p. 95; Léon (1976), p. 57; *DPF*, p. 554.

17. Pluchon (1985b), p. 93; see also Léon (1976), p. 59; Brau, pp. 65, 70, 71.

18. Pluchon (1985b), p. 103; N. L. Bourgeois, p. 451. Most Saint Domingue surgeons deserted from the merchant marine, and many or most of these must have slipped into the untamed worlds of plantation surgery.

19. On this legislation, see Joubert, pp. 27–32; Brau, pp. 59, 64–71, 77–78, 91, 105; Pluchon (1985b), pp. 101–4; and Léon (1976), pp. 58–59; *DPF*, p. 556.

20. Pluchon (1985b), p. 102; Brau, pp. 67–68, 82; see also Joubert, p. 32.

21. Brau, pp. 69, 71, 81; Léon (1976), p. 58.

22. *AA*, PauP Supplement, 1789, no. 23 (March 19); see also *AA*, PauP Supplement, 1777, no. 40 (October 4); *AA*, Cap, 1789, no. 74 (October 24); Léon (1976), p. 61.

23. On apothecaries and pharmacy legislation in the colony, see Pluchon (1985b), pp. 107–12; Joubert, p. 30; Brau, p. 55; Léon (1976), pp. 56, 60.

24. Pluchon (1985b), pp. 108, 110, and (1987), p. 147; see also Joubert, p. 21.

25. Pluchon (1985b), p. 111, mentions three colonial apothecaries (Chatard, Sénéchal, and Fusée Aublet) who became botanistes du roi and who published.

26. See Brau, pp. 90, 92, 111; Léon (1976), pp. 60–61; *DPF*, p. 388.

27. *DPF*, pp. 489, 1102; Cercle des Philadelphes (1788e), pp. 19–20, 40; Cauna (1987), p. 88. Authorities first consulted Alfort veterinarians in 1777 about an epizootic outbreak; see *DPF*, p. 291.

28. Pluchon (1985b), pp. 92, 103–4; see also Joubert, pp. 24–25, 29; N. L. Bourgeois, pp. 451–52, 458–59; Duvivier 2:172. Joubert also counts priests, nuns, and missionaries as illegal medical practitioners.

29. Cauna (1984), p. 51, and (1987), p. 87; Pluchon (1985b), p. 107.

30. Lafosse, p. 230. See also Léon (1976), pp. 40–46; (1952), p. 42; and similar, but later manuals by Ducoeur-Joly and Descourtilz.

31. See Joubert, pp. 23, 33; Pluchon (1987), pp. 143, 145–46, 155, and (1985b), pp. 109–10.

32. Pluchon (1987), p. 19, and (1985b), p. 111. Note the attack on *kaperlatas* in *AA*, PauP Supplement, 1789, no. 23 (March 19).

33. Pluchon (1987), p. 18, and (1985b), p. 111; Cornevin, pp. 97–99; Joubert, pp. 33–34; Davis, pp. 197–98; Cauna (1987), p. 128. Moreau (*DPF*, p. 520) calls the practice of medicine among slaves "a means of assassination."

34. Joubert, p. 33; see also Dunn, p. 309.

35. See Pouppée-Desportes, 1:1–15; 2:326–29, 330–33; 3:454n; Lacroix, 3:52; 4:137, 157; Pouzet; Brau, pp. 83–88; Duvivier, 2:169–70; Léon (1976), pp. 17–29; Botherel-Blanchet; Institut de France, p. 74; and listing, *DPF*, p. 1539. With several spellings, Pouppée-Desportes's name is sometimes alphabetized as Desportes. For the sequence of French physicians in Saint Domingue and their works, see also Debien (1943), pp. 25–42, and Loker (1981).

36. Gros, p. 345.

37. Pouzet, pp. 380, 382; Pouppée-Desportes, 1:2; vol. 3, "Avertissement" and "Approbation." Pouppée-Desportes resided in Cap François. His fellow médecin du roi in Léogane, Jean Damien-Chevalier, likewise published on colonial medicine. Damien-Chevalier's medical and botanical *Lettres à M. De Jean* appeared in 1752; his *Sur les Fièvres de l'île de St. Domingue* appeared in 1763. See Pluchon (1985b), p. 128; Léon (1933), pp. 26–29, and (1976), pp. 35–38; Loker (1981), no. 43.

38. On Arthaud, see esp. Hamy and Lautour; see also Maurel and Taillemite in *DPF*, p. x and index listing on p. 1445, where he is mistakenly styled Jean Arthaud. On Arthaud's medical training, see Hamy, p. 295; Arthaud (1771).

39. Arthaud (1776); *AA*, Cap, 1776, no. 2 (January 13); PauP, 1777, no. 21 (May 21); Feuille du Cap, no. 24 (June 14). Revival of the drowned by introducing tobacco smoke anally was a current fad in France. The full story of how Arthaud came to revive this female slave just two weeks after the equipment was installed might well speak volumes about eighteenth-century French medicine.

40. See Arthaud's exchanges over inoculation and venereal diseases, *AA*, Cap, 1774, nos. 35 (September 3), 42 (October 22); 1775, nos. 11 (March 18), 13 (April 1).

41. See Arthaud, "Discours prononcé dans l'assemblée provinciale . . . ," December 4, 1789; AN, Colonies F³ 194; Hamy, pp. 296–97; Pluchon (1985b), p. 91; Maurel (1961), p. 253; Fouchard (1955b), pp. 60–61. Arthaud can also be traced among personnel listed in the annual *Almanach de Saint-Domingue*.

42. Broussolle and Masson, in Pluchon, ed. (1985), pp. 69–87. This excellent study by Broussolle and Masson provides the basis for the present remarks.

43. The naval medical-surgical colleges were substantial medical schools, essentially university centers, with upward of two hundred students. The school at Brest officially became the Collège Royal de Chirurgie de la Marine in 1775; see Broussolle and Masson, pp. 73–75, 82–85; Launay, p. 7.

44. The naval medical corps was rife with internal tensions, especially between the physicians who headed the service and the top rank of staff surgeons; see Broussolle and Masson, pp. 72, 77–78, 85–86; Brau, p. 110.

45. See Brau, pp. 79, 89, 107–8; Broussolle and Masson, p. 75; Pluchon (1985b), pp. 113–14; Pouzet, p. 57.

46. Pluchon (1985b), pp. 113–14; Brau, pp. 101–5; *DPF*, p. 555; Broussolle

and Masson, p. 77; Pouzet, p. 37; Tanguy, p. 19; *AA,* 1776, Cap, no. 28 (July 13).

47. Brau, pp. 107, 110.

48. On Duchemin, see *DPF,* p. 497; Pluchon (1985b), p. 93; Brau, p. 108; Bissainthe, p. 756; Pouzet, p. 57; see also Pluchon (1987), p. 237.

49. Nothing like these exceptionally long letters (upward of five printed pages) ordinarily appeared in the *Affiches,* and they bespeak a conspiracy with the editor; see *AA,* PauP, 1778, no. 47 (December 1); PauP Supplement, no. 50 (December 22); 1779; nos. 2 (January 12), 3 (January 19), 4 (January 26).

50. Pluchon (1985b), pp. 113–14.

51. On Dazille, see Duvivier, 2:171–72; Léon (1976), pp. 38–40; Dazille (1792), pp. 3–6; *DPF,* p. 250; Pluchon (1985b), p. 95.

52. Dazille (1792), pp. 3–4; Duvivier, quoting Jourdan 2:171. The work on plantation medicine by Lafosse quoted above appeared in 1787, a decade after the first edition of Dazille's *Maladies des Nègres.*

53. Dazille (1776), p. 3.

54. Dazille (1785), pp. ix, 254–56; (1788), pp. 426–42; and (1792), p. 6; also Léon (1976), p. 39. Dazille's work on tetanus represents one part of a larger tetanus project sponsored by the naval ministry; see Duvivier, 2:198; Bissainthe, no. 7502; and chapter 12 in the present work. Regarding colonial medical works, one might mention the "Mémoires sur les maladies de Saint-Domingue," published as part of N. L. Bourgeois's *Voyages Intéressans* in 1787; see N. L. Bourgeois, pp. 409–504, and Léon (1976), pp. 46–49.

55. Dazille (1792), p. 8; (1785), pp. 31–38; and (1788), pp. 9, 14, 18.

56. See, for example, AN, C^{9B} 38, letters dated May 31 and August 7, 1788.

57. On the Eaux de Boynes, see *DPF,* pp. 764–73.

58. See *DPF,* pp. 768–70; Cercle des Philadelphes (1788b), passim and pp. 26, 97, 100, 235; Tippenhauer, pp. 118–21.

59. *DPF,* pp. 765–66, 770.

60. *DPF,* p. 767 and listing on p. 1492. Dazille had earlier inspected the Eaux de Boynes; see Dazille (1785), pp. 87–112; Léon (1928), p. 79, and (1976), pp. 62–66; Cabon [1928–33], 2:439.

61. See *DPF,* pp. 966–68, 1360–61, 1373–74. Pouppée-Desportes analyzed another hot spring in the remote Mirebalais in the 1730s; Arthaud later also investigated this spring; see *DPF,* pp. 917–19.

62. *DPF,* p. 521; Cornevin, p. 99. On inoculation in the colony, see also Cauna (1984), p. 41; (1987), p. 123; Pluchon (1985b), p. 93; Girod (1972), p. 80; Pouzet, p. 32; *DPF,* pp. 224–25, 250.

63. On Joubert de la Motte, see Léon (1976), p. 16; *DPF,* p. 521 and listing on pp. 1501–2; Pluchon (1985a), p. 182; Loker (1981), no. 46; and chapter 9 in the present work.

64. Pluchon (1985b), p. 97; *DPF,* pp. 225, 1490, 1538.

65. *DPF,* pp. 250–51, 521–22, 1559; Maurel (1961), p. 266; Pluchon (1985a), p. 183; also Loker (1981), no. 51.

66. On the Société Royale de Médicine, see Hannaway (1972) and (1974); Gillespie (1980); J. Meyer in Desaive et al.; McClellan (1985), pp. 221–22; see also W. B. Cohen, pp. 73–76.

67. The *Affiches Américaines* excerpted this paper, *AA*, PauP, 1787, no. 26 (March 31).
68. Lafosse, pp. 9, 11n, 138–44.
69. *DPF*, pp. 888–89.
70. Neufchateau, pp. 57–58.
71. Pouppée-Desportes, 1:15–30, 316, 319.
72. N. L. Bourgeois, pp. 414–16, and Nougaret in N. L. Bourgeois, p. 411.
73. Pluchon (1987) provides these details, pp. 259–60, 263–67.
74. *AA*, PauP Supplement, 1778, no. 50 (December 22).
75. Arthaud (1787), pp. 425–26.
76. Brau, p. 94; Pluchon (1987), p. 146.

CHAPTER 9: Economic Botany and Animal Economy

1. Brockway, pp. 8, 36–60, speaks to these points.
2. Thus the French, for example, made the export of indigo seeds a capital crime, even after it was pointless to do so; Brockway, pp. 55, 57; on the general point, see also Lacroix, 3:191.
3. Brockway, p. 72; A. W. Hill, pp. 191–92; De Candolle, pp. 165–70. Medicobotanical gardens appeared in Pisa, Florence, and Padua before 1545, Leyden in 1577, Montpellier in 1593, Nuremberg in 1625, Oxford in 1640, and so on. The garden of the London Society of Apothecaries, founded in 1663, represents a late example of this type. Brockway notes that European botanical gardens may have been modeled after pre-Columbian Aztec gardens.
4. Brockway, pp. 2, 72–74; see also De Candolle, pp. 174–78; Lacroix, 4:89. On the Jardin du Roi in Paris, see Laissus; Broc, pp. 20–21. Osborne, pp. 6–9, discusses French menageries at Versailles and elsewhere, emphasizing the developing role of zoos as "symbols of colonial conquest."
5. See Brockway, p. 74; Lacroix, 3:ix–x; 4:89–90; A. W. Hill, pp. 185, 210; De Candole, p. 178.
6. See Brockway, pp. 58, 75, and appendix; A. W. Hill, pp. 210–15; Howard, pp. 382–83; Hindle (1956), p. 277.
7. Botherel-Blanchet, pp. 23–30; Auvigne and Kernes, pp. 7–13; Laissus, p. 293.
8. Lacroix, 3:25–29; see also Brockway, p. 51; Institut de France, pp. 115, 147; and chapter 4 in the present work.
9. Lacroix, 3:25–26. Regarding the coffee trees, Bignon reminds Lignon that "utility consists in multiplication."
10. One might signal in particular J.-A. Peyssonnel (1694–1759), first royal physician in Guadeloupe and correspondent of the Paris Academy, the Marseille Academy, and the Royal Society of London; see Lacroix, 3:23–24, 39–45, 83; Institut de France, pp. 13, 93.
11. See De Candolle, p. 170; Brockway, p. 58; Barnwell and Toussaint, pp. 55, 66, 73, 83; Chardon, p. 68; Lacroix, 3:58, 200–203; 4:39; Lafforgue, p. 157; see also Cornu; Lokke, pp. 25–27, 109.
12. On Poivre, see Lacroix, 3:191–207; Lokke, pp. 49–54; Brockway, pp. 50–51; Broc, pp. 278–80.
13. See Lacroix, 3:203–7; Toussaint, p. 22; Brockway, p. 50; Tessier, pp. 585,

589, 594, 597; Lokke, pp. 50, 57; Lacroix 3:207; 4:90; Chaia, p. 133.

14. Lacroix makes the point, 3:4.

15. Lacroix, 3:32; Broc, pp. 123–24; Institut de France, p. 14; Barrère, front matter, and passim. Crouse (1977), p. 273, notes the early voyage (1652) to Guiana and a subsequent book by the French priest, A. Biet.

16. See Lacroix, 3:49–51; Chaia, pp. 130, 134–35; Martinière, p. 389; Devèze, p. 308; Institut de France, p. 10. Artur's manuscript history of Guiana never seems to have been published.

17. Lacroix, 3:58; 4:200–201, 208; Chardon, pp. 67–68; Pluchon (1985b), citing Chaia, p. 129; Fournier, p. 58; *DPF,* p. 761.

18. See Chardon, p. 71; Lacroix, 3:80–81, 90; Institut de France, pp. 136–37; De Candolle, p. 170.

19. Chaia, pp. 142–43, and Lacroix, 3:71–72, note other informants in South America.

20. Lacroix, 4:187–89.

21. Donkin, pp. 11, 14; Ros, p. 68. Donkin's remarkable study provides the starting point for any discussion of cochineal. Ross describes the traditional Amerindian dyeing process using cochineal; see also Cutbush.

22. On production figures, see Donkin, pp. 13, 21, 27–29.

23. See Donkin, pp. 25, 37; Ross, p. 69; Devèze, p. 135.

24. Donkin, pp. 37–38; Edelstein, p. 3. The Dutch cloth industry was a major consumer; the English used approximately 170,000 pounds in the late eighteenth century; and some Spanish cochineal, greatly adulterated and inflated in price, made its way to Saint Petersburg in Russia.

25. See memoir reprinted in Thierry de Menonville, 1:xl–li.

26. The expression is Moreau's; see *DPF,* p. 1019. See also *DPF,* p. 1552; Edelstein, p. 1; Donkin, pp. 46, 52; Mozard in *AA,* PauP, 1787, no. 84 (October 20).

27. Thierry de Menonville's "Voyage à Guaxaca" forms part of his *Traité de la Culture du Nopal et de l'Éducation de la Cochenille;* Edelstein retells the story, pp. 3–8. Having paid for his cochineal, Thierry de Menonville later resented charges that he stole the product!

28. *DPF,* p. 1020; Arthaud, "Éloge de Thierry de Menonville," in Thierry de Menonville, 1:cx–cxii; also Corvington, p. 99; Cabon [1928–33], 2:442–44.

29. Colonists also ate nopal. Moreau de Saint-Méry pronounced the plant "very agreeable" when cooked in a white sauce; *DPF,* p. 1071.

30. So says Thierry de Menonville, 1:cxix–cxx.

31. Thierry de Menonville, 2:377–81, 430–32, 435. See similar thinking by Gauché in Cercle des Philadelphes (1788b), p. 155.

32. Arthaud, "Éloge de Thierry de Menonville," in Thierry de Menonville, 1:xxiii.

33. Moreau de Saint-Méry (1784–90), 6:485; *DPF,* pp. 1020, 1084.

34. Several sources tell this story; see *DPF,* p. 1021; Moreau de Saint-Méry, (1784–90), 6:483–84; Arthaud in CA-HdC, fol. 165; *AA,* PauP Supplement, 1785, no. 39 (September 24); Corvington, pp. 140–41; Donkin, p. 46. Arthaud reprints Joubert's "Histoire" and attacks Joubert in the notes in Thierry de Menonville, 1:lix–lxvi.

35. On the Jardin du Roi, see Corvington, pp. 140–41, 168; *DPF*, p. 1021; Arthaud, *Notice sur la ville de Port-au-Prince*, p. 7. See also letter from De La Luzerne to Marbois, dated PauP, June 2, 1787; BN, Mss., n.a.f. 20278, vol. 2, fol. 55; letter from de Peiner and Prony, dated PauP, September 14, 1790; BN, Mss., n.a.f. 9545, fol. 40.

36. See *DPF*, pp. 274–75, 1459; Arthaud in Thierry de Menonville, 1:lxxiv; *AA*, PauP, 1787, nos. 44 (June 2), 69 (August 30); 1788, no. 1 (January 3). Brulley was also the author of a pamphlet on cochineal husbandry published in Paris in 1795; see citation in Donkin, p. 65.

37. The Cercle later supplemented its stock from the Jardin du Roi in Port-au-Prince; see *AA*, PauP Supplement, 1785, no. 46 (November 12); Arthaud in Thierry de Menonville, 1:xxv, lxxxiv.

38. These swatches are pinned to documents in AN, Colonies F^3 151.

39. From the subscription circular printed by the Cercle des Philadelphes for Thierry de Menonville's *Traité*; Cercle des Philadelphes (1785c); see also Brulley in *AA*, PauP Supplement, 1785, no. 46 (November 12).

40. At least one other commercial nopalry existed in Saint Domingue from 1785, that of one Genton at the Môle; see *AA*, PauP Supplement, 1786, no. 1 (January 7). Moreau signals Port-à-Piment and Port-de-Paix as areas for potential cochineal development; *DPF*, pp. 683, 779.

41. See Arthaud, *Notice sur la ville de Port-au-Prince*, pp. 7–8; Corvington, pp. 154–55, 193; *DPF*, pp. 292, 1011, 1071, 1159; Cabon [1928–33], 2:444.

42. See memoir dated October 3, 1786; AN, C^{9B} 36 (1); a ministerial addendum stated that, with Richard attached to the finance ministry, he could not be transferred to Saint Domingue without its permission.

43. *DPF*, pp. 1021–22; BN, Mss., n.a.f. 9545, "Correspondance et papiers d'Hippolyte Nectoux," fol. 7. Nectoux studied at the Jardin du Roi in Paris; surprisingly, the Cercle des Philadelphes never made Nectoux a member; Nectoux later went to Egypt with Napoleon. See Maurel (1961); Pluchon (1985a); Gillispie (1987), p. 6; Lokke, p. 240.

44. AN, Colonies, F^3 89, fols. 39–41. This and a companion volume (vol. 92) of the "Collection Moreau de Saint-Méry" in the Archives Nationales are given over almost entirely to materials concerning plant transfers among French colonial stations; see catalog by May, and Maurel in *DPF*, pp. xli–xlii.

45. See *DPF*, p. 1021; CA-HdC, fols. 176–176v; *AA*, PauP, 1788, no. 20 (March 8).

46. The clove tree (*Eugenia aromatica*, or *giroflier* in French) is the plant in question, although allspice (*Pimenta dioica*) may also have been involved; see Devèze, p. 207.

47. See *AA*, PauP, 1787, nos. 30 (April 14), 62 (August 4).

48. *AA*, PauP, 1789, no. 25 (April 29); AN, Colonies F^3 89, fols. 110–11; see related memoranda, fols. 42–128.

49. According to a printed notice selling cloves; AN, Colonies F^3 89, fols. 122–24.

50. Arthaud, *Notice sur la ville de Port-au-Prince*, pp. 7–8. Curiously, while the *Alexandre* winded its way to Saint Domingue in 1787 and 1788, Captain William Bligh carried the breadfruit tree to the Caribbean for the English aboard the

Bounty. The mutiny aboard the *Bounty* incidentally secured a victory for the French in the race to get breadfruit to the West Indian colonies first, but that meant little, for the indomitable Bligh succeeded in bringing breadfruit to Saint Vincent in 1793 on his second voyage from the Pacific; see Howard, pp. 381, 384.

51. *AA*, PauP, 1788, no. 84 (October 18).

52. See *DPF*, p. 575; Arthaud (1792), p. 4; *AA*, PauP, 1788, no. 84 (October 18); *AA*, Cap, 1788, no. 47 (November 22); Pluchon (1985b), p. 111.

53. *DPF*, pp. 631, 1450; *AA*, PauP, 1789, no. 10 (January 31); AN, Colonies F³ 92, fol. 200.

54. Arthaud, *Notice sur la ville de Port-au-Prince*, p. 8; *DPF*, p. 1021; *AA*, PauP, 1789, no. 53 (July 1). The Société Royale d'Agriculture published Nectoux's paper, "On the preparation and handling of plants and trees for shipment to America from the East Indies," in 1791; see *DPF*, p. 1022; AN, Colonies F³ 89, fols. 111–16; and a similar English pamphlet of 1789, "Directions for taking care of the cochineal insects while at sea," translated into French the same year; APS, Pam., v. 59 and v. 1097. Nectoux submitted other memoirs on coffee and spice trees to the Société d'Agriculture, BN, Mss., n.a.f. 9545, fol. 22v.

55. Howard, pp. 384–85; *DPF*, p. 1021; *AA*, PauP, 1788, no. 20 (March 8); 1789, no. 20 (March 7); AN, Colonies F³ 92, fols. 192–94, 201.

56. *AA*, PauP, 1788, no. 20 (March 8).

57. See Laissus, p. 335; Institut de France, p. 217; letter from Thouin to Nectoux, "Paris, X^bre 1789"; BN, Mss., n.a.f. 9545, fols. 18–20.

58. See Palisot certificate from "cy-devant académie;" PA, Adrien de Jussieu collection. Microfilm of these and related Palisot papers from the Bibliothèque Royale Albert in Brussels have been deposited at the American Philosophical Society (APS Microfilms 1350–52). The author thanks Charles Gillispie for alerting him to these papers as well as for arranging for the microfilm to be produced. For more on Palisot, see Institut de France, p. 181; Lacroix, 3:131; 4:209–21; Chardon, pp. 176–77.

59. See letter, Palisot de Beauvois to A.-L. de Jussieu, dated "Au Cap franc. isle St Domingue, le 24 juillet 1788"; APS, Microfilm 1350. See also Palisot letters to A.-L. de Jussieu: December 15, 1788; June 1, October 9, December 23, 1789; de la Haye to A.-L. de Jussieu, December 19, 1789; APS, Microfilm 1350. See also Laissus, pp. 289, 317, 324. Unfortunately, Palisot's extensive botanical and natural history collections from Saint Domingue were later lost at sea.

60. See Brockway, pp. 103–12; Headrick, chap. 3.

61. *AA*, PauP, 1787, no. 87 (November 1); 1788, nos. 84 (October 18), 100 (December 13); Marbois and de Vincent letter of September 14, 1788, and Thouin letter of December 1789 in Nectoux papers, BN, Mss., n.a.f. 9545, fols. 18–20, 38.

62. *AA*, PauP, 1789, no. 20 (March 7); Cercle des Philadelphes (1789a), p. 3.

63. Wimpffen, p. 210. Possibly reflecting on the previous administration of the garden, Wimpffen goes on to say, "It is much to be lamented that the fate of a useful institution should thus depend on the inclination, the caprice, or the ignorance of an individual."

64. *AA*, PauP, 1788, no. 20 (March 8); see also *DPF*, p. 1021.

65. *DPF*, pp. 224, 1011. Bamboo was imported as a building material and perhaps for fodder; Moreau noted the increasing attacks of insects on imported bam-

boo. The pecan tree project is noted in *AA*, PauP, 1787, no. 26 (March 31).

66. Letter from Hesse to Marbois, dated Paris, February 16, 1788; BN, Mss., n.a.f. 20277, fols. 259–60. Letter: La Luzerne to Nectoux, dated Paris, October 31, 1789; BN, Mss., n.a.f. 9545, fol. 7; *DPF*, pp. 1021–22.

67. See letters to Nectoux from the administrators, dated Port-au-Prince, September 14, 1790, and from Thouin, dated Paris, August 15, 1791; BN, Mss., n.a.f. 9545, fols. 22v, 40. See also Corvington, p. 194.

68. See extracts of the registers of the Provincial Assembly, dated January 23, 1792, in Nectoux papers, BN, Mss., n.a.f. 9545, fol. 46.

CHAPTER 10: Meteorology and Popular Science

1. J. Meyer in Desaive et al., pp. 9–20, 29, 139; McClellan (1985), pp. 220–22.
2. McClellan (1985), pp. 223–27.
3. *Gazette de St. Domingue*, 1764, no. 5 (February 29), an early number of the first colonial newspaper. The message of meteorology was thus one of the first to come through the medium of the newspaper.
4. *AA*, PauP, 1767, no. 25 (June 24).
5. *AA*, PauP, 1780, no. 3; Thierry de Menonville, 1:xxi–xxii.
6. *AA*, PauP, 1784, no. 38 (September 18).
7. Ibid.
8. *AA*, PauP, 1784, no. 41 (October 9).
9. *AA*, 1784, no. 45 (November 6); AN, Colonies F^3 152, fols. 168–71. Not only a conduit for meteorological observations, Lalande used his informants to solicit information about Saint Domingue, notably about tides; see *DPF*, p. 775; and the haughty letter from Lalande, *AA*, PauP, 1785, no. 27 (July 2).
10. Talman, pp. 72–73; Arthaud (1786a), p. 6. Reports of two other colonists, Chabaud in Cap François and Thomas in Camp-de-Louise, also appeared in Paris. The sudden interest in meteorology in Saint Domingue in the middle 1780s precipitated a small push for declination observations; see *AA*, PauP, 1787, no. 13 (September 13); McClellan (1985), p. 363n.
11. *AA*, PauP, 1787, no. 6 (January 20).
12. *AA*, PauP, 1788, no. 62 (August 2).
13. See statements to this effect and meteorological reports from Île de France in *AA*, PauP Supplement, 1785, no. 35 (August 7); PauP, 1787, no. 30 (April 14).
14. Joubert and Odelucq were two notable reporters; *AA*, PauP, 1786, nos. 2 (January 14), 6 (February 11), 19 (May 13); etc.; *DPF*, pp. 286–88.
15. Mozard repeated that "these things are doubtless very useful, and enlightened agriculturalists gain great advantages from them"; *AA*, PauP, 1787, no. 73 (September 13).
16. *AA*, PauP, 1788, no. 30 (April 12); Mozard again notes the utility of meteorological observations for agriculturalists, travelers, and sailors.
17. See *AA*, PauP, 1789, no. 15 (February 19); *AA*, Cap. 1789, nos. 6 (February 7); 12 (March 21); see also *AA*, Cap, 1788, no. 1 (January 5).
18. Talman, pp. 64, 72.
19. Cercle des Philadelphes (1787a), part I, article 49; Anglade (1973a), p. 36.
20. *AA*, PauP, 1786, no. 9 (March 4).
21. Charles Gillispie (1983), pp. 25, 67–68, 98–99; Richard Gillespie,

pp. 253–59. Charles Gillispie's (1983) monograph provides a wonderful account of the invention of aviation in France; Richard Gillespie's article on early ballooning is also a valuable source, as is Darnton (1968), pp. 18–23.

22. See *AA*, PauP, 1783, nos. 49 (December 3), 50 (December 10); Cap, 1784, nos. 5 (February 4), 11 (March 17); PauP, 1784, nos. 13 (March 31), 15 (April 14).

23. Mozard describes the Vaudreuil balloon, *AA*, PauP, no. 14 (April 7). The first balloon flight in the United States occurred in Philadelphia on January 9, 1793; see Gillispie (1985); Bellegarde (1953), p. 49.

24. *DPF*, pp. 289–90; *AA*, PauP, 1784, no. 16 (April 21).

25. *DPF*, p. 290; also Castonnet des Fosses (1884), p. 31, and (1893), p. 22n.

26. *AA*, PauP, 1784, no. 16 (April 21).

27. Castonnet des Fosses (1884), p. 31; (1893), p. 22n; *DPF*, p. 460; and Map 4.

28. Letter dated March 20, 1784; reprinted in Debien (1972), p. 432.

29. Letter dated "Au Fond," September 2, 1784; Debien (1972), p. 434, and report of one of these flights in *AA*, PauP, 1784, no. 33 (August 14). Mme Millet obviously confuses inflammable air (hydrogen) with rarefied atmospheric air.

30. Toussainte and Adolphe, ref. A-100, p. 18, list of pamphlet describing the flight on Île de France of a balloon 32 feet in diameter.

31. *AA*, PauP, 1784, no. 35 (August 28); *AA*, Cap, 1788, no. 15 (April 12).

32. See *AA*, Cap, 1787, no. 22 (June 2); 1788, nos. 9 (March 1); 50 (December 13); 1789, nos. 2 (January 10); 22 (April 25); *Gazette de Saint-Domingue*, 1791, no. 46 (June 8); *DPF*, p. 512; Cabon [1928–33], 2:437. The number of sessions per course did vary; one had sixteen sessions, whereas those on the road only four.

33. This quotation is a composite from Millon's announcements in *AA*, Cap, 1788, no. 9 (March 1) and 1789, no. 22 (April 25).

34. On Nollet and his famous course on experimental physics, see Torlais; see also McClellan (1979), pp. 437–38; (1985), p. 18.

35. *AA*, Cap, 1787, no. 18 (May 5).

36. See *AA*, Cap, 1774, no. 17 (April 30); *AA*, PauP, 1785, no. 15 (April 9); Fouchard (1955b), pp. 160–63; on policing of these shows, see *DPF*, p. 478.

37. *DPF*, pp. 511–12; Cabon [1928–33], 2:437.

38. *DPF*, p. 1060; *AA*, PauP, 1788, nos. 63 (August 9); 66 (August 16). Lightning hit the Intendancy in Port-au-Prince in 1788, a strike that proved to Mozard the superiority of pointed rods over blunt ones. See also *Gazette de Saint-Domingue*, 1791, no. 46 (June 8), where Primat claims to have demonstrated that electricity from an electrostatic generator is identical to lightning!

39. *DPF*, pp. 239, 272.

40. See *DPF*, pp. 510–11; BN, Mss., n.a.f. 22086, Papiers de Baudry des Lozières, fol. 169; CA-HdC, fol. 172v; *AA*, Cap, 1778, no. 50 (December 22). References to Baudry's and Dubourg's instruments may or may not be to the same equipment. Two other scientific "cabinets" show up in estate sales; see *AA*, PauP, 1772, no. 30 (July 22); 1777, no. 52 (December 23).

41. *AA*, PauP, 1788, no. 40 (May 17). Moreau de Saint-Méry held that earthquakes resulted from an explosion of air, water, and "inflammable puritic matter" in underground caverns that laced Saint Domingue and surrounding waters.

"This theory is conformable to the principles of a sane physics," said Moreau, in *DPF*, p. 719; for like views, see *AA*, PauP, 1789, no. 20 (March 7).

42. *DPF*, pp. 509–10; Shapin and Schaffer.

43. *AA*, PauP, 1770, no. 32 (August 1); see initial notice in *AA*, Cap, 1770, no. 24 (June 16), and another response, *AA*, PauP, 1770, no. 30 (July 18).

44. *AA*, PauP, 1770, no. 42 (October 10).

45. See, e.g., *AA*, PauP, 1772, no. 3 (January 15); 1782, no. 33 (August 14).

46. On Euler's three-body problem, see *AA*, PauP, 1769, no. 23 (June 7).

47. *AA*, PauP, 1777, no. 52 (December 23).

48. See Darnton (1968), passim and here pp. 3–10, 47–52; see also the major section devoted to mesmerism in Gillispie (1980), pp. 261–89.

49. See Darnton (1968), pp. 62–71; Gillispie (1980), pp. 264–82.

50. Maurel (1961), pp. 234–38; Pluchon (1985a), pp. 158–59; *DPF*, p. 345; Cabon [1928–33], 2:451–52; Darnton (1968), p. 58; Gillispie (1980), pp. 284–88. Back in France naval authorities received Puységur cooly for his shipboard experiments with mesmerism.

51. See *DPF*, p. 345; Pluchon (1985a), p. 158; Maurel (1961), p. 239; Debien (1955), 17n; Puységur (1787b); and Duvivier, 1:90.

52. Debien (1972) reprints this letter, p. 433. This remarkable colonial episode confirms the deep sexual and moral fears raised by mesmerism in official quarters; see Gillispie (1980), pp. 282–83.

53. *DPF*, p. 1322.

54. Reprinted in Debien (1955), pp. 16–17; Trembley, himself Swiss and a lesser member of the great Genevan family that included Abraham Trembley, was quite familiar with serious and advanced discussions of mesmerism in Europe, and he speaks of them at length in this letter.

55. Stein (1988), p. 50, has uncovered other instances where slave owners discussed the beneficial application of mesmerism to their slave property. According to Stein, "Letters exchanged between absentee owners and their colonial representatives were full of references to this matter."

56. Moreau tells this story in *DPF*, pp. 275–76; see also Maurel (1961), pp. 237–38. Pluchon (1987), pp. 66–69, now provides the most complete account.

57. See *AA*, Cap, 1784, no. 23 (June 9); PauP, 1784, nos. 24 (June 16); 25 (June 23); 26 (June 30); PauP Supplement, 1785, no. 33 (August 13); PauP, no. 36 (September 3); PauP Supplement, no. 37 (September 10).

58. *AA*, PauP, 1784, no. 25 (June 23).

59. See *DPF*, p. 345; Maurel (1961), pp. 235, 237; Pluchon (1985a), p. 158; Arthaud (1785), p. 2. Arthaud's brother-in-law, Baudry des Lozières, might possibly have been the otherwise unnamed third party in these trials.

60. *DPF*, p. 345; Maurel (1961), pp. 234–36; Pluchon (1985a), pp. 158–59.

61. CA-HdC, fol. 150; partially quoted in Pluchon (1985a), p. 159.

CHAPTER 11: Origins: Science or Freemasonry?

1. Anglade (1973b), p. 129, for example, calls the Mesmer affair the "crystallizing event" for the Cercle. See further chapter 10; CA-HdC, fols. 149v–150; *DPF*, pp. 345, 349; Pluchon (1985a), pp. 158–59; Maurel (1961), pp. 234–36; Arthaud (1785), p. 2.

2. See [Tramond] (1927), p. 603; Maurel (1938), writing in French in the *Franco-American Review*. Maurel suggests that the Cercle received funds from Masonic lodges, but no evidence exists that the Cercle des Philadelphes ever got a sou from Masonic sources; Maurel (1961), pp. 239–43. The texts of Maurel (1938) and (1961) are identical; citations are to the 1961 imprint.

3. See Faÿ (1932), pp. 257–59, 266; see also Maurel (1938) and (1961), p. 240, where she cites Faÿ. See also Faÿ (1942) [*sic*], the collaboration by Faÿ, Maurel, and Equy, and citations of Faÿ by Taillemite in *DPF,* p. xv.

4. Maurel (1961) in the *Revue française d'histoire d'outre-mer* adds a valuable new annotated appendix listing members of the Cercle des Phildealphes, but otherwise her 1961 article is word for word that published in 1938.

5. For example, Gabriel Debien (1938), p. 71: "Apparently created a combat Mesmerism, the Cercle in reality was of Masonic inspiration." Martineau, p. 148: "All the members of the Cercle were Freemasons." Desfeuilles (1957), p. 364, calls the Cercle a "para-masonic assembly." Even Georges Anglade (1973b), pp. 129–30, the scholar who comes closest to seeing the Philadelphes as part of the distinct scientific-society movement still viewed the Cercle as animated by a "Masonic flame." The noted historian of Freemasonry, Alain Le Bihan (1974), p. 41, similarly linked the Cercle to Masonic activity in Saint Domingue. See also Pressoir et al. (1953), p. 97; Taillemite (1958) in *DPF,* p. xiv; Julien (1977), p. 127.

6. Pluchon (1985a), pp. 158–60, 163, 166.

7. See Pluchon (1985a), pp. 161–66, 174–75. Note Pluchon's section headings (pp. 161, 163) labeling the Cercle a "politicized institution" and a "political machine." See also Maurel (1961), p. 241, and very early statements by Vaissière (1909), p. 332, that stress an active political program supposedly embodied by the Philadelphes.

8. Pluchon (1985a), pp. 159–60, 163, 165–66, 168.

9. Roche, in particular, distinguishes strongly between the Masonic and scientific society movements in eighteenth-century France; see Roche (1978), 1:257–80; also McClellan (1985), p. 137; Cobban, p. 135.

10. Jacob, esp. pp. 120, 245–49.

11. Roche (1978); McClellan (1985), pp. 6, 89–99, 133–37 and appendices.

12. Le Bihan (1967), pp. 291–95, 389–91, 393; *DPF,* p. 540; [Tramond] (1927), p. 605. The possibility of a third lodge in Cap François, Amitié, arises in Moreau (*DPF,* p. 427), but Moreau seems to have confused the name with the Verité lodge.

13. Le Bihan (1974), p. 46.

14. The proportion of Mason members of the Cercle des Philadelphes is not known. Le Bihan (1974), p. 41, signals several Mason Philadelphes, as does Pluchon (1985a); Maurel (1961), p. 240 and appendices.

15. Taillemite in *DPF,* pp. xiv–xv; Maurel in *DPF,* p. xxxviii; Moreau, *DPF,* p. 7.

16. Le Bihan (1967), pp. 406–7; Bissainthe, p. 381; Depréaux, esp. p. 12.

17. Maurel (1961), p. 240; Léon (1976), p. 53, gives Arthaud the title.

18. Maurel (1961), p. 253; Taillemite in *DPF,* p. x.

19. Jean Fouchard discovered this exchange, and he reprints major portions

with comments in Fouchard (1955b), pp. 52–55. Despite the implication, I find no evidence of Lerond's original proposal in the *Affiches Américaines*.

20. See Delile letters in *AA*, Avis du Cap, 1769, nos. 5 (February 1), 8 (February 22).

21. *AA*, Avis du Cap, 1769, no. 10 (March 8); Fouchard (1955b), pp. 52–53.

22. On this little-known publication, see *DPF*, pp. 496–97; Fouchard (1955b), p. 98; Bissainthe, p. 757.

23. Notices about the *Journal de Saint-Domingue* in the *Affiches Américaines* for 1766 listed articles about cotton, coffee, sugar, and ginger; *AA*, PauP, 1766, nos. 10 (March 5), 31 (July 30), 42 (October 15), and 53 (December 31).

24. See *AA*, Avis du Cap, 1769, no. 11 (March 15) for Toussaint's letter; Fouchard (1955b), pp. 53–55 reprints a long extract; Fouchard, too, leaves open the possibility that it is genuine. The biographical facts in the letter do not exactly match what is known of Toussaint Louverture, but the latter was literate, twenty-six years old in 1769, and likewise from the northern plain. He might have written the letter in question; consult Dorsainvil, pp. 81–83.

25. A noted proposer of civic schemes, the abbé Bernardin de Saint-Pierre (1737–1814) also commented on colonial affairs, having spent some time in the Indian Ocean colonies. He likewise wrote the early romance, *Paul et Virginie* (1785), set in an idyllic island colony; see Lokke, pp. 54–58, 86–87; Hahn, p. 105.

26. The editor of the *Affiches Américaines* took issue with this point, referring in a footnote to treatises on coffee, cotton, and sugar that appeared in the defunct *Journal de Saint-Domingue*.

27. *AA*, Avis du Cap, 1769, no. 11 (March 15); Fouchard (1955b), p. 55.

28. *AA*, Cap, 1769, no. 44 (November 6); Fouchard (1955b), p. 55.

29. See Arthaud (1791), p. 10; see also Léon (1933), p. 40, and (1976), p. 49.

30. *DPF*, p. 497. Moreau refers to the physician, Duchemin de l'Étang, and his abortive "Gazette de Médecine et d'Hippiatrique"; see above chapter 8.

31. *DPF*, p. 345. See also CA-HdC, fol. 150, for an allusion to other previous plans to found a learned society in Saint Domingue.

32. For background, see, again, Roche (1978); McClellan (1985). The only other French-style learned institution to spread to North America in the eighteenth-century was the curious and short-lived *Académie des Sciences et Beaux-Arts des États-Unis de l'Amérique* (Richmond, Virginia, 1786–89); notably the Richmond academy came after the Cercle des Philadelphes and never received the slightest token of government support; McClellan (1985), pp. 144–45.

33. On these proposals, see Lacroix, 4:11; Le Bihan (1974), p. 41; Toussaint (1966), pp. 22, 39; see also Barnwell and Toussaint, p. 87.

34. Hindle (1956), p. 277.

35. Hindle (1956), p. 277, mentions the Cercle des Philadelphes, but misses the essence of the institution as an official, indeed royal scientific society. Still, for Hindle, the Cercle was an unusually "active intellectual group [that] published a surprising number of papers on topics of the sort then engaging the American philosophical societies." On eighteenth-century "academies" and "societies," see McClellan (1985), pp. 13–34. The Cercle des Philadelphes, of course, was organized as a typical provincial "academy."

36. Arthaud, "Discours pour la Séance du 29 août," AN, Colonies F³ 152, fols. 220v–221r. See also Arthaud's speech of December 3, 1786, "Une académie peut-elle être utile à St. Domingue?"; AN, Colonies F³ 152, fols. 234–42; and also F³ 81, fols. 119–26.

37. Quoted in Pluchon (1985a), p. 159. See also Maurel (1961), p. 238.

38. CA-HdC, fol. 150. Similarly with regard to Arthaud's expressions, "new order of things" and "revolution" quoted by Pluchon (1985a), p. 165, note that Arthaud refers to a learned "company" supported by the government.

39. CA-HdC, fol. 151. Pluchon (1985a), pp. 160–61, quotes and paraphrases this passage; Maurel (1961), pp. 242–43, does the same.

40. Pluchon (1985a), p. 160; no oath appears in the detailed *Statuts* of the Cercle prepared in the fall of 1784; Cercle des Philadelphes (1985b).

41. Arthaud (1785), pp. 2–3; see also Maurel (1961), pp. 235–36; Pluchon (1985a), pp. 161–62. The passage is from a speech delivered at the first public session of the Cercle on May 11, 1785. The reference to this Venetian academy seems entirely a cover for the reference to Philadelphia.

42. APS Archives, 506.7294: C33.1 (no. 1). Baudry was a refugee in Philadelphia from 1792 to 1796.

43. See the petition to the administrators dated August 22, 1784, reprinted in Moreau de Saint-Méry (1784–90), 6: 559.

44. Arthaud (1785), p. 3. Arthaud uses the terms *savans* and *Amateurs des sciences* for "scientists and lovers of science" in the translation.

45. See Figure 18 heading part III; later membership certificates of the Cercle in APS Archives, 506.7294.C33.1 (nos. 3/4); and description of seal in Cercle des Philadelphes (1785a), *Statuts*, article xxxiv.

46. Moreau, *DPF*, p. 349, confirms the nine founders. Note Baudry's references to his qualities as "Fondateur" and "Membre Fondateur" of the Cercle des Philadelphes/Société Royale; APS Archives, 506. 7294. C33.1 (nos. 5, 6, 7).

47. See undated Baudry speech, APS Archives 506.7294.C33.1 (no. 6). Le Bihan (1967), pp. 406–7, makes plain Baudry's strong commitment to Freemasonry. See other Baudry Masonic materials in BN, Mss. n.a.f. 22084, e.g., at fol. 392.

48. See CA-HdC, fol. 152; Cercle des Philadelphes *Statuts* [1785a].

49. See Table 3. Table 3 derives from *DPF*, p. 349; the "Index des Noms de Personnes" by Maurel and Taillemite in *DPF*, pp. 1441–1559; and Maurel (1961), pp. 261–66; "1784 Tableau" reprinted in Pluchon (1985a), pp. 181–83; the printed "Tableau" in *AA*, 1785, no. 22 (28 May); and the undated 1785 manuscript list, BN, Mss., n.a.f. 22085, fols. 136–37.

50. See Arthaud letter to Minister, de Castries, dated April 9, 1785; AN, C⁹ᴮ 35.

51. Trained in Paris, Cosme d'Angerville was, according to Maurel, a "physician-surgeon" and "correspondent" of the Paris Faculty of Medicine. He did become a *correspondant* of the Academy of Surgery and the Royal Society of Medicine in Paris. Arthaud later calls d'Angerville the "doyen" of the Cercle des Philadelphes, which reinforces the notion that d'Angerville and not Baudry des Lozières was the third of the original three founders, along with Arthaud and Dubourg. On d'Angerville, see CA-HdC, fols. 154v, 165; Maurel in *DPF*,

p. 1470, and Maurel (1961), p. 263; *DPF*, p. 349; Pluchon (1985a), p. 182.

52. See *DPF*, pp. 528–29; and chapter 12 in the present work.

53. See Table 3; Pluchon (1985a), pp. 182–83; Pluchon (1985a), p. 158, suggests Dubourg was an apothecary.

54. On Dubourg, see CA-HdC, fols. 171r–173r; *DPF*, pp. 349, 528–29; Maurel in *DPF*, p. 1477. According to Moreau, Dubourg supposedly first developed a love for science from hearing Le Cat's lectures in Rouen.

55. See Pluchon (1985a), p. 182, and compare ms. "Tableau du Cercle des Philadelphes," BN Mss., n.a.f. 22085, fols. 136–37.

56. See Maurel in *DPF*, p. 1446; *DPF*, p. 349.

57. On Baudry des Lozières, see Maurel in *DPF*, p. 1449; Dépréaux. Baudry, who should have known, labels Poulet a lawyer, but still the initial connection to the legal establishment is weak; see BN Mss., n.a.f. 22095, fols. 136–37.

58. See Pluchon (1985a), p. 172; Maurel (1961), p. 253; and the fuller discussion in chapter 14.

59. See undated ms. of Baudry des Lozières, APS Archives, 506.7294.C33.1 (no. 6).

60. Ambiguities and crosscurrents do remain. Having made the distinction between the temple of the Masons and the Cercle des Philadelphes, in a speech to the Cercle a week later Baudry still let his personal Masonic colors show when, with outlandish imagery, he interjected the following. (His use of "science" in the singular is either incredibly obtuse or he means to evoke the secret knowledge of the Masons.) "La science est absolument nécessaire. C'est la verité universelle qui doit éclairer tous les Esprits comme le soleil éclaire tous les Corps et sans le quel on ne distingue rien Celui qui n'est point impregé de ce fluide des Savans [!], de cette lumiere pure, est aveugle comme un aveugle né. . . . En effet quel joye ne sent on pas quand le nuage de l'ignorance se diffuse, et qu'on apperçoit par degrés ses yeux s'ouvrir à la Lumière de la Science"; Baudry "Discours prononcé dans la Société des Philadelphes, le Dimanche 22 août 1784," APS Archives, 506.7294.C33.1 (no. 2). Members of the Cercle may well have shared Baudry's enthusiasms, but the Cercle itself did not become thereby an institution of Freemasonry.

61. See Baudry ms., APS Archives, 506.7294.C33.1 (no. 6), pp. 2–3. See also related remarks in Baudry ms., "Discours prononcé dans la Société des Philadelphes, le Dimanche 22 août 1784;" APS Archives, 506.7294.C33.1 (no. 2).

62. And, indeed, as Roche points out, episodes of revelry and licentiousness were not unknown in the background histories of eighteenth-century scientific societies; see Roche (1978), 1:24–29, 45–48; McClellan (1985), p. 91.

63. Undated Baudry ms., APS Archives, 506.7294.C33.1 (no. 6), pp. 4–5.

CHAPTER 12: Milestones on the Road to Recognition

1. By contrast, the academy in Châlons-sur-Marne took twenty-five years to earn letters patent in 1775; on this process, see Roche (1978), 1:1, and graph no. 1; McClellan (1985), pp. 14, 91, 134, and appendices.

2. Three new recruits joined between August 15 and August 22: Barré de Saint-Venant, a rich plantation owner, militia officer, and member of the Chamber of Agriculture; François de Chaumont, chief royal copyist for the navy in Cap

François; and Edme Ducatel, master of pharmacy and proprietor of a drugstore in Cap François; see Maurel in *DPF*, pp. 1447, 1478.

3. Moreau de Saint-Méry (1784–90), 6:559–60, reprints the petition of August 22, 1784 and the administrators' reply; Blanche Maurel presents excerpts in *DPF*, p. xxxvii. See also Arthaud speech of August 22, [1784]; AN, Colonies F³ 152, fol. 232; CA-HdC, fol. 151v; and *DPF*, p. 346.

4. See letters dated, "Le 22 7bre 1784" and "du 30 7bre 1784," CCP, letters 3 and 4, fols. 185–185v; see also CA-HdC, fol. 151v.

5. See copies: AN, Colonies F³ 81; AN, SdOM, "Bibliothèque de Moreau de Saint-Méry," vol. 147; APS Library. Figure 20 reproduces the first page from the APS copy; the interpolated date, 1786 (not seen), notes the document's receipt at the APS. See also *DPF*, p. 346; Maurel (1961), pp. 260–61.

6. Charles Arthaud enunciated this program in several speeches: August 22, [1784]; August 29, [1784]; December 3, 1786; AN, Colonies F³ 152, fols. 220–223, 232, 234–42; F³ 81, fols. 119–26. In all these speeches Arthaud emphasized the utilitarian, saying at one point, "Our establishment can only be well looked upon if we attach ourselves to the interests of colonists."

7. Arthaud said the Cercle shared Moreau's views regarding "Tableaux des Recherches." Moreau joined the Cercle on April 18, 1785, donating his *Lois et Constitutions* to the Cercle's library; see CA-HdC, fol. 154.

8. See CA-HdC, fols. 151v–152, and letter from Coustard, dated "14 9bre 1784," CCP, letter 11, fols. 187–187v.

9. *AA*, PauP, 1784, no. 48 (November 27). Mozard chauvinistically and incorrectly claimed more academies for France than all other nations combined.

10. CA-HdC, fol. 150v; and Arthaud, AN, Colonies F³ 152, fols. 190–91; see also Moreau in *DPF*, p. 346; Pluchon (1985a), p. 169; Maurel (1961), p. 244.

11. CA-HdC, fol. 152; and reply, "Le 9 X^{bre} 1784"; CCP, letter 5, fol. 185v.

12. See copies of Cercle des Philadelphes (1785a): PA, Pochette de séance for February 25, 1789; AN, Colonies F³ 81; AN, SdOM, "Bibliothèque de Moreau de Saint-Méry," vol. 147; APS Library. See also Puchon (1985a), p. 177.

13. Maurel (1961), p. 243. By 1787, the Cercle met weekly; see *DPF*, p. 348, and Trembley in Debien (1955), p. 8.

14. See Cercle des Philadelphes (1785a), articles xxxvi, lvi–lvii, lx, and annotations on copy in AN, Colonies F³ 81. Collecting dues proved onerous for the other officers, so the Cercle appointed a treasurer in early February 1785; see CA-HdC, fol. 153; see also *DPF*, p. 347.

15. The new associates: Prévost, a lawyer; Deschamps, a lawyer and later navy comptroller in Port-au-Prince; Verret, a royal engineer; Odelucq, a plantation owner and member of the Chamber of Agriculture; Vatable, a military officer in Guadeloupe; and Jacques-François Dutrosne (1749–1814), better known as Dutrône la Couture, the physician who also wrote on sugar-cane manufacture; see listings in Pluchon (1985a), appendices; Maurel (1961); CA-HdC, fols. 152, 159v; *DPF*, p. 1294; Vastey.

16. Dubourg stepped aside as president to become director of the (as yet nonexistent) botanical garden, with Dr. Peyré named assistant director. Arthaud took over as president. Poulet remained the nominal secretary, with Prévost added as assistant secretary. Dubourg doubled as librarian. The dozen or so other

Philadelphes became simple (resident) associates; on this reorganization, see CA-HdC, fol. 152v; remarks by Arthaud, AN, Colonies F³ 152, fols. 192–94 (speech of October 31, 1784); and chapter 11 in the present work.

17. See Mss., "Discours pour regler des rangs des associés du cercle"; "Discours prononcé dans la séance du 28 9bre 1784"; "Discours prononcé dans la séance du 15 août en presentans un plan pour les status de Cercle"; AN, Colonies F³ 152, fols. 188–89; fols. 195–203 (here 203); fol. 228.

18. See ms., "Discours prononcé dans la séance du 28 9bre 1784," AN, Colonies F³ 152, fols. 195–203; quotes on fols. 198–99, 202.

19. Letter dated "au Cap, 21 Février 1785"; AN, C⁹ᴮ 35. See also CA-HdC, fol. 152v; Pluchon (1985a), p. 169. Seventeen associates signed this letter.

20. See letters to "Monseigneur," dated April 1, and May 12, 1785; AN, C⁹ᴮ 35.

21. See ministry documents dated "Versailles, le 7 Juillet 1785"; and response from administrators dated "1 Xbre 1785, Port-au-Prince"; AN, C⁹ᴮ 35.

22. See CA-HdC, fols. 152v–153v; Maurel (1961), p. 266. François de Neufchateau later was president of the French Legislative Assembly and later still president of the Imperial Senate; see Maurel in *DPF,* p. 1490; Fouchard (1955b), pp. 57–58; Gillispie (1980), pp. 20, 371; Pluchon (1987), p. 228. The name, Neufchateau, appears with and without a circumflex.

23. CA-HdC, fols. 155v, 156v; *DPF,* pp. 652–53; Maurel in *DPF,* p. 1514. Elected after the first public meeting of the Cercle on May 11, 1785, Le Gras held off joining the Cercle precisely because he felt the organization might be merely "meteoric"; see Baudry papers, *éloge* of Le Gras, BN, Mss., n.a.f., vol. 22086 at fol. 137bis. In the short period until his death later in 1785, however, Le Gras became an enthusiastic member, donating 1,000 colonial livres for a prize and books for the Cercle's library.

24. See CA-HdC, fol. 153; Cercle des Philadelphes (1787b), and listings in Maurel (1961) and Pluchon (1985a); see also Institut de France, p. 93.

25. See letter to the Martinique administrators, AN, Colonies F³ 81, fol. 116.

26. De Galvez's letter dated Havana, April 30, 1785, in CCP, letter 6, fol. 186; see also CA-HdC, fols. 156v–157, 162v; Cercle des Philadelphes (1787b).

27. See ms. "Tableau du Cercle des Philadelphes," in the papers of Baudry des Lozières, BN, Mss., n.a.f. 22085, fols. 136–37. Listing de Trouillet as an honorary, but not François de Neufchateau, the Baudry ms. thus dates between February 13 and May 1, 1785. Pierre Pluchon (1985a), appendix 4, reprints another ms. "Tableau" from AN, Colonies F³ 81; Pluchon mistakes the date, which must be 1785, since the ms. lists an honorary class. Pluchon's (1985a) list closely resembles a third list (that distinguishes resident and colonial associates) in the *Affiches Américaines* for May 28, 1785; see *AA,* PauP, 1785, no. 22 (May 28). By May 1785, Poulet, the Cercle's original secretary, and Vatable, a militia major in Guadeloupe, had dropped out.

28. The new associates: Guyot (or Guiot), a surgeon attached to the militia; Jean Gelin, a veterinarian and royal instructor of veterinary surgery; the chevalier de Correjolles, a retired army officer and a rich proprietor. See previous note and Maurel in *DPF,* p. 1470; Maurel (1961), pp. 263, 264.

29. See appropriate entries in Maurel in *DPF,* and Maurel (1961). In addition

to Joubert, Monnier was médecin du roi at Saint-Marc; Roland, an army physician, became médecin du roi in 1786 or 1787. The fourth new physician, Pescaye from Port Margot, also dropped from the rolls before May 1785; he may or may not have been the same Pescaye *fils,* later a resident associate.

30. On Deshayes, see *DPF,* pp. 1408–9; Arthaud (1786a), esp. pp. 6–7; Maurel in *DPF,* p. 1513; Debien (1937), no. 118; Stehlé (1970), p. 92; Chardon, p. 170.

31. On Paul Belin de Villeneuve, see membership lists cited previously; CA-HdC, fols. 168–69; Maurel (1961), p. 248; Maurel in *DPF,* p. 1450; see also Moreau de Saint-Méry (1791), pls. 30, 31.

32. Arthaud, CA-HdC, fols. 154–155v, and Mozard, *AA,* PauP, 1785, no. 22 (May 28), give detailed descriptions; also *DPF,* pp. 346–47. The meeting took place in the Cercle's rooms, but not the ones it eventually let on the rue Vaudreuil.

33. Arthaud (1785). Later in 1785 Arthaud published this speech and the accompanying *Description de la ville du Cap* in Paris; Arthaud styles himself "Médecin du Roi au Cap, Président du Cercle," in this work clearly intended to promote awareness of Cap François and the Cercle in Paris and elsewhere.

34. Arthaud provides these details, CA-HdC, fols. 154v–155. Baudry, too, emphasized political docility, saying "that the man who devotes himself to the sciences never causes trouble, the savant never has the time"; see Baudry speech, BN, Mss., n.a.f. 22085, fols. 132–51, the quotation here from fol. 142v.

35. Arthaud says the papers were given, but Mozard is more believable in saying that time ran out; see CA-HdC, fols. 155–55v; *AA,* PauP, 1785, no. 22 (May 28).

36. *AA,* PauP, 1785, no. 22 (May 28).

37. See Roche (1978), 1:324–55 and graphs 10–15; McClellan (1985), pp. 298–99. AN, Colonies F³ 81, and AN, SdOm, BMSM, vol. 147, contain Cercle des Philadelphes (1785b). See also Arthaud's comments, CA-HdC, fol. 153v.

38. The Cercle noted that "it is up to colonists, rich proprietors, and all those inspired by humanity and patriotism to propose and fund prizes"; Cercle des Philadelphes (1785b), p. 4; see also Pluchon (1985a), p. 171.

39. *DPF,* p. 377. For similar vivid accounts, see Arthaud in Cercle des Philadelphes (1788a), pp. 10, 16–19; CA-HdC, fol. 154; Trembley quoted in Debien (1955), no. 114, p. 9; also Debien (1974), p. 121.

40. See *DPF,* pp. 377, 1017, 1098; Roussier, pp. 246, 248; Ménier (1978), p. 125. This "Dépôt des papiers publics" arose primarily so that parties in France could more easily obtain information about persons in the colonies, but establishing a colonial record free from insects was also a factor.

41. *AA,* PauP, 1781, no. 21 (May 22); see also Desfeuilles, no. 179, p. 366. Arthaud notes the attempt to have slaves pick out bugs; Cercle des Philadephes (1788b), p. 23.

42. Cercle des Philadelphes (1788a), pp. 14, 20, and (1788d), p. 1; CA-HdC, fol. 165. On government request, the Cercle tested samples from the Reveillon factory in Paris, but while deserving of nationalistic commendations, Reveillon paper did not prove immune from attacks. Later, as minister of the navy, La Luzerne pressed the Cercle to reopen its investigations into this matter; see letter,

La Luzerne to Arthaud, dated "Versailles, 19 mars 1789," AN, C⁹ᴮ 40. Paper samples from Saint Domingue supposedly invulnerable to insects are preserved in AN, Colonies F³ 267, fol. 175.

43. See CA-HdC, fol. 157; *AA,* 1786, Cap Supplement, nos. 5 (February 1), 8 (February 22), 21 (May 27); PauP Supplement, 1786, no. 21 (May 27); see also Barré de Saint-Venant, p. 293.

44. CA-HdC, fol. 158; see also *AA,* Cap. 1786, no. 8 (February 22).

45. On the Paris Musée (also known as the Lycée), see Taillemite in *DPF,* pp. xiv, xv; Gillispie (1980), pp. 181, 190–91. On the Cercle's "correspondence" with the Paris Musée, see CA-HdC, fol. 157r–157v; Maurel (1961), p. 241.

46. Arthaud proclaimed this "engagement sacré" in a speech delivered on October 31, 1784; AN, Colonies F³ 152, fol. 192; see also the Cercle's prospectus.

47. See chapter 9 for the general background to these efforts, and *AA,* PauP Supplement, 1785, no. 45 (November 5); Thierry de Menonville, 1:front matter; *DPF,* p. 347.

48. Arthaud in Thierry de Menonville, 1:xxvi; letter from administrators, "du 21 Juillet 1785," letter 9 in CCP, fols. 186v–187.

49. Arthaud in Thierry de Menonville, 1:vi–vii, xxvii, lxxxiv–xcviii. Three harvests were made through November 1785. Rain, ants, and insect galls on the nopal cacti proved problems.

50. Letter dated "du 6 aoust 1785," letter 7, CCP, fols. 186–186v. See also second letter from the administrators, dated "21 Aoust 1785," letter 8 in CCP, fol. 186v; and CA-HdC, fol. 157v. Moreau de Saint-Méry (1784–90), 6:560, reprints the ordinance of November 21, 1785, granting the Cercle des Philadelphes two *carreaux* on the Morne de Cap. See also Maurel (1961), p. 251, and further discussion in this chapter.

51. CA-HdC, fols. 157v–158; and Arthaud's "Discours pour la séance du 15 aoust 1785," AN, Colonies F³ 152, fols. 208–13, whence the expression, "élever un azile aux Sciences." Arthaud justified the Cercle's "repas philosophique," by mentioning the "Joie decente et une Liberté vraiement Philosophique" that united members on such occasions. See also Maurel (1961), p. 249.

52. Arthaud's *éloge* appeared as a separate publication, in the *AA* [1785, PauP Supplement, no. 46 (November 12)], and later again in Thierry de Menonville, 1:cx–cxviii; see also CA-HdC, fol. 158v. Mozard chimed in by *reprinting* a 1765 memoir on cochineal by the Cap Chamber of Agriculture, *AA,* PauP Supplement, 1785, no. 39 (September 24).

53. Letter from Coustard and Bongars, dated "13 8ᵇʳᵉ 1785," letter 10, CCP, fol. 187; see also CA-HdC, fols. 160, 166; Pluchon (1985a), p. 169.

54. A copy of the prospectus and samples of felt dyed with colonial cochineal can be found in AN, Colonies F³ 151; the Cercle reprinted the prospectus and a list of 145 subscribers in Thierry de Menonville 1:vi–xx.

55. *AA,* 1785, PauP Supplement, no. 45 (November 5).

56. The Cercle sent the ms. to Bergeret's press in Bordeaux, whence it was forwarded to Versailles and the navy ministry for censoring and a license to print. The navy ministry sent the ms. to the keeper of the seals (*garde des sceaux*), who in turn directed it to the director-general of the book trade (*directeur-général de la librairie*), who then passed it on to one M. de Machy, a royal censor. De Machy

approved the work and returned it through the same channels to the director-general of the book trade, the keeper of the seals, the navy ministry, and finally back to Bergeret in Bordeaux. For the actual privilege, assigned to the widow Herbault in Cap François, Bergeret had to apply separately to the General Directorate of the book trade and pay a fee to a designated royal secretary. These incredible details emerge from the relevant correspondence in AN, C⁹ᴮ 36 (2). See also Pluchon (1985a), p. 169.

57. Local administrators already fully controlled the press in Saint Domingue, but metropolitan authorities later explicitly transferred censorship of the Cercle's publications to the local administrators; see ministerial letter, "A Vᶦˡˡᵉˢ, Le 29 Xᵇʳᵉ 1786" in AN, C⁹ᴮ 36 (2). Moreau recognized the problems of publishing abroad, but still criticized the Cercle for spending so much to have its works printed in the colony; see DPF, pp. 353–54.

58. See AN, C⁹ᴮ 36 (2), items dated "Port-au-Prince, 29 Avril 1786" and "Versailles, le 10 aout 1786." The edition included reports on the Cercle's nopalry and a heavily annotated version of Joubert's plagiarized "Histoire Abrégée de la cochenille"; see further above chapter 9; Cutbush, p. 272.

59. AA, PauP, 1787, nos. 84 (October 20) and 85 (October 25).

60. Note Baudry des Lozières' "Mémoire sur la découverte du cotton animal," read to the Cercle on October 12, 1785. Baudry proposed having slaves harvest the cocoons of certain wasps that parasitized a particularly nasty worm that itself attacked valuable indigo and manioc plants. Baudry's "animal cotton" reportedly made wonderful stockings, and the whole enterprise represented a natural extension of the colonial economy. Notably, Baudry distinguished his "cotton" from other silks "through the trial of the electrical machine." See document, BN, Mss., n.a.f. 22085, fols. 235–40; and DPF, p. 1263.

61. See previous discussion in this chapter and administrators' letters dated "3 9ᵇʳᵉ 1785" and "1 9ᵇʳᵉ 1785," CCP, letters 20 and 21, fols. 190v–191; Moreau de Saint-Méry (1784–90), 6:560. See also Maurel (1961), pp. 246, 251.

62. A detailed announcement appears in AA, 1785, Cap, no. 51 (December 21); see also Arthaud's remarks in CA-HdC, fols. 159v–160. The Linnean orientation precipitated at least one colonial response in favor of Tournefort's system; see AA, 1785, PauP Supplement, no. 53 (December 31).

63. So say Moreau and Arthaud: DPF, p. 528; CA-HdC, fols. 160–160v.

64. Another project of the Cercle from the first half of 1786 ought not to be overlooked: the inquiry into a plant disease affecting the bâtard variety of indigo; see AA, Cap, 1786, no. 7 (March 1); AA, PauP, 1788, no. 62 (August 2), and relevant papers and letters, AN, Colonies F³ 152, fols. 116–41.

65. See letter from Arthaud and Prévost, dated "Au Cap, le 8 8ᵇʳᵉ 1785," AN, C⁹ᴮ 35; also, Pluchon (1985a), p. 169.

66. See 1786 items dated January 13, March 1, June 16, June 30, and September 1 in archival volume, AN, C⁹ᴮ 36 (2). As further evidence of its merit and good works, the Cercle mentioned its volume on tetanus in preparation.

67. CA-HdC, fols. 159–159v. On Barbé de Marbois, see Maurel in DPF, p. 1447, and DPF, p. 19; also Ott, p. 186n. See also Barbé's published accounts as intendant; Barbé de Marbois (1788, 1789, 1790). Barbé had previously served as French consul general in Philadelphia.

68. "Discours du 31 8bre 1785 en présence de Mr L'Intendant," AN, Colonies F³ 152, fols. 204–6.

69. CA-HdC, fol. 159.

70. See report, AN, C⁹ᴮ 35, and previous discussion in this chapter.

71. See ministerial document dated "23 Mars 1786," AN, C⁹ᴮ 36 (2). The administrators in Saint Domingue acknowledged this ministerial decision in their report dated, "St. Domingue, le 17 Juin 1786," AN, C⁹ᴮ 36 (2).

72. CA-HdC, fol. 160v; Maurel in DPF, p. 1507; Lacroix, 3:107.

73. La Luzerne letter dated "Le 22 Avril 1786," letter 14 in CCP, fols. 188–188v.

74. See letter dated "Le 1 Juin 1786" and "Reflexions envoiés à Mʳ de la Luzerne," AN, Colonies, F³ 152, fols. 214–16; also CA-HdC, 160v.

75. La Luzerne to Arthaud, "du 5 Juin 1786"; CCP, letter 15, fols. 188v–189v.

76. Symmetry with the first public meeting in 1785 demanded that the second occur on May 11, 1786, and so Arthaud reports it in his history of the Cercle; see CA-HdC, fol. 161. Notices in the Affiches Américaines confirm the June 20 date; AA, Cap, 1786, nos. 24 (June 14) and 27 (July 8). Moreau, DPF, p. 653, mistakes May 1, 1786, as the date of this meeting.

77. Moreau describes the new quarters, DPF, p. 349; see also AA, PauP, 1786, no. 27 (July 8). The Cercle first took rooms on or near the rue du Conseil. In early 1785 the authorities refused the Cercle permission to meet in the main administration building, le Gouvernement, and Arthaud much regretted the burden of rent thus imposed, and rightly so since many a provincial academy in France met for free at the local Hôtel de Ville.

78. Donations eventually created a small library of perhaps a few hundred volumes; Maurel (1961), pp. 241, 246; DPF, pp. 653, 1409; CA-HdC, fols. 153, 154; Cercle des Philadelphes (1785a), articles xxv, xxix.

79. With an official guardian, the Cercle's cabinet came to include various natural history specimens, mineralogical samples, and ethnographic artifacts; see DPF, pp. 149, 234, 349; AA, Cap, 1786, no. 7 (February 15); Cercle des Philadelphes (1785a), article xxiv; Pluchon (1985a), pp. 169, 178.

80. On this meeting, see CA-HdC, fol. 161; AA, PauP, 1786, no. 27 (July 8); letter from de Marbois "du 11 Juin 1786"; CCP, letter 16, fol. 189v.

81. "Education des nègres, tant physique que morale." See the original ms. of Baudry's éloge, BN, Mss., n.a.f. 22086, fol. 137bis, 1–11.

82. Arthaud (1786a); see also CA-HdC, fol. 160; Maurel in DPF, p. 1513.

83. CA-HdC, fol. 161v; AA, PauP, 1787, no. 27 (July 8); Cercle des Philadelphes (1786b).

84. The Cercle received a curt "thank you" in reply; letter "du 29 Juin 1786"; CCP, letter 17, fol. 189v. The Cercle made no mention of responses to the Neufchateau paper prize contest at this point.

85. See various lists in AA, PauP, 1785, no. 22 (May 28); AA, Cap, 1786, nos. 5 (February 1), 6 (February 8), 7 (February 15), 8 (February 22), 9 (March 1), 28 (July 22).

86. Figures reported according to the contemporary categories grouping papers in the newspaper; see AA, Cap, 1786, no. 5 (February 1).

87. Letter, Arthaud to Franklin, APS Mss. collections. I am grateful to Mme

Claude Anne Lopez of the Papers of Benjamin Franklin for orienting me to the Franklin–Arthaud correspondence. Arthaud sent Franklin his *Description* of Cap François, his discourse from the Cercle's public meeting of 1785, and probably the Cercle's prospectus, statutes, and first prize program.

88. Letter, Franklin to Arthaud, dated "Philadelphia, July 9, 1786"; Library of Congress, Manuscript Division. A contemporary French translation of this letter exists in CCP, as letter 18, fol. 190.

89. On these points, see Lingelbach; McClellan (1985), pp. 142, 327. Commented Moreau: "Il est permis d'espérer que le suffrage du philosophe américain pour la première société littéraire formée dans une colonie française est un augure qui s'accomplira"; *DPF*, p. 348.

90. See Arthaud letters to Franklin, March 15, 1786; August 25, 1786; October 13, 1787; APS, Franklin letters; letters of July 28, 1788; November 21, 1788; January 29, 1789; July 27, 1789; APS Archives. Note references to the Cercle des Philadelphes in APS, "Early Proceedings," pp. 146 (November 3, 1786), 163 (September 19, 1788), 168 (January 16, 1789), 171 (March 20, 1789). As a result of this "exchange," the American Philosophical Society preserves a rich store of material related to the Cercle des Philadelphes.

91. See letter, Franklin to Arthaud, dated Philadelphia, December 11, 1787, Library of Congress, Manuscript Division; APS, "Early Proceedings," p. 168 (January 16, 1789); APS ms. minutes, entry for January 16, 1795; Taillemite in *DPF*, p. xxix.

92. CA-HdC, fols. 161v–162. Mozard announced Franklin as an honorary associate of the Cercle later in October of 1786; *AA*, PauP, 1786, no. 40 (October 7).

93. See letter "du 27 Juillet 1786"; CCP, letter 19, fol. 190v; Cercle des Philadelphes (1786a), p. 14n; Dazille (1788), p. 425; N. L. Bourgeois, p. 411. Tetanus is an acute infectious disease characterized by spasm of voluntary muscles, especially of the jaw, caused by a specific toxin of the bacillus, *Clostridium tetani*.

94. See Cercle des Philadelphes (1786a), passim and p. 104; CA-HdC, fol. 162. The Cercle printed the volume at its own expense.

95. See previous discussion, and CA-HdC, fols. 162–162v; letter from Coustard and Barbé de Marbois, "du 6 Avril 1786"; CCP, letter 22, fols. 191–191v.

96. Administrators' letter, "du 20 Avril 1786"; CCP, letter 23, fol. 191v.

97. *DPF*, p. 347; see also Maurel in *DPF*, p. 1355.

98. The Cercle's new garden was located along the Haut du Cap River, near the town abattoir and the cours Villeverd. Arthaud characterized it as a swamp flooded by strong tides; Moreau said it turned into a mire at the least rain; see CA-HdC, fol. 162v; *DPF*, p. 548. A horrible *bidonville* presently occupies the site of the Cercle's concession.

99. On this meeting, see CA-HdC, fols. 163–163v; *AA*, PauP, 1786, no. 57 (December 21); Arthaud "Discours prononcé . . . Le 11 X 1786"; AN, Colonies F³ 81.

100. The Cercle commissioned the bust locally for 100 pistoles donated by M. de Larche; see Figure 18 and descriptions in Maurel (1961), pp. 250–51; Pluchon (1985a), p. 170; CA-HdC, fols. 162v–163; AN, Colonies F³ 152, fol. 266.

101. CA-HdC, fol. 163.

102. CA-HdC, fol. 163v.

103. See "Discours prononcé . . . Le 11 X 1786"; AN, Colonies F³ 81, fols. 119–26v; and related remarks in AN, Colonies F³ 152, fols. 234–39, and CA-HdC, fol. 163v. Arthaud admitted his earlier ideas about public education were mistaken, but declared a colonial academy of science a separate issue.

104. See Arthaud, "Discours," AN, Colonies F³ 81, fols. 124v–125. Presented publicly before the colony's administrators, Arthaud's sense here of a "nouvel ordre de choses" based on science and an improved colonial economy is a far cry from the Masonic spin given to these words by others.

105. Letter, "Mrs. de la Luzerne et du Marbois, Le 29 X^bre 1786," AN, C⁹ᴮ 36 (2).

106. See letter to the Cercle from de Marbois and de la Luzerne, "du 5 avril 1787," CCP, letter 24, fols. 192–192v. Note Arthaud's comment apropos of provisional royal recognition: "Le Cercle est arrivé à une epoque Satisfaisante et glorieuse"; CA-HdC, fol. 167 and further 167v.

CHAPTER 13: On to Letters Patent

1. See CA-HdC, fol. 168v; Cercle des Philadelphes (1788e), p. 56; AA, PauP Supplement, 1778, no. 37 (September 19).

2. See Cercle des Philadelphes (1788e), pp. 240–41, 87, 123–35, 240–41; also, CA-HdC, fols. 167v–168, 192v, 194. For other reports of glanders inoculations, see AA, PauP, 1787, no. 68 (August 25); 1788, no. 30 (April 12).

3. Cercle des Philadelphes (1788e); see also Pluchon (1987), p. 254.

4. AA, PauP, 1787, no. 74 (September 13). Note Moreau's praise, DPF, p. 292.

5. So says Arthaud, AN, Colonies F³ 152, fol. 274v.

6. CA-HdC, fol. 169v.

7. On this meeting see CA-HdC, fols. 170–71; AA, PauP, 1787, no. 68 (August 25). The Cercle had been in contact with the abbé Jacques-François Dicquemare or Dicquemar (1733–89) at least since 1786, when Lefebvre-Deshayes sent Dicquemare a colonial coffee brandy that won the praise of the Rouen academy; see AA, PauP, 1786, no. 47 (November 16).

8. The Cercle also commissioned a bust of Deshayes and later, in 1789, added a bust of Thierry de Menonville, for a total of four in its collection, including the king's and Dicquemare's. About the Menonville bust Mozard commented, "C'est une nouvelle preuve que le Cercle cherche par tous les moyens d'honorer ceux qui cultivent ici les sciences utiles"; see AA, PauP, 1789, no. 10 (January 31); DPF, pp. 165, 1409.

9. CA-HdC, fol. 170v; AA, PauP, 1787, no. 68 (August 25). The Cercle made Baumès a national associate and sent his winning memoir to France to be printed. Dufresnoy, a royal physician in Valenciennes in France, submitted his memoir late, but, impressed, the Cercle awarded Dufresnoy a silver medal and elected him a national associate. Dufresnoy's memoir dealt with the narcissus plant as a specific in treating tetanus, and in a civic-minded gesture of the highest order, the Cercle then made narcissus available gratis to the public through Ducatel's pharmacy in Cap François; see AA, 1787, Cap, no. 43 (October 27); CA-HdC, fol. 171.

10. Cercle des Philadelphes (1987b); *Almanach de Saint-Domingue,* 1789, 1790, 1791; CCP, letter 26, fols. 192v–193; Pluchon (1985a), appendices.

11. Notably including Siméon Worlock, plantation manager, smallpox inoculator, and *correspondant* of the Paris Royal Society of Medicine. On his election, Worlock funded a prize contest on using lye in sugar processing; see CA-HdC, fol. 169; Maurel in *DPF,* p. 1559.

12. On this singular correspondence network, see McClellan (1985), pp. 185–86.

13. Mozard's words, *AA,* PauP, 1787, no. 68 (August 25); on Chabanon, see also Maurel in *DPF,* p. 1464, and *DPF,* p. 222.

14. A few eighteenth-century scientific societies elected a few accomplished women to their ranks, including Mme du Châtelet, who became an associate of the Bolognese Academy; on a lesser scale, Mlle le Masson le Golft played a similar role in the Cercle. See Ehrman p. 39; and McClellan (1985), p. 297n.

15. Konvitz, pp. 125–26, and references, *AA,* PauP, 1784, no. 51 (December 18); *AA,* Cap, 1786, no. 28 (22 July); *AA,* Cap, 1786, no. 7 (February 15).

16. *AA,* PauP Supplement, 1786, no. 1 (January 7).

17. Lefebvre-Deshayes, p. 379n; *AA,* PauP, 1786, no. 27 (July 8); Maurel (1961), p. 246, where she mistakes Mlle le Golft as the author of the ode.

18. CA-HdC, fol. 164.

19. CA-HdC, fol. 173v. On A.-H. Tessier (1741–1837) as a reformer in French agriculture, see Gillispie (1980), pp. 335–36; Lacroix, 4:74.

20. See Cercle des Phildelphes (1787a); Hamy, p. 298. Anglade (1973b) presents an extended analysis of the Cercle's agricultural survey; Anglade (1973a), a work that is itself rare, reprints these *Questions,* pp. 30–40; see originals AN, Colonies F³ 127; AN, SdOM, BMSM, vol. 147, at fol. 271.

21. CA-HdC, fol. 173v; *AA,* PauP, 1787, no. 83 (October 18); Anglade (1973b), p. 131.

22. La Luzerne to Arthaud, "du 30 7bre 1787"; CCP, letter 28, fol. 193v; see also CCP, letter 29, fols. 193v–194; and CA-HdC, fol. 173v.

23. Anglade (1973a), pp. 30–36; Anglade (1973b), pp. 133–36.

24. *AA,* Cap, 1789, no. 72 (October 17). Typical titles: "Fermentation of Indigo," "Fabrication of Sugar," "Irrigation of the Artibonite."

25. Anglade (1973b), pp. 132–33; Taillemite in *DPF,* p. xvi; and agricultural memoirs preserved in Moreau's papers in the Archives Nationales via May, e.g., Colonies F³ 87.

26. Without realizing it, Debien (1981) reprints one such response.

27. CA-HdC, fol. 164, part of a comparatively lengthy presentation of d'Ingrandes and his work; see also *DPF,* pp. 1201–2; Maurel in *DPF,* p. 1500.

28. CA-HdC, fols. 164–164v.

29. CA-HdC, fols. 181–181v; see also reference to Arthaud's own "Les moyens de conserver les nègres" in *AA,* Cap Supplement, 1786, no. 6 (February 8).

30. Letter dated "au Cap. 30 mai 1786"; BN, Mss., n.a.f. 22086, fol. 412; and CA-HdC, fol. 161. Baudry des Lozières wrote an *Égaremens du négrophilisme* (1802), and the marquis d'Aussigné, a colonial associate, likewise compared slaves and peasants; see Cercle des Philadelphes (1788c).

31. Stein (1979), pp. 196, 238; Pluchon (1987), p. 168; Girod (1970), p. 101.

For related attitudes toward Amerindians, see Crouse (1977), pp. 218–19.

32. Arthaud (1789b), pp. 274–75.

33. Broc (esp. pp. 219–29, 445–59) and W. B. Cohen (esp. chaps. 3 and 8) provide the most recent authoritative accounts; see also enlightening discussions of these issues in Miller, chap. 7, and Greene, chaps. 6–8; and related paper by Palisot de Beauvois, "Les Races de l'homme suivant les auteurs," APS microfilm 1352.

34. See CA-HdC, fols. 173v, 180; La Luzerne to Arthaud, "du 20 7bre 1787"; CCP, letter 27, fol. 193; and ministry memo dated "16 Janvier 1788," AN, C^{9B} 38.

35. "Versailles, le 17 Janv. 1788"; AN, C^{9B} 38; copy in CCP, fol. 194, letter 31; see also CA-HdC, fols. 180, 181v; Pluchon (1985a), p. 170. Still, La Luzerne warned the Cercle not to address him directly, but to work through the colonial administrators, a warning repeated from "Versailles, 6 Mars 1788"; see copies, AN, C^{9B} 38; CCP, letter 33, fol. 194v.

36. See CA-HdC, fols. 174, 181v; AN, Colonies F^3 152, fol. 280; Marbois memo, "le 28 janv. 1789"; AN, C^{9A} 163, Intendance, no. 477; ministry note, "20 Mars 1788"; letter, "Versailles, 27 mars 1788"; AN, C^{9B} 38; copy, CCP, letter 35, fol. 195; Maurel (1961), p. 244; Pluchon (1985a), p. 170.

37. See Figure 18; CA-HdC, fols. 180–181; DPF, p. 348; Maurel (1961), p. 243. A mint silver *jeton* exists in the Cabinet des Medailles of the Bibliothèque Nationale; I thank M. Giard of the Cabinet des Medailles for helping me find the Cercle's medal in its collections. The legend, "Simon, G[raveur]. du Roi," appears chiseled under Louis's shoulder. Zay, p. 234, illustrates this coin without identifying it. French law forbade minting in the colonies.

38. Taillemite in DPF, p. xiv; Maurel (1961), p. 243. Gébelin's was not the Musée at the Palais Royal with which the Cercle was connected thru Moreau.

39. See CA-HdC, fol. 180; and administrators' letter, "du 20 Avril 1788"; CCP, letter 34, fol. 195. The Cercle struck some gold *jetons*, and it owned a seal with the same beehive emblem; see AA, PauP, 1786, no. 27 (July 8); Maurel (1961), p. 257; DPF, p. 348; for the seal, see Baudry des Lozières papers, "Séance du 26 fevrier 1787," in APS, Mss. collecitons.

40. See Cercle to "Monseigneur," "17 Avril 1788," and administrators to La Luzerne, "25 Juillet 1788," AN, C^{9B} 38. Having been warned again, the Cercle also wrote to the local administrators, who similarly reported to La Luzerne, with the Cercle's promise to stay within bureaucratic bounds.

41. Letter to Cercle, dated "Versailles, 2 8bre 1788"; AN, C^{9B} 38.

42. Letters to Marbois, "Versailles, 2 8bre 1788"; and from Marbois, "Port-au-Prince, le 28 fev. 1789," AN, C^{9B} 38.

43. Gauché in Cercle des Philadelphes (1788b), pp. 73, 183; CA-HdC, fol. 174.

44. See Cercle des Philadelphes (1788b), passim and p. 167. The fourteen other authors included Benjamin Rush, whose paper on the Harrogate baths outside of Philadelphia was translated.

45. AA, PauP, 1789, no. 65 (August 12).

46. CA-HdC, fols. 174v–175.

47. See relevant procès-verbaux, AN, C^{9B} 40; memoir, AN, Colonies F^3 152,

fols. 66–75; *AA,* PauP, 1788, no. 30 (April 12), and CA-HdC, fol. 175. (Mozard's suggestion: soak sacks of flour in seawater.)

48. See Cercle des Philadelphes (1788d); *AA,* PauP, 1789, no. 20 (March 7); Cercle des Philadelphes (1789a), p. 4.

49. The Cercle received some thirty clove trees; see CA-HdC, fols. 176–176v, where Arthaud gushes over colonial spice culture; and see chapter 9 in the present work.

50. See report, *AA,* Cap. 1788, no. 22 (May 31).

51. CA-HdC, fol. 180; Fouchard (1955b), p. 58; *DPF,* pp. 270–71; Maurel in *DPF,* p. 1473. Palisot de Beauvois wrote of de la Haye to A.-L. de Jussieu: "He is the only man in the whole colony from whom one can get both exact information and decent specimens. He is currently engaged in a botanical work which I think will please you"; letter dated "Du Cap le 15 X^bre 1788"; Adrien de Jussieu papers, APS microfilm, 1350.

52. See CA-HdC, fols. 178–80, and prospectus in BN, SdOM, BMSM, 147 at fol. 275; nineteen individuals took twenty-nine subscriptions, including La Luzerne who signed for ten; the resulting 3,500 livres ought to have been sufficient to publish the edition; see *AA,* Cap, 1789, no. 38, (June 20); and La Luzerne letter, dated "Versailles, 19 mars 1789"; AN, Colonies C^9B 40.

53. See Palisot letter, "Du Cap le 9 8bre 1789," APS microfilm, no. 1350. Debien (1937), no. 98, notes de la Haye's ms. of 202 pages and 206 plates, many in color, in the "Bibliothèque du Ministère de la Marine." Duvivier, 1:136, erroneously lists de la Haye's *Florindie* as published in 1788.

54. *AA,* PauP, 1788, no. 74 (September 13); and printed "Notice sur la séance publique du Cercle des Philadelphes, tenue le 15 Août 1788," Cercle des Philadelphes (1788c): AN, Colonies F^3 81; AN, SdOM, BMSM, 147 at fol. 272.

55. So suggests Castonnet des Fosses (1884), p. 30, in a passage that is otherwise pure fantasy: "Le 15 août de chaque année, il y avait une séance générale [du Cercle des Philadelphes] et, pour toute la ville, c'était une fête que les dames rehaussaient de leur présence. Les nègres tiraient des pétards sous les fenêtres de la salle et les cris joyeux montraient qu'ils n'étaient pas indifférents à ces tournois d'un nouveau genre."

56. Cercle des Philadelphes (1788d).

57. On Barré, see Thésée and Debien, here pp. 357–67; Maurel in *DPF,* p. 1447.

58. Letter dated "au qtier Morin le 25 mars 1788," BN, Mss., n.a.f. 20277 (I), fols. 15–16. Barré left for France also to represent the *doléances* of the Chamber of Agriculture and planter interests before the upcoming Estates General; see Thésée and Debien, p. 395; Pluchon (1985a), p. 170; Maurel (1961), p. 250; CA-HdC, fols. 176v–177.

59. See letters dated "17 Avril 1788" and "Au Cap français le 12 9bre 1788," AN, Colonies C^9B 38; "7 aoust 1788," AN, Colonies C^9A 162.

60. M. de Vaivre, colonial superintendent within the naval ministry, agitated to secure letters patent for the Cercle, and he may also have been at the meeting with La Luzerne; Thésée and Debien mention intervention by Condorcet as well; see *DPF,* pp. 347–48; Thésée and Debien, pp. 389–90, 398. Pluchon (1985a), pp. 170–71; Maurel in *DPF,* p. 1482; Maurel (1961), p. 252.

61. See Arthaud, "Discours . . . pour les status du Cercle," AN, Colonies F³ 152, fols. 188–89; ms. statutes dated "22 Sept. 1788" and administrators' letter dated "Port-au-Prince le 1 9ᵇʳᵉ 1788," both in AN, C⁹ᴬ 162; undated letter signed by thirteen associates of the Cercle, AN, Colonies C⁹ᴮ 38.

62. "Leur donner des lettres patentes"; see memorandum dated "11 Xᵇʳᵉ 1788," and related items, AN, Colonies C⁹ᴬ 162.

63. Pluchon (1985a), pp. 170–71. These two drafts of the letters patent and related materials are preserved, AN, Colonies C⁹ᴬ 162.

64. See memorandum dated "17 Avril 1789," noting the Cercle's "Demandes," the administrators' recommendations, and the minister's responses; AN, Colonies C⁹ᴬ 162. In the meantime, the Cercle again asked La Luzerne to expedite the letters patent. La Luzerne replied cordially, but without indicating that letters patent were on their way; see three items related to this overture, all dated "Versailles, le 19 Mars 1789"; AN, Colonies C⁹ᴮ 40.

65. See unsigned ministry memorandum dated "16 Mai 1789," AN, C⁹ᴬ 162; DPF, p. 347; Pluchon (1985a), p. 171. Contemporary and modern sources give as the offical title both "Société royale des sciences et des arts du Cap François" and "Société royale des sciences et arts du Cap François."

66. See Pluchon (1985a), p. 171; Maurel (1961), pp. 252–53; see also relevant documents, AN, Colonies C⁹ᴬ 162.

67. AN, SdOM, BMSM, 143, fol. 281; also AA, PauP, 1789, no. 75 (September 16); and Maurel (1961), p. 252. Moreau cataloged an eleven-page edition of the letters patent issued at Cap François; Duvivier, 1:179, lists a thirteen-page version published in Port-au-Prince.

68. See Cercle's letter dated "1 7ᵇʳᵉ 1789," administrators' letter to La Luzerne dated, "Port-au-Prince, le 7 8ᵇʳᵉ 1789," and La Luzerne letter to administrators dated "9 Janvier 1790," AN, Colonies C⁹ᴬ 162; see also Pluchon (1985a), pp. 171, 178; and the Almanach de Saint-Domingue.

69. Barré de Saint-Venant, p. 323, seems to take sole responsibility for the contact with the Paris Academy, and Moreau seconds Barré's claim, in DPF, p. 348. Barré had been in contact with Paris Academy of Sciences previously in the 1760s; see Thésée and Debien, pp. 362–63, and PA-PV, vol. 88 (1769), pp. 281–85; Pochettes de séances for July 23 and 29, 1769.

70. See Thésée and Debien, p. 398, and Maurel (1961), pp. 252, 262, neither entirely accurate accounts of this meeting. See also DPF, p. 348, and minutes of this meeting, PA-PV, vol. 108 (1789), p. 54.

71. "Discours prononcé à l'Académie royale des Sciences, le mercredy 25 Fev. 1789 par Mr. Barré de Sᵗᵉ Venant Président du Cercle des Philadelphes du Cap François"; PA, Pochette de séance, February 25, 1789.

72. See Barré, "Discours," PA, Pochette de séance, February 25, 1789, p. 9. Barré's portrait of Saint Domingue is utterly fantastic: substantial housing not required (the colony never got cold), the heat not excessive and always moderated by mild breezes, and clothing practically superfluous. No wild animals or venomous creatures disturbed this garden of Eden. Even the bread tasted better than in France! Barré went on to note that "the wholesomeness of Saint Domingue's climate has been proved by experiments the Cercle performed on the quality of atmospheric air."

73. Barré, "Discours," PA, Pochette de séance, February 25, 1789, p. 11; and Geggus (1989), pp. 1306–7, on the changed political climate. Regarding Barré's strikingly political call for free trade, with the government poised to grant letters patent to the Cercle, one can question whether Barré's personal politics was entirely congruent with institutional realities and the contemporary politics of science in old Saint Domingue.

74. Barré, "Discours," PA, Pochette de séance, February 25, 1789, pp. 5–7; on the barometric phenomenon, see also *DPF*, pp. 288–89.

75. This tie was a formal one, written into the statutes of the institutions from 1706. The special quality of being "one and the same body" allowed members of the two academies reciprocal rights and privileges not available to other societies with which the Paris Academy maintained relations. See McClellan (1985), pp. 96, 366; Hahn, p. 105.

76. PA-PV, vol. 108 (1789), p. 54. Loménie de Brienne, the archbishop of Sens, president of the Academy in 1789, had fallen as chief minister by the time of the meeting, and probably did not chair the meeting.

77. See original, signed committee report, PA, Pochette de séance for March 28, 1789; see also transcription into the Academy's minutes, PA-PV, vol. 108 (1789), pp. 87–92. Condorcet's precise role in this affair remains unclear; Thésée and Debien, p. 399; credit Condorcet generally with an active role in promoting the Cercle des Philadelphes at this time.

78. Committee report, PA-PV, vol. 108 (1789), pp. 87–88, 89. See also annotated "Statuts du Cercle des Philadelphes" with Moreau's endorsement, PA, Pochette de séance, February 25, 1789.

79. Report in PA-PV, vol. 108 (1789), p. 91.

80. See undated "Supplément au rapport précédent"; PA-PV, vol. 108 (1789), pp. 91–92; this supplement also proposed that the Academy exchange publications with the Cercle and send its own *Mémoires*, the *Savants Étrangers* series, and the *Connaissance des temps*.

81. PA-PV, vol. 108 (1789), pp. 92–93.

82. See report of this meeting, *AA*, PauP, 1789, no. 20 (March 7); and Arthaud, "Discours pour La Séance . . . du 21 février 1789"; AN, Colonies F³ 81, fols. 134–36, and copy, Colonies F³ 152, fol. 224v.

83. Arthaud quoted in *AA*, PauP, 1789, no. 20 (March 7).

84. See report, "Séance publique du Cercle des Philadelphes du Cap," in Cercle des Philadelphes (1789a), pp. 3–4. A minor note marred preparations for the August celebration. The Cercle had taken it upon itself to award a medal to the ex-slave, Jean Jasmin, for funding the hospice for free people of color in Cap François. The impulse was generous and liberal but contrary to the realities of racial division in Saint Domingue, and the local administrators rejected the idea of any award for Jasmin. Moreau tells the story in some detail, *DPF*, p. 414.

85. AN, Colonies C⁹ᴬ 162.

86. Compare Cercle des Philadelphes (1789a), AN, Colonies F³ 81, with Cercle des Philadelphes (1789b), AN, SdOM, BMSM, 147, pp. 276–80.

87. *DPF*, p. 349.

CHAPTER 14: Profile of an Institution

1. Pluchon (1985a), p. 185.

2. Three 1785 membership lists are described in chapter 12; see also Cercle des Philadelphes (1787b); Pluchon (1985a), appendices; Maurel (1961), pp. 261–66. Pluchon reprints several of these lists, including those in the *Almanach de Saint-Domingue* for 1790 and 1791. The *Almanach* for 1789 (BN, Lc^{32} 18) gives another list of Cercle associates.

3. The author compiled an electronic database of information contained in the sources just described. Maurel's annotated index of names in *DPF,* pp. 1441–1559, likewise proved a valuable source.

4. Compare Pluchon's estimate of 120–130 associates; Pluchon (1985a), p. 172. Figures are reliable through 1790, i.e., counting names on the *Almanach* list for 1791. Not counted are six unaffiliated individuals whose names appear on Maurel (1961) but nowhere else. On the lone female member, Mlle le Masson le Golft, see previous discussion.

5. On these comparative statistics, see Roche (1978), table 18.

6. Twelve came from Port-au-Prince, forty-four from elsewhere in Saint Domingue, including five from the Artibonite Valley. The Cercle elected twelve additional colonial associates from Guadeloupe, Martinique, and Saint Lucia.

7. Data for Graph 3 derive from the computer database mentioned in n. 3.

8. See McClellan (1981), pp. 556–58; Roche (1978), table 18.

9. Maurel (1961), p. 253 (note her punctuation); see also pp. 246–48.

10. Pluchon (1985a), pp. 172, 185; on *grands blancs,* see chapter 3 in the present work.

11. Graph 4 derives from the database mentioned in n. 3. Percentages sum to more than 100 percent with members counted in more than one category where appropriate.

12. Maurel (1961), pp. 246–47; Pluchon (1985a), p. 172; (1985b), p. 92.

13. Tarrade, p. 145; see also Geggus (1989), pp. 1301, 1307; Stein (1979), p. 152, provides a nuanced definition of contemporary *négociants*.

14. Pluchon (1985a), p. 185.

15. Pluchon and Maurel both spotlight Cercle associates in the legal profession; see Maurel (1961), p. 248; Pluchon (1985a), p. 185.

16. So notes Maurel (1961), p. 248; see also McClellan (1985), pp. 247–50.

17. Not to be confused with better-known Swiss relatives, this Jean Trembley wended his way from Switzerland to Saint Domingue in the 1750s, where he ultimately made good; see Debien (1955), p. 14; Debien (1956), pp. 47–50.

18. Debien (1955) reprints this and several other letters of Trembley to Bonnet; see here, pp. 8–9. Surprisingly, Bonnet's name does not show up on any of the membership lists of the Cercle des Philadelphes/Société Royale.

19. Letter dated "Du Cap le 15 X^{bre} 1788," PA, Collection Adrien de Jussieu— Don de M. Ramond-Gontaud; APS microfilm, no. 1350.

20. Surpisingly, the Cercle did not elect viscount P.-M.-F. de Pagès, captain in the royal navy, author of his world travels, and *correspondant* of the Paris Academy of Sciences. De Pagès resided on the colony's southern arm not far from Lefebvre-Deshayes, and by all lights should have joined the society in Cap François; see *DPF,* p. 1226; Institut de France, pp. 180–81.

21. Debien (1937), nos. 127, 128; *AA*, PauP, 1788, no. 38 (May 10).

22. The Bordeaux Academy followed up on the Tessier inquiry; see *AA*, PauP, 1789, no. 43 (May 27); BN, SdOM, BMSM, 37; on Barré's bridge, see Thésée and Debien, pp. 399–400; *DPF*, p. 221; and related documents, AN, Colonies C⁹ᴮ 40.

23. Pluchon (1985a), p. 180; PA-PV, vol. 108 (1789), p. 218.

24. By an index of common members, the Cercle affected close, albeit informal, connections to the Royal Society of Medicine. Seven members of the Cercle, including Arthaud and Worlock, became *correspondants* of the Paris medical society. Vicq d'Azyr, permanent secretary of the Royal Society of Medicine, pronounced the *éloges* of three Philadelphes: Lefebvre-Deshayes, Joubert de la Motte, and Cosme d'Angerville; see AN, Colonies F³ 267, fol. 176.

25. See *DPF*, p. 348; Maurel (1961), p. 241; *AA*, PauP Supplement, 1786, no. 21 (May 27); *AA*, PauP, 1787, no. 68 (August 25); also Stein (1979), p. 188.

26. See *AA*, Cap Supplement, 1786, nos. 5 (February 1), 8 (February 22), 21 (May 27); *AA*, PauP, 1789, no. 20 (March 7); *AA*, Cap, 1789, no. 3 (January 17); Pluchon (1985a), p. 159; Maurel (1961), p. 241; regarding the Dijon Academy, see also *AA*, PauP, 1786, no. 53 (December 7).

27. The Cercle and the Marine Academy at Brest also shared indirect, "structural" ties through common members at the Paris Academy of Sciences. Arthaud (1792), p. 3, implies his election to the Royal Society in Montpellier; on possible connections to the Orleans Academy, see Debien (1937), no. 12.

28. At least forty-five individuals (28 percent of the Cercle's membership) possessed at least one additional learned-society affiliation. One national associate of the Cercle, another Gauché, was not a Fellow of the Royal Society of London, per a contemporary claim. The Royal Society of London did recieve a mule skin from Saint Domingue, and one report from the *Philosophical Transactions* filtered into the colony; see Pluchon (1985a), appendices; Maurel (1961), p. 261; *DPF*, p. 291; *AA*, PauP, 1789, no. 18 (February 28). On the role of common members, see McClellan (1985), passim and pp. 180–82, 340 n.144.

29. Arthaud, "Compliment . . . du 23 avril 1787"; AN, Colonies F³ 152, fol. 217.

30. Nicolson, p. 365.

31. Arthaud (1786b) as reprinted in Hamy, p. 307; see also *DPF*, p. 349, where Moreau laments "those unfortunate whose race would still exist if the original conquerors enjoyed the peaceable inclination to study and the soul of the Philadelphes."

32. See Arthaud (1789a), who explains head deformities based on climatological determinism; see also the timid reply by Amic; and Hamy, p. 301.

33. See Arthaud 1790 ms., "Dissertation sur les phallus des naturels du païs," in Hamy, pp. 310–14, here p. 311; and Arthaud (1789a), p. 254, quoting La Condamine on the nobility of the American savage; see also Knight, p. 12.

34. Arthaud (1789a), pp. 250, 251; see also CA-HdC, fol. 177.

35. According to Hamy, p. 299; see also *AA*, Cap, 1786, no. 28 (July 22).

36. CA-HdC, fol. 177; Hamy, pp. 299, 305; *AA*, Cap, 1786, no. 7 (February 15).

37. See *AA*, Cap, 1786, no. 28 (July 22), announcing the Cercle's proposed

"Tableau minéralogique de la Colonie"; see also *AA*, PauP, 1786, no. 57 (December 21); Cercle des Philadelphes (1785a), article xxx; (1789a and b); Gauché in Cercle des Philadelphes (1788b), pp. 178–79.

38. Gauché in Cercle des Philadelphes (1788b), p. 182.

39. See Genton, p. 176; the first steam engine came to Saint Domingue in 1786; Hispaniola would seem to lack significant amounts of coal or copper.

40. CA-HdC, fols. 177–78; see also *DPF*, p. 111.

41. At least seven colonial articles appeared in Rozier's Journal, written by Arthaud (1787, 1789a, 1789b, 1790), Amic, Lefebvre-Deshayes, and Genton; also Hamy, pp. 298–99. On Rozier's Journal, see McClellan (1979) and (1985), pp. 191–93.

42. *AA*, PauP, 1788, no. 4 (January 12).

43. Lefebvre-Deshayes, pp. 374, 377.

44. Arthaud (1786a), p. 11.

45. "Le Cercle . . . présente un bilan négatif"; Pluchon (1985a), p. 174; see also p. 175, where Pluchon speaks of the Cercle's "faillite."

CHAPTER 15: The Fall of the Société Royale of Cap François

1. Moreau explicitly closes his account of Saint Domingue as of October 18, 1789, the day he says revolution broke out in Saint Domingue; see original title page of the *Description de la Partie Française de l'Isle de Saint-Domingue*, reproduced in *DPF*, p. 1, and Moreau in *DPF*, pp. 7, 9, 10.

2. The standard sources in English are by Ott, James (1963), and Geggus (1982). See also Dorsainvil; Cornevin; Frostin (1975), chap. 7; Lokke, pp. 119–23; Cauna (1987), chaps. 7–8.

3. Maurel (1961), pp. 253, 256–57; Pluchon (1985a), pp. 173–74. On the rumors that inflamed the colony, see further, Ott, pp. 37, 43.

4. Palisot letter to A.-L. de Jussieu, "Du Cap, le 23 X^{bre} 1789," APS Archives, microfilm, no. 135; Palisot's emphasis. See other reports in Arthaud to Moreau, "19 8bre 1789," AN, Colonies F^3 194; Governor-General, De Peinier, to La Luzerne, "Port au Prince le 24 8bre 1789," AN, Colonies C^{9B} 40, item 57. Also, Maurel (1961), pp. 256–57; Taillemite in *DPF*, p. xxi; Pluchon (1985a), pp. 173–74; Fouchard (1955b), pp. 60–61.

5. "Discours prononcé dans l'assemblée provinciale de la partie du nord, le 4 Décembre 1789," AN, Colonies F^3 194; see also printed "Extrait des Registres de l'Assemblée provinciale" AN, Colonies C^{9B} 39; and Arthaud's (presumably later) *Réfutation* (s.d.), in which he defends himself as a "citoyen français" against the calumnies of his enemies by citing Rousseau!

6. Both Maurel and Pluchon say that Arthaud was banished from the Royal Society for up to two years, during which time he supposedly went to France. But Arthaud remained on the published lists of members, and he definitely reappeared as secretary of the Royal Society by August 1790. More probably, Arthaud was restored to his positions in the fall of 1789 when he made clear his support for slavery and *grand blanc* policies. See Maurel (1961), pp. 257–58; Pluchon (1985a), p. 174. See also Arthaud ms., AN, Colonies F^3 152, fols. 96–115; Cercle des Philadelphes (1790).

7. Pluchon (1985a), p. 174.

8. Maurel (1961), p. 257, quotes Auvray's "Discours"; see original, AN, Colonies F³ 194. On the provincial assemblies, see Ott, p. 33.

9. Bussan, quoted in Maurel (1961), p. 258; Bussan was seneschal at the Cap.

10. Maurel (1961), pp. 257–58.

11. On the political situation during this period, see Ott, pp. 31–35; James (1963), pp. 66–72; Lokke, pp. 123–25, 130; Dorsainvil, pp. 55–56.

12. See *DPF*, p. 401, and Moreau de Saint-Méry (1790), p. 29; Arthaud ms. dated "le 29 juillet 1790," AN, Colonies F³ 152, fols. 96–115.

13. See eight-page Cercle des Philadelphes (1790), AN, SdOM, Recueil Colonies, 37, BMSM 39 (no. 26). This curious document resembles other printed reports of Society meetings, except for the grammatical past conditional replacing the conventional past tense in the text.

14. See Cercle des Philadelphes (1790), pp. 7–8. It seems inconceivable that the Cercle did not elect Bonnet, but his name does not appear on membership lists for 1790 or 1791, which it ought if he joined in 1789 or 1790.

15. On the Ogé revolt and the "mulatto question," see Ott, pp. 36–37; James (1963), pp. 73–75; Dorsainvil, pp. 60–62; Geggus (1989), p. 1302; Taillemite in *DPF*, p. xxiv; Lokke, pp. 133–39.

16. Arthaud (1791), p. 88n establishes the date of this meeting and Blanchelande's presence; on Blanchelande, see Ott, p. 37; Dorsainvil, p. 57.

17. Arthaud (1791), 114-page *Observations,* according to the full title, "Adressées au Comité de Salubrité de l'Assemblée Nationale." On the debates in France, see detailed discussion in Hahn, chap. 7.

18. Arthaud (1791), pp. 90–95. About the Society's garden Arthaud said, "We have been thwarted in our plans, but the administration of Saint Domingue will some day take this question into consideration."

19. Arthaud (1791), p. 97.

20. *Gazette de Saint-Domingue,* no. 55, July 9, 1791. See also *DPF,* pp. 401 and 401n, where Moreau, discussing these commemorative efforts, expresses outrage at the "ridiculous custom of using Latin to express French virtues."

21. See Thouin letter of this date, BN, Mss., n.a.f. 9545, fols. 22v–23.

22. See accounts in Ott, pp. 47–48; James (1963), pp. 85–88; Pluchon (1987), pp. 118, 301–4; Dorsainvil, pp. 65–67; Cornevin, pp. 42–43; Doucet, pp. 181–85; Davis, pp. 201–2. James and Ott say that the Boukman ceremony occurred on August 22, 1791, with the revolt breaking out immediately. Other authorities, followed here, date the Boukman ceremony to August 14–15, with the revolt following on August 22. See also Frostin (1975), chap. 7, on black and white revolts; and remarks by Hugo, p. 83.

23. See *AA,* Cap, 1791, nos. 62 (August 3) and 64 (August 10); Charles Mozard, longtime editor, quit Saint Domingue for France on August 3, 1791.

24. A four-page printed notice of this meeting, Cercle des Philadelphes (1791), appeared within a day or two; Moreau de Saint-Méry saved a copy, now preserved within the AN, SdOM, Recueil Colonies, 2ᵉ Série, 18; BMSM, 18.

25. Letter from Prieux to Arthaud, dated "Au Cap, 20 aout 1791," AN, Colonies F³ 152, fol. 248.

26. The Cercle printed this "Programme" as part of the notice of the meeting of August 16, 1791; see Cercle des Philadelphes (1791).

27. Cercle des Philadephes (1791), pp. 2–3.

28. On the slave revolt in Saint Domingue and related points touched on here, see Ott, pp. 48–49; James (1963); pp. 85–90; Pluchon (1987), pp. 126–132; Davis, pp. 202–4; Dorsainvil, pp. 67–69; Cornevin, p. 43.

29. See undated four-page ms. address in Arthaud's hand, "À l'assemblée Générale de la partie française de St. Domingue," AN, Colonies F³ 152, fols. 246–47.

30. On the First Civil Commission and circumstances in Saint Domingue, see Ott, pp. 55–59, 62 n.44; Dorsainvil, pp. 69–71; James (1963), p. 103; Stein (1979), p. 44. The French Constitution of 1791 pointedly did not apply to the colonies and left them to decide their own internal affairs.

31. Regarding this meeting, see Arthaud (1792), passim and esp. pp. 3, 14n; Maurel (1961), pp. 258–59. See also Debien (1943), p. 40.

32. As is clear from Geanty's letter to Arthaud dated "Le 30 Juin 1792"; AN, Colonies F³ 152, fols. 93–94. Arthaud's fate is not known, but Hamy is surely wrong to say that Arthaud died in 1791. If Arthaud remained in Paris, he may well have perished at the guillotine. Despite Maurel's claim, *DPF*, p. 1445, Arthaud was not a member of the Club Mass   see Debien (1953a). Arthaud's brother, Jean Artaud, escaped to Cuba; see Devèze correspondence, Girard papers, APS microfilm.

33. Letter from Geanty to Arthaud, "Le 30 Juin 1792"; AN, Colonies F³ 152, fols. 93–94. The annotation in Arthaud's hand: "recue le 1 9bre."

34. On Sonthronax and the Second Civil Commission, see Ott, pp. 65–68; Dorsainvil, pp. 71–73; James (1963), pp. 121–29; Lokke, pp. 139–40.

35. Ott, pp. 69–71; James (1963), pp. 126–28; Basket, pp. 87, 96.

36. Hahn, pp. 238–40; McClellan (1985), p. 253.

37. On the Third Civil Commission and efforts of the Directory to govern in Saint Domingue, see Dorsainvil, p. 89; Cornevin, pp. 48–49.

38. On Giroud, see Lacroix, 3 : 109–111; on Raimond, see Cook; Maurel in *DPF*, p. 1540; *DPF*, p. 1237; Ott, pp. 27, 30; Lokke, p. 137.

39. Giroud letter of June 28, 1796, quoted in Lacroix, 3 : 113.

40. On the tortuous developments in the period from 1793 to 1804, see Ott, chaps. 5–9; Doucet, pp. 203–11; James (1963), pp. 145–377; Geggus (1982); Cornevin, pp. 44–54; Dorsainvil, pp. 81–135; see also Knight, p. 157.

41. Léon (1952), p. 40; see also Duvivier, 1 : 133; Descourtilz, title page, and 2 : 412, 457. Descourtilz describes the botanical garden at the Cap hospital and the Eaux de Boynes spa in 1803, where Gauché still could be found as "proprietor" of the spa.

42. Barré de Saint-Venant, pp. 441–42.

43. On the Leclerc expedition, see Ott, chaps. 8–9; James (1963), chaps. 12–13; Dorsainvil, chaps. 13–16; Cauna (1987), pp. 241–45; Cornevin, pp. 50–54; Stein (1979), pp. 44–45; Lokke, pp. 233–34; Davis, pp. 65–67.

44. Ott, p. 152. An earthquake in 1842 leveled Cap François yet again; see Cabon [1928–33], 2 : 564. Comparatively few traces of the colonial world remain in modern Haiti.

45. Ott, p. 182, puts total French losses of the Leclerc expedition—medical and otherwise—at 40,000. See also Bonnette, p. 471; Houdaille, pp. 869–70;

Gentilini, p. 57; and ms., "Relation historique, Topographique & Médicale de l'Expédition de St.-Domingue en 1802," by army surgeon major, C. S. Cuynat in the John Kobler/Haitian Revolution Collection, NYPL-SC, original in the Bibliothèque Publique of Dijon.

46. Ott, p. 182. Demanding heavy indemnities, the French recognized Haiti only in 1825; the threat that the French might invade again remained real for many years. France abolished slavery entirely only in 1848; the United States did not recognize Haiti until 1862. See Cornevin, pp. 40–58; Dorsainvil, pp. 189–90. On the world-historical significance of the Haitian revolution, see also Fox-Genovese and Genovese, esp. p. 404.

CONCLUSIONS: Science and Colonial Development

1. Gillispie (1980), chap. 1, emphasizes rational state administration as another reason why the French state called upon science in the late eighteenth-century.

2. The Société Royale de Médecine appears at several junctures here. Its archives, not consulted for this study, doubtless hold many treasures regarding the history of colonial science; see Gillispie (1980), p. 195.

3. See Blerald, pp. 12–18; Tarrade, pp. 531–47, 598–608, 639–41, 700, 782; Fox-Genovese and Genovese, p. 6; and above, Introduction and chapter 4.

4. Mintz (1974), p. 60; Pluchon, ed. (1980), p. 32.

5. See remarks to this effect by Tarrade, p. 782; Cauna (1987), pp. 20, 197; Knight, pp. 138–45, 152; Kolchin, pp. 360, 362. Fox-Genovese and Genovese, esp. chaps. 1–2 and epilogue, analyze related tensions in the transition from merchant capitalism to industrial capitalism; compare Williams, who links plantation economies with the emergence of the Industrial Revolution.

6. See Tarrade, p. 146, and above, chapter 2. By the same token, Stein warns of reading too much modernity into the eighteenth century. About the slave trade in particular, Stein sees "the tradition of medieval merchants more than of modern industrial capitalists," and he remarks on "the curious amalgam of traditional and modern features" in the sugar trade at the end of the Old Regime; see Stein (1979), pp. 201–2 and (1988), pp. 168–69, 174.

7. See Knight, pp. 128–29; Lewis, pp. 121–23; Geggus (1989), pp. 1292–93; Fox-Genovese and Genovese, p. 397; Williams.

8. Chambers, esp. pp. 299, 306–10.

9. Chambers, p. 309, likewise gives up on the traditional notion of the Enlightenment for analyzing colonial science, preferring instead what he calls "scientific Enlightenment," as the more fruitful analytical concept.

10. Gillispie (1980), p. 549; McClellan (1985), p. 25.

11. See DPF, pp. 888–89; CA-HdC, fol. 158; AA, Cap, 1786, no. 8 (February 22), and above, chapters 8 and 7.

12. On French science and Kulturpolitik, see Stroup, p. 53; Hahn, chap. 2; McClellan (1985), pp. 250–51. These points were sharpened in conversation with my colleague Harold Dorn.

13. See Basalla; MacLeod, pp. 225–29. See also I. B. Cohen, and Basalla's view foreshadowed in Fleming, p. 180. No mention is made of Saint Domingue by Basalla or anyone else in the literature under discussion here.

14. Surprisingly, Basalla (and others) neglect to consider the history of colo-

nial medicine in thinking about the history of colonial science.

15. For discussions of these various models, see Reingold and Rothenberg, pp. xi–xii; MacLeod, pp. 222–29; Chambers, pp. 305–15; Jarrell; and Meinig, esp. pp. 65–66, 258, 260–61, for general models of colonial development. Pyenson (1989) presents a suggestive three-dimensional model for colonial science along a functionary-bureaucratic axis, a research axis, and a mercantilist-business axis. Like French colonial science in later periods, the functionary-bureaucratic aptly characterizes science in colonial Saint Domingue.

16. Colonial histories of science in British North America evidence this point. Historians of American science have dealt primarily with the institutionalization and professionalizing of American science, the development of an independent national science tradition, and American divergence from European examples. The function of science for colonization has not been a special concern; see remarks to this effect in Hollinger, and Reingold (1976a, 1976b, and 1978); see also Oleson and Brown, pp. 343, 345. The critical bibliography compiled by Rothenberg (1982) makes evident an essential absence of work on science and colonialism per se in America. Struik (1962) is somewhat the exception here.

BIBLIOGRAPHY

Contemporary Writings

Abeille. 1790. *Apperçu Rapide sur les Colonies*. Paris: n.p.

Almanach Historique et Chronologique de Saint-Domingue. 1765–91. Cap François and Port-au-Prince.

American Philosophical Society. 1885. "Early Proceedings from the Manuscript Minutes of Its Meetings, 1744–1838." *Proceedings of the American Philosophical Society* 22:1–711.

Amic. 1791. "Lettre de M. Amic, docteur en Médecine." *Journal de physique* 39:132–36 and plate.

Arthaud, Charles. *See also* Cercle des Philadelphes.

———. 1771. *Dissertation sur la dilation des artères*, etc. Paris: P. G. Cavelier.

———. 1776. *Traité des Pians*. N.p.

———. 1785. *Discours prononcé à l'ouverture de la première séance publique du Cercle des Philadelphes, Tenue au Cap-François le 11 mai 1785, avec une Description de la ville du Cap, pour servir à l'histoire des maladies que l'on y observe*. Paris: n.p.

———. 1786a. *Précis historique sur Monsieur le chevalier Lefebure-Deshayes*. [Cap-François]: Imprimerie Royale du Cap.

———. 1786b. *Recherches sur la Constitution des naturels du pays, sur leurs arts, leur industrie et les moyens de leur subsistance*. Au Cap: Imprimerie Royale. [Reprinted in Hamy, "Arthaud."]

———. 1787. "Observations sur les effets de la piqûre de l'Araignée-crabe des Antilles, et Description de la Bête à mille pieds de Saint-Domingue." *Journal de physique* 30:422–26.

———. 1789a. "Dissertation sur la conformation de la Tête des Caraïbes." *Journal de physique* 34:250–55.

———. 1789b. "Observation sur les Albinos." *Journal de physique* 35:274–78.

———. 1790. "Observations sur une Fièvre." *Journal de physique* 36:379–82 and plate.

———. 1791. *Observations sur les lois concernant la Médecine et la Chirurgie dans la Colonie de St.-Domingue*. Au Cap-Français: Rians.

———. 1792. *Description de l'Hôpital Général du Cap*. [Au Cap.]

————. N.d. *Réfutation de la pièce justificative du septième chef de la première denonciation solennelle d'un ministre, fait à l'Assemblée nationale, en la personne du Comte de la Luzerne.* Cap-Français: Imprimerie royale.

————. N.d. *Notice sur la ville du Port-au-Prince.* N.p.

Barbé de Marbois, François. 1788. *État des Finances de Saint-Domingue* [1785– 1787]. Port-au-Prince: Mozard.

————. 1789. *État des Finances de Saint-Domingue* [1788]. Port-au-Prince: Mozard.

————. 1790. *Observations personnelles à l'Intendant de Saint-Domingue.* Paris: Knapen.

Barré de Saint-Venant, Jean. 1802. *Des Colonies Modernes sous la Zone Torride et particulièrement de celle de Saint-Domingue.* Paris: Brochot.

Barrère, Pierre. 1741. *Essai sur l'histoire naturelle de la France equinoxiale.* Paris: Piget.

Bellin, Jacques-Nicolas. 1764. *Le Petit Atlas Maritime,* tome 1: *Amérique Septentrionale et les Isles Antilles.* [Plates.] N.p.

————. 1807. *Hydrographie françoise. Recueil des Cartes Marines dressées au Dépôt des cartes, plans et journaux, par M. Bellin et autres.* 2 vols. Paris: n.p.

Bourgeois, Nicolas Louis. 1787. *Voyages Intéressans dans différentes colonies françaises, espagnoles, anglaises, &c. . . . et un Mémoire sur les Maladies les plus communes à Saint-Domingue.* London: Bastien. [Papers, memoirs of N. L. Bourgeois (1710–76), published by (his nephew) P. J. B. Nougaret.]

Bouguer, Pierre. 1738. "Extrait d'une lettre de M. Bouguer écrite à M. de Réaumur, du Petit Goave dans l'Isle de Saint-Domingue, le 26 Octobre 1735, sur la Longueur du Pendule." *HMARS,* 1735, Mémoires section, pp. 522–28.

Cercle des Philadelphes. *See also* entries under Arthaud, Charles.

————. 1784. *Prospectus du Cercle des Philadelphes établie au Cap.* Au Cap: Imprimerie royale.

————. 1785a. *Statuts du Cercle des Philadelphes.* Au Cap: Imprimerie royale.

————. 1785b. *Programme de prix proposés par le Cercle des Philadelphes, à son Assemblée publique du 11 mai 1785.* Au Cap: Imprimerie Royale.

————. 1785c. *Souscription proposée par le Cercle des Philadelphes pour l'Édition des Ouvrages de feu M. Thiery de Menonville.* Cap-François: Imprimerie Royale. [AN, Colonies F³ 151.]

————. 1786a. *Dissertation et Observations sur le Tétanos, Publiées par le Cercle des Philadelphes au Cap-François.* Au Cap-François: Rians.

————. 1786b. *Programme de prix proposés par le Cercle des Philadelphes, à son Assemblée publique du 20 juin 1786.* Au Cap: Imprimerie royale.

————. 1787a. *Questions relatives à l'Agriculture de Saint-Domingue.* Au Cap-François: Imprimerie Royale.

————. 1787b. *Tableau du Cercle des Philadelphes Établi au Cap-François.* Cap-François: Imprimerie royale.

————. 1788a. *Dissertation sur le papier, dans laquelle on a rassemblé tous les essais qui ont été examinées par le Cercle des Philadelphes, sur les moyens de préserver le papier de la piqûre des insectes.* Presented by Charles Arthaud. Port-au-Prince: Mozard.

————. 1788b. *Mémoires du Cercle des Philadelphes. Tome Premier.* Port-au-Prince: Mozard.

————. 1788c. *Notice sur la Séance publique du Cercle des Philadelphes, Tenue le 15 Août 1788*. Port-au-Prince: Mozard. [AN, SdOM, BMSM 147; also, AN, Colonies F³ 81.]

————. 1788d. *Programme des prix proposés par le Cercle des Philadelphes*. Port-au-Prince: Mozard. [AN, Colonies F³ 81; AN, SdOM, BMSM 147.]

————. 1788e. *Recherches, Mémoires et Observations sur les Maladies Épizootiques de Saint-Domingue*. Au Cap-François: Imprimerie Royale.

————. 1789a. *Programme des prix proposés par le Cercle des Philadelphes du Cap, à son Assemblée du 17 août 1789*. Au Cap-François: Imprimerie royale. [AN, Colonies F³ 81.]

————. 1789b. *Programme des prix proposés par la Société royale des sciences et des arts du Cap-François, dans sa séance du 17 août 1789*. Au Cap-François, de l'Imprimerie Royale. [AN, SdOM, BMSM 147.]

————. 1790. *Prospectus des travaux que la Société Royale des Sciences et des Arts du Cap-François se proposoit de présenter dans la séance publique qui devoit avoir lieu le 17 août 1790*. Au Cap: Imprimerie Royale. [AN, SdOM, Recueil Colonies, 37, BMSM 39 (no. 26).]

————. [1791]. *Lectures qui ont été faites à la séance publique de la Société Royale des Sciences et Arts de la Ville du Cap-Français, le 16 août 1791*. Au Cap-François: Imprimerie Royale. [AN, SdOM, Recueil Colonies, 2ᵉ Série, 18, BMSM 18.]

Chabert, J. -B. Marquis de. 1786. "Mémoire sur l'usage des horloges marines." *HMARS*, 1783, Mémoires section, pp. 49–66.

Chambre d'Agriculture, [Cap-François,] Saint-Domingue. N.d. *Lettre bien Importante de la Chambre d'Argiculture de Saint Domingue, Adressée aux Membres du comité colonial, séant à Paris*. N.p.

Chastonnet-Desterre, Gabriel. 1796. *Considerations sur l'État présent de la Colonie Française de Saint-Domingue*. N.p.

D'Alembert and Diderot. 1751–65. *Encyclopédie, ou Dictionnaire raisonné des Sciences, des Arts et des Métiers*, 17 vols. Paris: Le Breton et al.

Damien-Chevalier, Jean. 1752. *Lettres à M. De Jean*. Paris.

————. 1763. *Sur les Fièvres de l'île de St. Domingue*. Paris.

Dazille, [Jean Barthélemy]. 1776, 1792. *Observations sur les Maladies des Nègres: Leurs causes, leurs traitemens et les moyens de les prévenir*. 1st ed., Paris: Didot. 2d ed., "considérablement augmentée," Paris: Croullebois.

————. 1785. *Observations Générales sur les Maladies des climats chauds, leurs causes, leurs traitements et les moyens de les prévenir*. Paris: Didot.

————. 1788. *Observations sur le Tétanos, précédées d'un discours sur les moyens de perfectionner la Médecine-Pratique sous la zone torride*, etc. Paris: Planche.

[Descahos.] 1781. *Lettre de M. Descahos, Habitant Riverain du Fleuve de l'Artibonite En l'Isle de Saint-Domingue*. London: n.p.

Descourtilz, Michel Étienne. 1809. *Voyages d'un naturaliste et ses observations faites sur les trois règnes de la Nature*. 3 vols. Paris: Dufart.

Deshayes. *See* Lefebvre-Deshayes.

Dubuisson, Paul Ulric. 1780. *Nouvelles considérations sur Saint-Domingue*. Paris: Cellot & Jombert.

Ducoeur-Joly, S. J. 1802. *Manuel des Habitants de Saint-Domingue*. Paris: Lenoir.

Dutertre, Jean-Baptiste. 1667–71. *Histoire générale des Antilles habitées par les Fran-çois.* 4 vols. Paris.

Dutrône la Couture, Jacques-François. 1790. *Précis sur la canne et sur les moyens d'en extraire le sel essentiel, suivi de plusieurs mémoires.* . . . Paris: Duplain. 2d ed., 1791; 3d ed., 1801.

Edwards, Bryan. 1797. *An Historical Survey of the French Colony in the Island of St. Domingo.* . . . London: John Stockdale.

Feuillée, Louis. 1714, 1725. *Journal des observations physiques, mathématiques et bota-niques, fait par l'ordre du Roi sur les côtes orientales de l'Amérique, depuis 1707 jusques en 1712,* 3 vols. Vols. 1 and 2, Paris: Giffard. Vol. 3, Paris: Mariette.

Fleurieu, C. P. Claret de, Comte d'Eveux. 1773. *Voyage fait par ordre du Roi en 1768 et 1769 à différentes parties du monde, pour éprouver en mer les horloges marines inventées par M. Berthoud.* 2 vols. Paris: Imprimerie royale.

Fontenelle, Bernard Le Bovier de. 1703. "Sur la Longueur du Pendule." *HMARS,* 1700, Histoire section, 114–16.

Fusée Aublet, Jean-Baptiste-Christophe. 1775. *Histoire des Plantes de la Guyane française,* 4 vols. Paris: Didot.

Gazette de S. Domingue. 1764. Cap François. [Weekly newspaper, January–May 1764; BN, 4° Lc¹² 17/19.]

Gazette de Saint-Domingue. 1791. Port-au-Prince. [Semiweekly newspaper, Janu-ary–November 1791; BN 4° Lc¹² 25.]

Genton, M. de. 1787. "Essai de minérologie de l'Isle de Saint-Domingue dans la partie Françoise." *Journal de physique* 31: 173–77.

Girod-Chantrans, [Justin]. 1785, 1786. *Voyage d'un Suisse dans différentes colonies d'Amérique pendant la dernière guerre, avec une table d'observations météorologiques faites à Saint-Domingue.* Neuchâtel: Société Typographique. London: Poinçot. [Modern, abridged edition with introduction by Pierre Pluchon. Paris: Tallan-dier, 1980.]

Godin, [Louis]. 1738. "La Longueur du pendule simple qui bat les secondes du temps moyens, Observée à Paris & au Petit Goave en l'Isle Saint-Domingue." *HMARS,* 1735, Mémoires section, pp. 505–21.

Hilliard d'Auberteuil. 1776. *Considérations sur l'état présent de la colonie française de Saint-Domingue.* Paris.

La Condamine, Charles-Marie de. 1738. "De la measure du Pendule à Saint-Domingue." *HMARS,* 1735, Mémoires section, pp. 529–44.

Lafosse, J. F. 1787. *Avis aux habitans des colonies, particulièrement à ceux de l'Isle de S. Domingue, sur les principales causes des maladies qu'on y éprouve le plus communé-ment & sur les moyens de les prévenir.* Paris: Royez.

La Hire, Phillippe de. 1707. "Comparaison de l'observation de l'Eclipse . . . faite dans l'Isle de S. Domingue." *HMARS,* 1706, Mémoires section, pp. 481–82.

Lefebvre-Deshayes, Étienne. 1785. "Notices sur l'Anémone de mer à plumes ou animal-fleur." *Journal de physique* 27: 373–81 and plate.

Lind, James. 1785. *Essai sur les maladies des Européens dans les pays chaudes . . . Tra-duit de l'Anglois par M. Thion de la Chaume* [from 3d ed., 1777]. Paris: Barrois.

Moreau de Saint-Méry, Méderic Louis Élie. 1784–90. *Lois et Constitutions des Colonies Françoises de l'Amérique Sous-le-Vent.* 6 vols. Paris: Quileau, Méquigon jeune, Moutard, etc.

————. 1790. *Éloges de M. Turc de Castelveyve et de M. Dolioules. . . .* Paris: Rochette.

————. 1791. *Recueil de Vues des Lieux Principaux de la Colonie Françoise de Saint-Domingue, gravées par les soins de M. Ponce,* etc. Paris: A.P.D.R.[?]

————. 1796a. *Description topographique et Politique de la partie espagnole de l'isle de Saint-Domingue,* 2 vols. Philadelphia: [by the author].

————. 1796b. *Topographical and Political Description of the Spanish Part of Saint-Domingo. . . .* Translated from the French by William Cobbett. Philadelphia: [by the author].

————. 1797–98. *Description Physique, Civile, Politique et Historique de la Partie Française de l'Isle de Saint-Domingue, Avec des Observations générales sur sa Population, sur le Caractère & les Mœurs de ses divers Habitans; sur son Climat, sa Culture, ses Productions, son Administration, &c. &c.* 2 vols. Philadelphia. [Modern edition in 3 vols., edited by Blanche Maurel and Étienne Taillemite. Paris: Société de l'Histoire des Colonies Françaises/Librairie Larose, 1958; Société Française d'Histoire d'Outre-Mer, 1984.] [Cited as *DPF.*]

————. 1985. *A Civilization That Perished: The Last Years of White Colonial Rule in Haiti,* Edited and translated by Ivor D. Spencer. Lanham, Md.: University Press of America.

Neufchateau, N. L. François, Comte de. [1786]. *Les Études du Magistrat, Discours prononcé à la Rentrée du Conseil Supérieur du Cap, le Jeudi 5 Octobre 1786.* Au Cap-Français: n.p.

Newton, Isaac. 1966. *Mathematical Principles of Natural Philosophy.* Translated by Andrew Motte and Florian Cajori. 2 vols. Berkeley: University of California Press.

Nicolson, Jean-Barthélemy-Maximilien. 1776. *Essai sur l'Histoire Naturelle de St. Domingue.* Paris: Gobreau.

Pingré, Alexandre-Guy. 1772. "Observations du passage de Vénus sur le disque du soleil faite au Cap François, isle de St.-Domingue, le 3 juin 1769." *HMARS,* 1769, Mémoires section, pp. 513–28.

————. 1773a. "Précis d'un voyage en Amérique, ou Essai Géographique sur la position de plusieurs Isles. . . ." *HMARS,* 1770, Mémoires section, pp. 487–513.

————. 1773b. "Rapport des Observations Faites sur mer pour la détermination des longitudes . . . Lues à la rentrée de L'Académie des Sciences [1773]." *Journal de physique* 2:1–11.

Poissonnier-Desperrières, Antoine. 1763. *Traité des fièvres de l'île de Saint-Domingue.* Paris.

————. 1767. *Traité sur les maladies des gens de mer.* Paris.

Pouppée-Desportes, [Jean-Baptiste-René]. 1770. *Histoire des Maladies de S. Domingue,* 3 vols. Paris: Lejay.

Puységur, A.-H.-A. Chastenet, Comte de. 1787a. *Le Pilote de l'Isle de Saint-Domingue et les Débouquemens de cette Isle . . . Publié par ordre du Roi.* Paris: De l'Imprimerie Royale.

————. 1787b. *Détail sur la navigation aux côtes de Saint-Domingue et dans ses Débouquemens.* Paris: De l'Imprimerie Royale.

Raynal, Guillaume-Thomas. 1781. *Histoire philosophique et politique des Établis-*

semens et du Commerce des Européens dans les deux Indes. 10 vols. Genève. [Many editions; original in 1770.]

————. 1785. *Essai sur l'Administration de St. Domingue.* N.p.

Société royale des sciences et des arts du Cap Français. *See* Cercle des Philadelphes.

Tessier [Henri-Alexandre]. 1793. "Mémoire sur l'importation et le progrès des arbres à épicerie, dans les Colonies Françoises." *HMARS*, 1789, Mémoires section, pp. 585–97.

Thierry de Menonville, N. J. 1787. *Traité de la Culture du Nopal et de l'Éducation de la Cochenille dans les Colonies Françaises de l'Amérique; Précédé d'un Voyage à Guaxaca*, 2 vols. Cap-François: Chez la veuve Herbault. Paris: Delalain. Bordeaux: Bergeret. [Published by the Cercle des Philadelphes.]

Ulloa, Don Antonio. 1752. *Voyage historique de l'Amerique Meridionale fait par ordre du roi d'Espagne.* 2 vols. Amsterdam and Leipzig: Arkstée & Merkus. [French translation by E. de Mauvillon of original Spanish edition of 1748; includes separate translation, *Observations astronomiques et physiques fait par ordre du roi d'Espagne*, of companion work by Don George [Jorge] Juan [y Santaclia], original Spanish edition likewise of 1748.]

Valverde, Antonio Sanchez. 1785. *Idea del Valor de la Isla Española, y utilidades que da ella puede sacar su Monarquia.* Madrid: Pedro Marin.

Wimpffen, Francis Alexander Stanislaus, Baron de. 1797. *A Voyage to Saint Domingo in the Years 1788, 1789, and 1790.* Translated by J. White. London: n.p.

Later Writings

Aiton, E. J. 1989. Review of *Science Reorganized: Scientific Socieites in the Eighteenth Century*, by James E. McClellan III. *The Heythrop Journal*, January: 105–7.

Anderson, Benedict. 1988. "Criollismo." *London Review of Books*, January 21, pp. 5–6.

Anglade, Georges. 1973a. "Document: Questions relatives à l'agriculture de Saint-Domingue, publiée par le Cercle des Philadelphes du Cap-François, 1787." *Les Cahiers du Centre Haïtien d'Investigations en Sciences Sociales* 7: 16–40.

————. 1973b. "Une enquête du Cercle des Philadelphes du Cap." *Conjonction* 120: 127–44.

————. 1982. *Atlas Critique d'Haïti.* Montreal: Erce & CRC.

Armas, Juan Ignacio. 1888. *La Zoología de Colón y de los Primeros Exploradores de América.* Havana: O'Reilly.

Armitage, Angers. 1953. "The Pilgrimage of Pingré: An Astronomer-Monk of Eighteenth-Century France." *Annals of Science* 9: 47–63.

Aubin, Eugène. 1910. *En Haïti: Planteurs d'autrefois, nègres d'aujourd'hui.* Paris: Armand Colin.

Aucoc, Léon. 1889. *L'Institut de France: Lois, statuts et règlements concernant les anciennes académies et l'Institut, de 1639 à 1829.* Paris: Imprimerie Nationale.

Auvigne, R., and J. P. Kernes. 1956. "Nantes, herbier des Isles: Ou le rôle joué par les botanistes nantais dans l'introduction en France des végétaux exotiques au 18ᵉ siècle." *Histoire de la médecine* 6, no. 11: 7–13.

Barnwell, P. J., and A. Toussaint. 1946. *A Short History of Mauritius.* London: Longmans.

Basalla, George. 1967. "The Spread of Western Science." *Science* 156:611–22.

Basket, Sir James. 1824. *History of the Island of St. Domingo: From Its First Discovery by Columbus to the Present Period.* New York: Mahlon Day. [Original edition, London: 1818.]

Bellegarde, Dantès. 1923. "La société française de St.-Domingue à la veille de la Révolution." *Revue de l'Amérique Latine* 4:104–14.

———. 1950. "La société française de Saint-Domingue avant 1789." *Conjonction* 25–26:8–20; 27:1–6.

———. 1953. *Histoire du peuple Haïtien (1492–1952).* Port-au-Prince: n.p.

Bernal, J. D. 1971. *Science in History.* 4 vols. Cambridge, Mass.: MIT Press.

Bissainthe, Max. 1951. *Dictionnaire de bibliographie Haïtienne.* Washington, D.C.: Scarecrow Press.

Blerald, Alain-Philippe. 1986. *Histoire économique de la Guadeloupe et de la Martinique du XVIIe siècle à nos jours.* Paris: Karthala.

Bohannan, Paul, and Philip Curtin. 1987. "The African Slave Trade." In *The Human Perspective: Readings in World Civilization II,* edited by Lynn H. Nelson, pp. 20–30. San Diego: Harcourt Brace Jovanovich.

Bonnette, P. 1938. "Le calvaire médical de Saint-Domingue et de la Guadeloupe (Ans X et XI)." *Hippocrate: Revue d'Humanisme Médicale* 6:470–90.

Bordier, A. 1884. *La colonisation scientifique et les colonies françaises.* Paris: Reinwald.

Botherel-Blanchet, Catherine. 1977–78. "Traité des plantes usuelles de Saint-Domingue du Docteur Pouppé Desportes, Médecin du Roi, 1770." Thèse de doctorat en médecin, Université de Nantes, no. 1965.

Bourgeois, Charles. 1970. "Le Père Louis Feuillée, Astronome & Botaniste du Roi." In *Comptes Rendus du 99e Congrès National des Sociétés Scientifiques (Pau, 1966),* Section des Sciences, 1:9–19. Ministère de l'Éducation Nationale, Comité des Travaux Historiques et Scientifiques. Paris: Bibliothèque Nationale, 1970.

Boxer, C. R. 1965. *The Dutch Seaborne Empire, 1600–1800.* London: Hutchinson.

Boyden, Hayne. 1936. "The Mysterious Hot-Air Engine of Haiti." *U.S. Naval Institute Proceedings* 62:86–96, 714.

Brau, Paul. 1931. *Trois siècles de médecine coloniale française.* Paris: Vigot Frères.

Broc, Numa. [1975]. *La géographie des philosophes: Géographes et voyageurs français au XVIIIe siècle.* Paris: Editions Ophrys.

Brockway, Lucile H. 1979. *Science and Colonial Expansion: The Role of the British Royal Botanic Gardens.* Studies in Social Discontinuity. New York: Academic Press.

Broussolle, B., and Ph. Masson. 1985. "La santé dans la Marine de l'Ancien Régime." In Pluchon, ed., *Histoire des Médecins,* pp. 69–87.

Burg, B. R. 1984. *Sodomy and the Pirate Tradition: English Sea Rovers in the Seventeenth-Century Caribbean.* New York: New York University Press.

Bush, Barbara. 1990. *Slave Women in Caribbean Society: 1650–1838.* Kingston and London: Heineman; Bloomington: Indiana University Press.

Butel, Paul. 1982. *Les Caraïbes au temps des flibustiers, XVIe–XVIIe siècles.* Paris: Aubier Montaigne.

Cabon, Adolphe. 1916a. "Contribution à l'étude de la géographie d'Haïti." *Bul-*

letin Semestriel de l'Observatoire Météorologique du Séminaire-Collège St.-Martial, January–June: 149–74.

———. 1916b. "Notes historiques sur la détermination de la position géographique d'Haïti." *Bulletin Semestriel de l'Observatoire Météorologique du Séminaire-Collège St.-Martial,* January–June: 51–68.

———. 1917. "Contribution à l'étude de la géographie d'Haïti." *Bulletin Semestriel de l'Observatoire Météorologique du Séminaire-Collège St.-Martial:* 93–128.

———. [1928–33.] *Histoire d'Haïti: cours professé au Petit Séminaire—Collège Saint-Martial.* 4 vols. Port-au-Prince: La Petite Revue.

Caillet, Robert. 1954. "Le remède universel du docteur Ailhaud." *Revue d'histoire de la pharmacie* 42:251–66.

Cale, John G. 1971. "French Secular Music in Saint-Domingue, 1750–1795." Ph.D. dissertation, Louisiana State University.

Canny, Nicholas, and Anthony Pagden, eds. 1987. *Colonial Identity in the Atlantic World: 1500–1800.* Princeton: Princeton University Press.

Carron, Claude-Henri. 1941. "L'Ankylostomse à Saint-Domingue." Thèse pour le Doctorat en Médecine, Faculté de Médecine de Paris.

Castonnet des Fosses, H[enri Louis]. 1884. *L'île de Saint-Domingue au XVIIIᵉ siècle.* Nantes: Mellinet.

———. 1886. *Saint Domingue sous Louis XV.* Angers: Lachèse et Dolbeau.

———. 1893. *La perte d'une colonie: La révolution de Saint-Domingue.* Paris: A. Faivre.

Cauna, Jacques. 1981. "Vestiges de sucreries dans la plaine du Cul-de-Sac." *Conjonction* 149:65–104.

———. 1984. "L'état sanitaire des esclaves sur une grande sucrerie (Habitation Fleuriau de Bellevue, 1777–1788." *Revue de la Société Haïtienne d'Histoire et Géographie* 42, no. 145:18–78.

———. 1985. "Vestiges de sucreries dans la plaine du Cul-de-Sac." *Conjonction* 165:31–58.

———. 1987. *Au temps des isles à sucre: Histoire d'une plantation de Saint-Domingue au XVIIIᵉ siècle.* Paris: Éditions Karthala.

Centre d'Études de Géographie Tropicale. 1985. *Atlas d'Haïti.* Bordeaux: C.N.R.S. et Université de Bordeaux, 3.

Chaia, Jean. 1979. "Science, médecine et état sanitaire en Guyane au XVIIIᵉ siècle." *Mondes et Cultures* 39:129–43.

Chailley-Bert. J. 1898. *Les compagnies de colonisation sous l'Ancien Régime.* Paris: Armand Colin.

Chambers, David Wade. 1987. "Period and Process in Colonial and National Science." In Reingold and Rothenberg, eds., *Scientific Colonialism,* 297–321.

Charles, A. 1953. "L'Acajou du XVe au XXe siècles." *Revue de 'La Porte Océane'* 9, May: 3–10.

Chardon, Carlos E. 1949. *Los Naturalistas en la America Latina.* Ciudad Trujillo, Republica Dominicana: Editora del Caribe.

Chavalier, G. A., G. Debien, et al. 1953. "Recherches collectives: Chronique documentaire pour une nouvelle histoire coloniale: Les papiers privés et l'Amérique française." *Revue d'histoire de l'Amérique française* 6:536–59; 7:88–109.

Cobban, Alfred. 1963. *A History of Modern France*, vol. 1: *1715–1789*. Baltimore: Penguin Books.

Cohen, I. B. 1961. "The New World as a Source of Science for Europe." In *Actes du IXe Congrès International d'Histoire des Sciences, Madrid (1959)*, pp. 96–130. Paris: Hermann.

Cohen, William B. 1980. *The French Encounter with Africans: The White Response to Blacks, 1530–1880*. Bloomington: Indiana University Press.

Cole, Charles Woolsey. 1939. *Colbert and a Century of French Mercantilism*. 2 vols. New York: Columbia University Press.

———. 1971. *French Mercantilism: 1683–1700*. New York: Octagon Books. [Original edition: New York: Columbia University Press, 1943.]

Cook, Mercer. 1941. "Julien Raimond." *Journal of Negro History* 26:139–70.

Coquery-Vidrovitch, Catherine. 1986. [Articles:] "Coloniale, Histoire," and "Impérialisme, Histoire de." In *Dictionnaire des Sciences Historiques*, edited by André Bugière, pp. 141–46, 350–53. Paris: Presses Universitaires de France.

Cornevin, Robert. 1982. *Haïti*. Que Sais-je? series, no. 1955. Paris: Presses Universitaires de France.

Cornu, Henri. [1975, 1980]. *Paris et Bourbon: La politique française dans l'Océan Indien de 1664 à 1815*. Académie des Sciences d'Outre-Mer, *Travaux et Mémoires*, n.s. no. 4 (1975). St. Denis: Anchaing, 1980.

Corvington, Georges. 1970. *Port-au-Prince au cours des ans: La ville coloniale, 1743–1789*. Port-au-Prince: Imprimerie Henri Deschampes.

Crosby, Alfred W. 1986. *Ecological Imperialism: The Biological Expansion of Europe, 900–1900*. Cambridge: Cambridge University Press.

Crouse, Nellis M. 1966. *The French Struggle for the West Indies, 1665–1713*. New York: Octagon Books. [Originally published in 1943.]

———. 1977. *French Pioneers in the West Indies, 1624–1664*. New York: Octagon Books. [Reprint of 1940 original.]

Curtin, Philip D. 1977. "The Black Experience of Colonialism and Imperialism." In Mintz, ed., *Slavery, Colonialism and Racism*, pp. 17–29.

Cutbush, Edward. [1811]. "On Cochineal." In *Archives of Useful Knowledge*, edited by James Mease, 1:257–273. [Philadelphia.]

Darby, William. 1823. *Darby's Edition of Brooke's Universal Gazeteer*. Philadelphia: Bennett & Walton.

Darnton, Robert. 1968. *Mesmerism and the End of the Enlightenment in France*. Cambridge: Harvard University Press.

———. 1982. *The Literary Underground of the Old Regime*. Cambridge: Harvard University Press.

———. 1985. *The Great Cat Massacre and Other Episodes in French Cultural History*. New York: Basic Books, 1984. Random House/Vintage, 1985.

Darnton, Robert, and Daniel Roche, eds. 1989. *Revolution in Print: The Press in France: 1775–1800*. Berkeley: University of California Press.

Darondel, Louis. 1939. *Légendes et traditions dans l'histoire de Saint-Domingue: Essai de critique*. Port-au-Prince: Compagnie Lithographique.

Davis, Wade. 1985. *The Serpent and the Rainbow*. New York: Simon and Schuster.

Debbasch, Yvan. 1977. "Le Maniel: Further Notes." In Price, ed., *Maroon Societies*, pp. 143–48.

Debien, Gabriel. 1937. "Les sources manuscrites de l'histoire et de la géographie de Saint-Domingue." Port-au-Prince: Valcin. Also in *Revue de la Société d'Histoire et de Géographie d'Haïti* 6: 13–62.

D[ebien], G[abriel]. 1938. Review of "Une Société de Pensée à Saint-Domingue," by Blanche Maurel. *Revue d'histoire des colonies* 27, no. 118: 71–72.

———. 1943. "Notes bibliographiques sur l'histoire de Saint-Domingue." *Revue de la Société d'Histoire et de Géographie d'Haïti* 14: 25–42.

———. 1949. "Une maison d'éducation à Saint-Domingue: 'Les Religieuses' du Cap." *Revue d'histoire de l'Amérique française* 2: 557–75.

———. 1953a. *Les colons de Saint-Domingue et la Révolution: Essai sur le Club Massiac (Août 1789–Août 1792)*. Paris: Armand Colin.

———. 1953b. "Réfugés de Saint-Domingue à Cuba, 1793–1815." *Revista de Indias* 13: 559–605.

———. 1954. "Réfugés de Saint-Domingue à Cuba, 1793–1815." *Revista de Indias* 14: 11–36.

———. 1955. "Profils de colons: Jean Trembley." *Revue de "La Porte Océane"* 11, no. 113: 14–19; no. 114: 8–10.

———. 1956. *Études Antillaises (XVIIIe Siècle)*. Paris: Armand Colin. [Cahiers des Annales.]

———. 1967. "Les cimetières à Saint-Domingue." *Conjonction* 105: 27–40.

———. 1972. "Une Nantaise à Saint-Domingue (1782–1786)." *Revue de Bas-Poitu et des Provinces de l'Ouest* 83: 413–36.

———. 1974. "Un Officier du regiment de Forez à Saint-Domingue en 1764." *Conjonction* 124: 115–39.

———. 1981. "Pour ameliorer les cases, les hopitaux et la nourriture des esclaves à Saint-Domingue à la fin du XVIIIe siècle." *Revue de la Société Haïtienne d'Histoire de Géographie et de Géologie* 39, no. 131: 1–17.

De Candolle, A. P. 1922. "Jardins de Botanique [Notice abrégée de l'histoire et l'administration]." *Dictionnaire des Sciences Naturelles*, 24: 165–81.

De Lattre. 1972 [Document:] "Considerations sur la ville du Cap français." *Revue de la Société Haïtienne d'Histoire de Géographie et de Géologie* 37: 44–56.

Depréaux, Albert. 1924. "Le Commandant Baudry des Lozières et la Phalange de Crête-Dragons (Saint-Domingue, 1789–1792)." *Revue de l'Histoire des Colonies Françaises* 12: 1–43.

Dermigny, Louis. 1954. "Saint-Domingue et le Languedoc au XVIIIe siècle." *Revue d'histoire des colonies* 42: 47–70.

Desaive, J. P., J. P. Goubert, et al. 1972. *Médecins, climat et épidémies à la fin du XVIIIe siècle*. Paris: CNRS.

Desfeuilles, André. 1957. "À propos de la ferme des Postes de Saint-Domingue." *Le Vieux Papier* no. 178: 329–36; no. 179: 361–72.

Devèze, Michel. 1977. *Antilles, Guyanes, La Mer des Caraïbes de 1492 à 1789*. Paris: S.E.D.E.S.

Donkin, R. A. 1977. *Spanish Red: An Ethnogeographical Study of Cochineal and the Opuntia Cactus*. Philadelphia: American Philosophical Society. [*Transactions of the American Philosophical Society* 67, part 5.]

Dorsainvil, J.-C. [1924, 1955]. *Manuel d'histoire d'Haïti*. Port-au-Prince: Henri Deschampes.

Doucet, Louis. 1981. *Quand les Français cherchaient fortune aux Caraïbes*. Paris: Fayard.

Dunn, Richard S. 1972. *Sugar and Slaves: The Rise of the Planter Class in the English West Indies*. Chapel Hill: University of North Carolina Press.

Durant, Will, and Ariel Durant. 1935–75. *The Story of Civilization*. 11 vols. New York: Simon and Shuster.

Duvivier, Ulrich. 1941. *Bibliographie générale et méthodique d'Haïti*. 2 vols. Port-au-Prince: Imprimerie de l'État.

Edelstein, Sydney M. 1958. "Spanish Red—Thiery de Menonville's Voyage to Guaxaca." *American Dyestuff Reporter* 47:1–8.

Ehrman, Ester. 1986. *Mme du Châtelet*. Leamington Spa, U.K.: Berg Publishers.

Elicona, Anthony Louis. 1934. *Un colonial sous la Révolution en France et en Amérique: Moreau de Saint-Méry*. Paris: Jouve.

Emerson, Rupert. 1969. "Colonialism." *Journal of Contemporary History* 4:3–16. [Special number devoted to colonialism and decolonialization.]

Ewen, Charles R. 1988. "The Short Unhappy Life of a Maverick Caribbean Colony." *Archaeology*, July–August: 41–47, 76.

Faÿ, Bernard. 1932. "Learned Societies in Europe and America in the Eighteenth Century." *American Historical Review* 37:255–67.

———. 1942. *La Franc-maçonnerie et la révolution intellectuelle du XVIIIᵉ siècle*. Paris: n.p.

Faÿ, Bernard, Blanche Maurel, and Jean Equy. 1942–43. *Histoire de France*. 2 vols. Paris: J. de Gigord.

Finley, M. I. 1976. "Colonies—An Attempt at a Typology." *Transactions of the Royal Historical Society* 26:167–88.

Fleming, Donald. [1964]. "Science in Australia, Canada, and the United States: Some Comparative Remarks." In *Proceedings of the 10th International Congress of the History of Science, 1962*, 1:179–96. Paris: Hermann.

Fouchard, Jean. 1954. "Les Joies de la lecture à Saint-Domingue." *Revue d'Histoire des Colonies* 41:103–11.

———. 1955a. "Les Joies de la table à Saint-Domingue." *Revue de la Société haïtienne d'histoire de géographie et de géologie* 27:59–63.

———. 1955b. *Plaisirs de Saint-Domingue: Notes sur sa vie sociale, littéraire et artistique*. Port-au-Prince: Imprimerie de l'État.

———. 1955c. *Artistes et répetoire des scènes de Saint-Domingue*. Port-au-Prince: Imprimerie de L'État.

———. 1955d. *Le Théatre à Saint-Domingue*. Port-au-Prince: Imprimerie de l'État.

———. 1972. *Les marrons de la liberté*. Paris.

Fournier, P. 1932. *Voyages et découvertes scientifiques des missionnaires naturalistes Français à travers le monde (XVᵉ à XXᵉ siècles)*. Paris: Paul Lechevalier.

Fox-Genovese, Elizabeth, and Eugene D. Genovese. 1983. *Fruits of Merchant Capital: Slavery and Bourgeois Property in the Rise and Expansion of Capitalism*. New York: Oxford University Press.

Frostin, Charles. 1962. "L'intervention britannique à Saint-Domingue en 1793." *Revue française d'histoire d'outre-mer* 49:293–365.

———. 1973. "Les 'Enfants Perdus de l'État' ou la condition militaire à Saint-

Domingue au XVIIIe siècle." *Annales de Bretagne* 80:317–43.

———. 1975. *Les révoltes blanches à Saint-Domingue aux XVIIe et XVIIIe siècles (Haïti avant 1789)*. Paris: L'École.

Furet, François. 1965. "La 'Librairie' du royaume de France au 18e siècle." In *Livre et Société dans la France du XVIIIe siècle*, edited by François Furet. La Haye: Mouton. pp. 3–30.

Geggus, David Patrick. 1982. *Slavery, War and Revolution: The British Occupation of Saint-Domingue, 1793–1798*. Oxford: Clarendon Press.

———. 1989. "Racial Equality, Slavery, and Colonial Secession during the Constituent Assembly." *American Historical Review* 94:1290–1308.

Genovese, Eugene D. 1981. *From Rebellion to Revolution: Afro-American Slave Revolts in the Making of the New World*. New York: Vintage Books. [Original publication: Baton Rouge: Louisiana State University Press, 1979.]

Gentilini, M., and J. -P. Nozais. 1985. "Expansion coloniale et santé." In Pluchon, ed., *Histoire des médecins*, pp. 45–65.

Gillespie, Richard. 1984. "Ballooning in France and Britain, 1783–1786: Aerostation and Adventurism." *Isis* 75:249–68.

Gillispie, Charles Coulston. 1980. *Science and Polity in France at the End of the Old Regime*. Princeton: Princeton University Press.

———. 1983. *The Montgolfier Brothers and the Invention of Aviation*. Princeton: Princeton University Press.

———. 1985. "U.S. Flight No. One: January 9, 1793." *Invention and Technology:* 63–64.

———. 1987. *Monuments of Egypt: The Napoleonic Edition*. Princeton: Princeton Architectural Press.

Girod, François. 1970. *Une fortune coloniale sous l'Ancien Régime: La famille Hecquet à Saint Domingue, 1724–1796*. Annales Littéraires de l'Université de Besançon, 115. Paris: Les Belles Lettres.

———. 1971. "Les villes dans la partie française de Saint-Domingue au XVIIIe siècle." Académie des Sciences, Belles-Lettres et Arts de Besançon, *Procès-Verbaux et Mémoires* 179:225–50.

———. 1972. *La vie quotidienne de la société créole: Saint-Domingue au XVIIIe siècle*. Paris: Hachette.

Gordon, Sadie C. B. 1937. "Geographic Influences in the History of Haiti." M.A. Thesis: Columbia University.

Goubert, Pierre. 1952. "Une belle enquête: Saint-Domingue au XVIIIe siècle." *Annales: Économies, Sociétés, Civilisations* 7:329–31.

———. 1974. *The Ancien Régime: French Society, 1600–1750*. Translated by Steve Cox. New York: Harper & Row.

Greene, Graham. 1984 [1966]. *The Comedians*. Middlesex, U.K.: Penguin Books.

Greene, John C. 1977. *The Death of Adam: Evolution and Its Impact on Western Thought*. Ames: Iowa State University Press.

Gros, le Dr. 1896 "Un médecin des colonies au XVIIIe siècle: Pouppée-Desportes." *Archives de Médecine Navale et Coloniale* 66:345–57.

Hagelberg, G. B. 1985. "Sugar in the Caribbean: Turning Sunshine into Money." In Mintz and Price, eds., *Caribbean Contours*. pp. 85–126.

Hahn, Roger. 1971, 1986. *Anatomy of a Scientific Institution: The Paris Academy of Sciences, 1666–1803.* Berkeley: University of California Press.

Hall, Gwendolyn Midlo. 1971. *Social Control in Slave Plantation Societies: A Comparison of St. Domingue and Cuba.* Baltimore: Johns Hopkins University Press.

Hamy, E. -T. 1908. "Charles Arthaud de Pont-à-Mousson (1748–1791)." *Bulletins et Mémoires de la Société d'Anthropologie de Paris,* ser. 5, 9:295–314.

Hannaway, Caroline C. F. 1972. "The Société Royale de Médecine and Epidemics in the *Ancien Régime.*" *Bulletin of the History of Medicine* 46:257–73.

———. 1974. "Medicine, Public Welfare and the State in 18th-Century France: The Société Royale de Médecine of Paris: 1776–1793." Ph.D. dissertation, Johns Hopkins University.

Headrick, Daniel R. 1981. *The Tools of Empire: Technology and European Imperialism in the Nineteenth Century.* New York: Oxford University Press.

Hector, Michel C., and Claude D. Moise. 1962. *Le régime colonial français à Saint-Domingue.* Mimeograph. Port-au-Prince.

Hessen, Boris. 1931. "Social and Economic Roots of Newton's 'Principia.'" In *Science at the Crossroads.* London: Kinga.

Hicks, John W. 1987. "Red Crabs on the March on Christmas Island." *National Geographic* 172:820–31.

Hill, Arthur W. 1915. "The History and Functions of Botanic Gardens." *Annals of the Missouri Botanical Garden* 2:185–240.

Hill, Christopher. 1987. "Success Story: Captain Kidd and the War against the Pirates." *New York Review of Books* 34, January 29, pp. 14–16.

Hindle, Brooke. 1956. *The Pursuit of Science in Revolutionary America: 1735–1789.* Chapel Hill: University of North Carolina Press.

———. 1981. *Emulation and Invention.* New York: New York University Press.

Hoetink, H. 1985. "'Race' and Color in the Caribbean." In Mintz and Price, eds., *Caribbean Contours,* pp. 55–84.

Hollinger, David A. 1980. "On Science and American Society." *ISIS* 71:478–80.

Home, R. W. 1987. "The Beginnings of an Australian Physics Community." In Reingold and Rothenberg, eds., *Scientific Colonialism,* pp. 3–34.

Houdaille, Jacques. 1978. "Quelques données sur la population de Saint-Domingue au XVIIIᵉ." *Population* 28:859–72.

Howard, Richard A. 1954. "A History of the Botanic Garden of St. Vincent, British West Indies." *Geographical Review* 44:381–93.

Hughes, Jonathan. 1983. *American Economic History.* Dallas: Scott, Foresman & Co.

Hugo, Victor. 1826. *Bug-Jargal.* [Modern edition: *Le dernier jour d'un condamné* and *Bug-Jargal,* presented by Roger Bouderie. Paris: Gallimard, 1970.].

Humphreys, R. A. 1965. "The Development of the American Communities outside British Rule." In *New Cambridge Modern History,* vol. 8: *The French and American Revolutions, 1763–1793,* edited by A. Goodwin, chap. 14. Cambridge.

Hyma, Albert. 1942. *The Dutch in the Far East: A History of the Dutch Commercial and Colonial Empire.* Ann Arbor, Mich.: George Wahr.

Institut de France, Académie des Sciences. *Les membres et les correspondants de l'Académie royale des sciences, 1666–1793.* Paris: Au Palais de l'Institut.

Jacob, Margaret C. 1981. *The Radical Enlightenment: Pantheists, Freemasons and Republicans*. London: George Allen & Unwin.

James. C. L. R. 1963. *The Black Jacobins: Toussaint L'Ouverture and the San Domingo Revolution*. 2d ed., rev. New York: Vintage.

———. 1974. "Haiti, History." In *The New Encyclopaedia Britannica*, 15th ed., Macropaedia, 8:550–52. Chicago: Encyclopedia Britannica.

Jarrell, Richard A. 1987. "Differential National Development and Science in the Nineteenth Century: The Problems of Quebec and Ireland." In Reingold and Rothenberg, eds., *Scientific Colonialism*, pp. 323–50.

Johnson, Jeffrey A. 1985. "German Science Abroad." *Science* 231:414–15.

Joubert, Louis. 1955. *Histoire de législation de la pharmacie aux antilles françaises*. Thèse, Université de Strasbourg, Faculté de Pharmacie.

Julien, Charles-André. 1976. *Les Français en Amérique au XVIIᵉ siècle*. Paris: S.E.D.E.S./C.D.U.

———. 1977. *Les Français en Amérique de 1713 à 1784*. Paris: S.E.D.E.S./C.D.U.

Klein, Herbert S. 1986. *African Slavery in Latin America and the Caribbean*. New York: Oxford University Press.

Knight, Franklin W. 1978. *The Caribbean: The Genesis of a Fragmented Nationalism*. New York: Oxford University Press.

Kolchin, Peter. 1987. *Unfree Labor: American Slavery and Russian Serfdom*. Cambridge: Belknap Press of Harvard University Press.

Konvitz, Josef W. 1987. *Cartography in France, 1660–1848: Science, Engineering and Statecraft*. Chicago: University of Chicago Press.

Lacombe, Robert. 1958. *Histoire Monétaire de Saint-Domingue et de la République d'Haïti jusqu'en 1874*. Paris: Éditions Larose.

Lacroix, Alfred. 1932–38. *Figures de Savants*. 4 vols. Paris: Gauthier-Villars. (Vols. 3–4 [1938] = *L'Académie des Sciences et l'étude de la France d'outre-mer de la fin du XVIIe siècle au début du XIX*.)

Lafforgue, Annie. 1980. "Le Jardin de l'État de Saint-Denis-de-la-Réunion." *Revue française d'histoire d'outre-mer* 67:157–60.

Laguerre, Michel S. 1982. *The Complete Haitiana: A Bibliographic Guide to the Scholarly Literature, 1900–1980*. 2 vols. Millwood, N.Y.: Kraus International Publications.

Laissus, Yves. 1964, 1986. "Le Jardin du Roi." In *Enseignement et diffusion des sciences en France au XVIIIᵉ siècle*, edited by René Taton, pp. 287–342.

Lanier, Clément. 1953. "Le lumière française aux iles Alizées: Quatre visages de missionnaires et de chercheurs." *Conjonction* 48:91–106.

Launay, Catherine. 1978–79. "Un Document: Observations sur les Maladies des Nègres: leurs causes, leurs traitements et les moyens de les prévenir de J. B. Dazille (1738–1812)." Thèse de doctorat en médecine, Université de Nantes. no. 2192.

Laurent, Gerard M. 1956. *Trois mois aux archives d'Espagne*. Port-au-Prince: Les Presses Libres.

Lautour, A. M. [1939]. "Arthaud, Charles." In *Dictionnaire de biographie française*, edited by J. Balteau et al., 3: cols. 1164–65. Paris: Letouzey.

Le Bihan, Alain. 1967. *Loges et Chapitres de la Grande Loge et du Grand Orient de France (2ᵉ Moitié du XVIIIe siècle)*. Paris: Bibliothèque Nationale. [Commission d'Histoire Économique et Sociale de la Revolution Française, *Mémoires et documents*, 20.]

————. 1974. "La Franc-Maçonnerie dans les colonies françaises du XVIIIe siècle." *Annals historiques de la Révolution Française* 46:39–62.

Lefebvre, Georges. 1947. *The Coming of the French Revolution*. Translated by R. R. Palmer. Princeton: Princeton University Press.

Legendre, Lucien Jean. 1958. "Catalogue de la Bibliothèque Haïtienne des Frères de l'Instruction Chrétienne." M.A. thesis: St. Michael's College, Winooski, Vt.

Léon, Rulx. 1928. *La Pratique médicale à Saint-Domingue*. Paris: Presses Modernes.

————. 1931. "Tableau des maladies dans l'ancienne colonie française de Saint-Domingue." *Bulletin de la Société de Médecine d'Haiti* 5:84–87.

————. 1933. *Notes bio-bibliographiques: Médecins et naturalistes de l'ancienne colonie française de Saint-Domingue*. Mimeograph, 89 pp. Port-au-Prince: Bibliothèque du Service d'Hygiène.

————. 1952. "Descourtilz." *Conjonction* 39:40–48.

————. 1953. "Une esquisse de l'histoire de la médecine en Haiti." *Conjonction* 47:5–17.

————. [1954]. *Les Maladies en Haïti*. Port-au-Prince: Imprimerie de l'État.

————. 1976. *Notes Bio-Bibliographiques: Médecins et naturalistes de l'ancienne colonie française de Saint-Domingue*. Rev. ed. Port-au-Prince: Imprimerie Panorama [102 pp.]

Lewis, Gordon K. 1983. *Main Currents in Caribbean Thought: The Historical Evolution of Caribbean Society in Its Ideological Aspects, 1492–1900*. Baltimore: Johns Hopkins University Press.

Lherisson, Camille. 1953. "L'oeuvre scientifique des missionnaires catholiques français à Saint-Domingue." *Conjonction* 43:5–13.

Ligue Féminine d'Action Sociale. [1954]. *Femmes Haitiennes*. Port-au-Prince: Deschamps.

Lingelbach, William E. 1956. "B. Franklin and the Scientific Societies." *Journal of the Franklin Institute* 261:9–31.

Liss, Peggy K. 1982. *Atlantic Empires: The Network of Trade and Revolution, 1713–1826*. Johns Hopkins Studies in Atlantic History and Culture. Baltimore: Johns Hopkins University Press.

Loker, Zvi. 1976a. "Un diplômé de Montpellier—Lopez de Pas, 'Médecin du Roi' à Saint-Domingue." *Revue d'histoire de la médecine hébraïque* 29:53–55.

————. 1976b. "Lopez de Paz: 'Médecin du Roi' au Petit Goave (Un Manuscrit inédit de l'époque coloniale)." *Le Nouvelliste* 15–16, May.

————. 1981. "Professionnels médicaux dans la colonie de St. Domingue au XVIIIᶜ siècle." *Revue de la Société Haïtienne d'Histoire et de Géographie* 39, no. 133:5–33.

Lokke, Carl Ludwig. 1976. *"France and the Colonial Question: A Study of French Opinion: 1763–1801."* Columbia University Studies in History, Economics and Public Law, no. 365. New York: Octagon Books. [Original edition: Columbia University Press, 1932.]

Lowenthal, Ira P., and Drexel G. Woodson, eds. 1974. *Catalogue de la Collection Mangonès*. Yale University, Antilles Research Program, Occasional Papers 2. New Haven: Antilles Research Program.

McClellan, James E., III. 1979. "The Scientific Press in Transition: Rozier's Journal and the Scientific Societies." *Annals of Science* 36:425–49.

———. 1981. "The Académie Royale des Sciences, 1699–1793: A Statistical Portrait." *ISIS* 72:541–67.

———. 1985. *Science Reorganized: Scientific Societies in the Eighteenth Century.* New York: Columbia University Press.

MacLeod, Roy. 1987. "On Visiting the 'Moving Metropolis': Reflections on the Architecture of Imperial Science." In Reingold and Rothenberg, eds., *Scientific Colonialism,* pp. 217–49.

Madiou, Thomas. 1847–1904. *Histoire d'Haïti.* 4 vols. Port-au-Prince: Courtois.

Manuel théorique et pratique de la flagellation des femmes esclaves, d'après le manuscrit inédit d'un planteur espagnol (XVIIIᵉ siècle). [1920s?] Paris: Librairie Franco-Anglaise.

Martineau, Alfred. 1938. Review of "Une Société à Saint-Domingue," by Blanche Maurel. *Revue d'histoire des colonies* 26:148–49.

Martineau, Alfred, et al. 1932. *Bibliographie d'histoire coloniale (1900–1930).* Paris: Société de l'Histoire des Colonies Françaises.

Martinière, Guy. [1981]. "L'Administration de la Guyane Française au XVIIIᵉ siècle." In *L'Historial Antillais,* edited by Roland Suvelor, pp. 361–405. Fort-de-France: Dajani.

Marx, Karl. n. d. *On Colonialism.* Moscow: Foreign Languages Publishing House. [Selections from Marx's writings, 1930?]

Mathieu, Joseph B. 1964. "Saint-Domingue dans les relations internationales (1744 à 1795)." Thèse du Diplôme ès sciences politiques, Institut Universitaire des Hautes Études Internationales de Genève. Typescript

Maurel, Blanche. 1935. "La Poste entre la France et les Isles de l'Amérique à la fin de l'ancien régime." *Revue de la Société d'Histoire et de Géographie d'Haïti* 6, no. 9:1–12.

———. 1938. "Une société de pensée à Saint-Domingue: Le 'Cercle des Philadelphes' au Cap Français." *Franco-American Review* 2:143–67.

———. 1961. "Une société de pensée à Saint-Domingue: Le 'Cercle des Philadelphes' au Cap Français." *Revue française d'histoire d'outre-mer* 47:234–66.

May, L. P. 1936. "Inventaire de la Collection Moreau de Saint-Méry." AN, Salle des Inventaires.

Meinig, D. W. 1986. *The Shaping of America: A Geographical Perspective on 500 Years of History,* Vol. 1: *Atlantic America, 1492–1800.* New Haven: Yale University Press.

Ménier, Marie Antoinette. 1978. "Les Sources de l'histoire de la partie française de l'Ile de Saint Domingue aux Archives Nationales de France." *Conjonction* 140:119–35.

Ménier, M. -A., and G. Debien. 1949. "Journaux de Saint-Domingue." *Revue d'histoire des colonies* 36:424–75.

Mercier, L. 1933. "La Vie au Cap Français en 1789." *Revue d'histoire des colonies* 21:101–30.

Mercier, Roger. 1969. "Les Français en Amérique du Sud au XVIIIe siècle: la mission de l'Académie des sciences (1735–1745)." *Revue française d'histoire d'outre-mer* 56:327–74.

Miller, John Chester. 1977. *The Wolf by the Ears: Thomas Jefferson and Slavery*. New York: Free Press.

Mintz, Sidney W. 1974. "The Caribbean Region." In Mintz, ed., *Slavery, Colonialism and Racism*, pp. 45–71.

———. 1985a. *Sweetness and Power: The Place of Sugar in Modern History*. New York: Viking.

———. 1985b. "From Plantations to Peasantries in the Caribbean." In Mintz and Price, eds., *Caribbean Contours*, pp. 127–53.

Mintz, Sidney W., ed. 1977. *Slavery, Colonialism and Racism*. New York: Norton. [Special number of *Daedalus: Journal of the American Academy of Arts and Sciences*, Spring, 1974.]

Mintz, Sidney W., and Sally Price, eds. 1985. *Caribbean Contours*. Baltimore: Johns Hopkins University Press.

Montague, Ludwell Lee. 1940, 1966. *Haiti and the United States, 1714–1938*. Durham, N.C.: Duke University Press, 1940. Reprint: New York: Russell & Russell, 1966.

Montpeyroux, A. B.-B. marquis de. 1944. *Pages d'histoire aux iles du vent*. Lyon: Nouvelle Revue Héraldique.

Negre, André. 1965, 1967. "Médecine et médecins aux Antilles au XVIIe siècle d'après le Réverend Père Labat." *Histoire de la Médecine* 15:2–17; 16:12–39.

Nicolas, Maurice. 1968. "À travers l'histoire: Les premiers Chambres de Commerce et d'Agriculture." *Bulletin de la Chambre de Commerce et d'Industrie de la Martinique* 25:52–56.

Oleson, Alexandra, and Sanborn C. Brown, eds. 1976. *The Pursuit of Science in the Early American Republic: American Scientific and Learned Societies from Colonial Times to the Civil War*. Baltimore: Johns Hopkins University Press.

Osborne, Michael A. 1987. "The Société Zoologique d'Acclimatation and the New French Empire: The Science and Political Economy of Economic Zoology during the Second Empire." Ph.D. dissertation: University of Wisconsin, Madison. [UM no. 8712432.]

Ott, Thomas O. 1973. *The Haitian Revolution, 1789–1804*. Knoxville: University of Tennessee Press.

Pacquet, Gilles, and Jean-Pierre Wallot. 1987. "Nouvelle-France/Québec/Canada: A World of Limited Identities." In Canny and Pagden, eds., *Colonial Identity in the Atlantic World*, pp. 95–114.

Parry, J[ohn] H[orace]. 1971. *Trade and Domination: The European Overseas Empires in the Eighteenth Century*. London: Weidenfeld and Nicolson.

———. 1981. *The Age of Reconnaissance*. Berkeley: University of California Press.

Pluchon, Pierre. 1984. *Nègres et Juifs au XVIIIᵉ siècle: Le racisme au siècle des Lumières*. Paris: Tallandier.

———. 1985a. "Le Cercle des Philadelphes du Cap-Français à Saint-Domingue: seule Académie coloniale de l'Ancien Régime." *Mondes et Cultures* 45:157–91.

———. 1985b. "La santé dans les colonies de l'Ancien Régime." In Pluchon, ed., *Histoire des Médecins* (1985), pp. 89–129.

———. 1987. *Vaudou sorciers empoisonneurs de Saint-Domingue à Haïti*. Paris: Karthala.

Pluchon, Pierre, ed. 1980. *Justin Girod-Chantrans, Voyage d'un Suisse dans les Colonies d'Amérique*. Paris: Tallandier.

———. 1985. *Histoire des médecins et pharmaciens de marine et des colonies*. Toulouse: Privat.

Pouzet, Pierrick. 1978–79. "Histoire des maladies de St. Domingue par Mr J. B. Pouppé-Desportes (1704–1748)." Thèse de doctorat en médecine, Université de Nantes, no. 2141.

Pressoir, Catts. 1950. *L'Enseignement de l'histoire en Haïti*. Mexico, D.F.: Instituto Panamericano de Geografía e Historia.

Pressoir, Catts, Ernst Trouillot, and Henock Trouillot. 1953. *Historiographie d'Haïti*. Mexico, D.F.: Instituto Panamericano de Geografía e Historia, publication no. 168.

Préval, Guerdy. 1985. *Proverbes haïtiens illustrés*. Ottawa: Canadian Centre for Folk Culture Studies, paper no. 59.

Price, Richard. 1983. *First-Time: The Historical Vision of an Afro-American People*. Johns Hopkins Studies in Atlantic History and Culture. Baltimore: Johns Hopkins University Press.

Price, Richard, ed. 1979. *Maroon Societies: Rebel Slave Communities in the Americas*. 2d ed. Baltimore: Johns Hopkins University Press.

Priestley, Herbert Ingram. 1939. *France Overseas through the Old Regime: A Study of European Expansion*. New York: D. Appleton-Century Co.

Pyenson, Lewis. 1982. "Cultural Imperialism and Exact Sciences: German Expansion Overseas 1900–1930." *History of Science* 20:1–43.

———. 1985. *Cultural Imperialism and Exact Sciences: German Expansion Overseas 1900–1930*. Studies in History and Culture, vol. 1. New York: Lang.

———. [1989]. "Pure Learning and Political Economy: Science and European Expansion in the Age of Imperialism." In *New Trends in the History of Science*, edited by R. P. W. Visser, et al., pp. 209–78. Amsterdam and Atlanta, Ga.: Rodopi.

Rebell, Hugues. 1927, 1966. *Les nuits chaudes du Cap Français*. Paris: Henri Jonquier. Paris: Jérôme Martineau.

Reingold, Nathan. 1976a. "Definitions and Speculations: The Professionalization of Science in America." In Oleson and Brown, eds., *The Pursuit of Knowledge in the Early American Republic*, pp. 33–69.

———. 1976b. "Reflections on 200 Years of Science in the United States." *Nature* 262:9–13.

———. 1978. "O Pioneers!" *Wilson Quarterly* 2:55–64.

———. 1987. "Graduate School and Doctoral Degree: European Models and American Realities." In Reingold and Rothenberg, eds., *Scientific Colonialism*, pp. 129–49.

Reingold, Nathan, and Marc Rothenberg, eds. 1987. *Scientific Colonialism: A Cross-Cultural Comparison*. Washington, D.C.: Smithsonian Institution Press.

Richard, Robert. 1954. "À propos de Saint-Domingue: la monnaie dans l'économie coloniale (1674–1803)." *Revue d'histoire des colonies* 42:22–46.

Roche, Daniel. 1978. *Le siècle des lumières en province: Académies et académiciens provinciaux, 1680–1789*. 2 vols. Paris: Mouton.

————. 1981. *Le Peuple de Paris: Essai sur la culture populaire au XVIIIᵉ siècle.* Paris: Aubier Montaigne.

Rodman, Selden. 1980. *Haiti: The Black Republic.* 5th ed., rev. Old Greenwich, Conn.: Devin-Adair. [Original edition, 1954.]

Ronan, Colin A. 1974. "Pingré, Alexandre-Gui." In *Dictionary of Scientific Biography,* edited by C. C. Gillispie, 10:614–16. [New York: Scribner's Sons].

Ross, Gary N. 1986. "The Bug in the Rug." *Natural History,* March: 67–73.

Rothenberg, Marc. 1982. *The History of Science and Technology in the United States: A Critical and Selective Bibliography.* New York: Garland.

Roussier, Paul. 1929. "Le dépôt des papiers publics des colonies." *Revue d'Histoire Moderne* 4:241–62.

Saintoyant, J. 1929. *La colonisation française sous l'ancien régime.* 2 vols. Paris: La Renaissance du Livre.

Saint-Vil, Jean. 1978. "Villes et bourgs de Saint Domingue au XVIIIe siècle: Essai de géographie historique." *Conjonction* 138:5–32.

Salone, Émile. N.d. *La Colonisation de la Nouvelle-France: Étude sur les origines de la nation canadienne française.* 3d ed. Paris: Guilmoto. Dubuque, Iowa: Wm. G. Brown Reprint Library.

Schofield, Robert E. 1963. "Histories of Scientific Societies: Needs and Opportunities for Research." *History of Science* 2:70–84.

Shannon, Magdaline W. 1979. "Bibliography of Saint-Domingue, Especially for the Period of 1700–1804." *Revue de la Société Haïtienne d'Histoire de géographie et de géologie* 37, no. 125:5–55.

Shapin, Steven, and Simon Schaffer. 1986. *Leviathan and the Air Pump: Hobbes, Boyle and the Experimental Life.* Princeton: Princeton University Press.

Spencer, Ivor D. *See* Moreau de Saint-Méry (1985).

Stearns, Raymond Phineas. 1970. *Science in the British Colonies of North America.* Urbana: University of Illinois Press.

Stehlé, Henri. 1966. "Quelques mises au points historiques relatives à l'introduction de végétaux économiques aux antilles françaises." *Bulletin de la Société d'histoire de la Guadeloupe* 5–6:27–37.

————. 1967. "Évolution de la connaissance botanique et biologique aux Antilles françaises." In *Comptes Rendus du 91ᵉ Congrès National des Sociétés Savantes (Rennes, 1966),* Section des Sciences, 1:275–90. [Ministère de l'Éducation Nationale; Comité des Travaux Scientifiques et Historiques. Paris: Gauthier-Villars et la Bibliothèque Nationale.

————. 1970. "La Contribution des savants français à l'étude des sciences naturelles aux Antilles." In *Comptes Rendus du 94ᵉ Congrès National des Sociétés Savantes (Pau, 1969),* Section des Sciences, 1:81–98. [Paris: Bibliothèque Nationale.]

Stein, Robert Louis. 1979. *The French Slave Trade in the Eighteenth Century: An Old Regime Business.* Madison: University of Wisconsin Press.

————. 1988. *The French Sugar Business in the Eighteenth Century.* Baton Rouge: Louisiana State University Press.

Stone, Carl. 1985. "A Political Profile of the Caribbean." In *Caribbean Contours,* edited by S. Mintz and S. Price, pp. 13–53.

Stroup, Alice. 1987. "Royal Funding of the Parisian Académie Royale des Sciences during the 1690s." *Transactions of the American Philosophical Society* 77, pt. 4.

Struik, Dirk J. 1962. *Yankee Science in the Making.* Rev. ed. New York: Collier Books.

———. 1984. "Early Colonial Science in North America and Mexico." *Quipu* 1:24–54.

Talman, C. Fitzhugh. 1906. "Climatology of Haiti in the Eighteenth Century." [U.S. Weather Bureau] *Monthly Weather Review and Annual Summary* 34:64–73.

Tanguy, Marc. 1977–78. "Le Traité des fièvres de l'île de Saint-Domingue." Thèse de doctorat en médecine, Université de Nantes.

Tarrade, Jean. 1972. *Le Commerce colonial de la France à la fin de l'Ancien Régime: L'évolution du régime de "l'exclusif" de 1763 à 1789.* 2 vols. Paris: Presses Universitaires de France.

Taton, René, ed. 1964, 1986. *Enseignement et diffusion des sciences en France au XVIIIᵉ siècle.* Paris: Hermann.

Taylor, E. G. R. 1951. *The Haven-Finding Art.* London: Hollis and Carter.

Taylor, Jean Gelman. 1983. *The Social World of Batavia: European and Eurasian in Dutch Asia.* Madison: University of Wisconsin Press.

Thésée, Françoise. 1974. Review of *Les marrons de la liberté,* by Jean Fouchard. *Revue française d'histoire d'outre-mer* 61:178–81.

Thésée, Françoise, and G. Debien. 1974. "Un colon niortais à Saint-Domingue: Jean Barré de Saint-Venant (1737–1810)." *Bulletin de la Société Historique et Scientifique des Deux Sèvres* 7:355–523.

Thompson, G. A., trans. and ed. 1970. *The Geographical and Historical Dictionary of America and the West Indies.* 5 vols. New York: Lenox Hill. Original edition, 1812.

Tippenhauer, L. Gentil. 1893. *Die Insel Haiti.* Leipzig: Brockhaus.

Tooley, Ronald Vere, complier. 1979. *Tooley's Dictionary of Mapmakers.* New York: Alan R. Liss.

Torlais, Jean. 1964, 1986. "La physique expérimentale." In René Taton, ed., *Enseignement et diffusion des sciences en France,* pp. 619–45.

Toussaint, Auguste. 1966. *Une Cité tropicale: Port-Louis de l'Ile Maurice.* Paris: Presses Universitaires.

Toussainte, A., and H. Adolphe. 1956. *Bibliography of Mauritius (1502–1954).* Port-Louis, Mauritius: Esclapon.

[Tramond, Joannès]. 1927. "Une Loge de Francs-Maçons à Saint-Domingue en 1740." *Revue de l'histoire des colonies françaises* 15:603–6.

Tramond, J[oannès]. [1928]. *Saint-Domingue en 1756 et 1757 d'après la correspondance de l'ordonnateur Lambert.* Paris: Société de l'Histoire des Colonies Française.

———. 1932. "Antilles et Guyane." In Martineau, *Bibliographie d'histoire coloniale,* pp. 374–390.

Trouillot, Enock. 1979. "L'Historiographie d'Haïti pendant ces vingt dernières années." *Revue de la Société Haïtienne d'Histoire de géographie et de géologie* 37, no. 125:71–79.

Trouillot, Henock. 1965. "Economie et finances de Saint-Domingue." *Revue de la Société Haïtienne d'Histoire de géographie et de géologie* 33, no. 110:1–139.

Vaissière, Pierre de. 1909. *Saint-Domingue (1629–1789): La Société et la vie créoles sous l'ancien régime*. Paris: Librairie Académique Perrin.

Vastey, Pompée Valentin, Baron de. 1814. *Le Système colonial dévoilé*. Au Cap-Henri [Cap Haïtien]: Roux.

Vignols, Léon. 1927. "Les cabarets et leurs grands protecteurs." *Revue d'Histoire Economique et Sociale* 15:359–65.

Vigoureux, G. 1921. *Jérémie: Son origine et son développement*. Port-au-Prince: Pierre Noël.

Vilaire, Patrick, Michèle Oriol, and Réginal Cohen. 1981. *Images d'Espanola et de St. Domingue*. Port-au-Prince: Henri Deschamps.

Watts, David. 1987. *The West Indies: Patterns of Development, Culture and Environmental Change since 1492*. Cambridge: Cambridge University Press.

Webster, Richard A., et al., 1974. "European Overseas Exploration and Empires." In *The New Encyclopaedia Britannica*, 15th ed., Macropaedia, 18:866–91. Chicago: Encyclopedia Britannica.

Weinberg, Gregorio. 1978. "Sobre la historia de la tradición cientifica latino-americana." *Interciencia* 3:72–78.

Williams, Eric. 1964. *Capitalism and Slavery*. London: André Deutsch.

Wilson, Charles. 1975. "The Economic Role and Mainsprings of Imperialism." In *Colonialism in Africa*, vol. 4: *The Economics of Colonialism*, edited by Peter Guigan and L. H. Gann, pp. 68–94. Cambridge: Cambridge University Press.

Woolf, Harry. 1959. *The Transits of Venus: A Study of Eighteenth-Century Science*. Princeton: Princeton University Press.

Zay, E. 1892. *Histoire monétaire des colonies françaises*. Paris: J. Montorier.

Zupko, Ronald Edward. 1978. *French Weights and Measures before the Revolution: A Dictionary of Provincial and Local Units*. Bloomington: Indiana University Press.

INDEX

❦

Académie française, 237
Académie Royale des Belles-Lettres
   (Arras), 237, 267
Académie Royale de Chirurgie, 138
Académie Royale de Marine (Brest), 4,
   124–25, 253, 326n.184, 355n.27
Académie Royale des Sciences (Paris):
   approves irrigation plans, 73; astro-
   nomical/cartographical work, 117–25;
   botanical enterprises, 149–50; and
   Cercle, 251–56, 353n.80; longitude
   work, 122–23; investigates mesmerism,
   176; mathematical expertise in, 294;
   membership, 255, 261; mineral water
   analyses, 244; publications, 102, 165;
   supervises balloon flights, 168; tests
   cochineal, 156
Académie Royale des Sciences et Belles-
   Lettres (Bordeaux), 219, 257, 267, 277
Académie Royale des Sciences, Belles-
   Lettres et Arts (Lyon), 169–70
Académie Royale des Sciences, Belles-
   Lettres et Arts (Rouen), 161, 267
Académie des Sciences, Arts et Belles-
   Lettres (Dijon), 144, 168–69, 267
Acadians (settlers), 62
Actors, 95, 321n.74. See also Theaters
Adanson, Michel, 152
Admiralty courts, 44, 86
Affiches Américaines: and Cercle, 238–39,
   279; and colonial theater, 321n.73;
   debate over literary society in, 188–91;
   described, 20, 98, 99–101, 306n.59,
   307n.26, 321–22; letter from prostitute
   in, 60; as meteorological clearing house,

165–66; reports on balloon flights in,
   169; science published in, 126, 171, 226.
   See also Mozard, Charles
Affranchis, 47–48, 51. See also Free people
   of color; Mulattoes
African-Americans. See Free blacks; Slaves
Agroindustries, 12, 14, 68, 316n.23
Alcohol, alcoholism, 30, 57
Alfort veterinary school, 135, 233–34
Almanach de Saint-Domingue, 98, 124, 236,
   251, 259, 318n.21, 321n.87, 327n.8
American Academy of Arts and Sciences
   (Boston), 5, 301n.10
American Philosophical Society (Phila-
   delphia), xiii, 5, 227, 268, 347n.90
Amsterdam Botanic Garden, 66
Anglade, Georges, 17, 239, 304n.46
"Animal magnetism." See Mesmerism
Apothecaries/apothecary shops, 102–3,
   134
Arawak Indians, 35, 308nn.5, 6, 355n.31;
   name Hayti, 15; studies of, 114, 268–69
Army. See Military
Artaud, Jean, 137, 358n.32
Arthaud, Charles (1748–93?): addresses
   assemblies, 275, 282; addresses Cercle,
   212, 215, 231, 247, 256, 341n.6; attacks
   medical opponents, 140–42, 178–80; on
   Cercle's origins, 179–80; contacts Franklin, 226; contra meteorology, 166; elected to APS, 227, 268; on flour stores,
   244–46; as founder and leader of Cercle,
   200, 212, 223, 343n.33; history of
   Cercle, 195, 247; on Indians, 269; life/
   fate, 137–38, 274–75, 284, 356n.6;

〖383〗

41; affected by earthquakes, 27–28; fire-fighting equipment, 28; founded, 81; hospital, 94, 129; Masonic lodges, 106; medical personnel in, 131; military barracks, 89; poorhouse, 320n.66; population, 75; port, 319n.33; theater, 94–95; water system, 88. *See also* Jardin du Roi; Jardin Royal

Port-de-Paix (town), 75, 81

Postal system, 81

Pouppée-Desportes, J. B. R., 137, 146

Press, the, 97–102. See also *Affiches Américaines;* Pornography

Print trades, 98, 101

Prostitutes/prostitution, 59–60, 90, 96, 321n.78

Providence des Hommes/Femmes. *See* Cap François, poorhouses/hospices

Provincial Assembly of Saint Domingue, 162, 275, 276, 278, 279. *See also* Colonial Assembly

Public health service, 129–31. *See also* Colonial medicine

Puységur, A.-H.-A. de Chastenet, count de: cartographical work, 125; proponent of mesmerism, 176–78

Pyenson, Lewis, 6, 8, 302n.14, 360n.15

Quinine, 113, 161

Race, theories of, 241–42, 311n.1

Race relations/tensions, 3, 44, 47, 51, 54

Racism: race prejudice, 58; in Saint Domingue, 80, 96, 136

Raimond, Julien, 286

Raynal, abbé G.-T., 305n.53

Religious orders, 38, 90, 91, 92, 309n.27

Reunion. *See* Île Bourbon

Reveillon paper factory, 343n.42

Richelieu, sends missionaries, 111–12

Richer, Jean, 118

Ricord, Joseph (colonist), 71

Roads, 78–80; royal, 80, 143

Roche, Daniel, 186

Rochefort, César de, 323n.4

Royal navy: as coloniing agent, 4; medical branch/corps, 31, 138–42, 228; presence in Saint Domingue, 41, 83; repository of charts and maps, 124; tests chronometers, 122

Royal Society of London, 151, 192, 355n.28

Royal Society of Sciences and Arts du Cap François. *See* Cercle des Philadelphes; Société Royale des Sciences et des Arts du Cap François

Rozier's Journal (*Observations sur la physique*), 146, 270–71, 277, 356n.41

Rubber, research concerning, 151

Rush, Benjamin, 350n.44

Sailing, transatlantic, 23–24

Saint Christopher (Saint Kitts) colony, 34

Saint Domingue (French colony): acclimatization, 30; as armed camp, 41; bells, 319n.35; as bilingual colony, 54; churches in, 38, 85; climate, 26, 352n.72; coasts, 26; cultivation in, 64; destroyed in Revolution, 5; distance from France, 23; dowsers in, 103; fauna in, 31, 33; feral animals in, 31–32; forests, 32; as French province, 9–10, 40, 44, 83, 295–96; French recollections of, 15; on frontier with nature, 30–32; goods and services available in, 102–3; historical development, 34–46; homeless in, 56, 91; importance, 2–5, 10, 12, 63; imported plants/animals, 32–33, 157, 162; industries (miscellaneous), 69, 70; isolated areas, 79; libraries in, 101; location, 304n.42; as "lost" colony, 14–19; maps of, 120, 124–26, 326; marine life, 31; material environment, 23–33; medical personnel in, 129–35; official meridian, 324n.8; mountains, 24, 55, 79–80; name, 15–16, 304n.43; overlooked historiographically, 17–19; pests in, 31; plains, 24–25; policed society; population, 3, 48–51; quicksand, 79; racial balance, 49; reading in, 101–2; reputation, 57–58, 301n.9; revolution in, 14–15, 273–87, 303n.38, 356n.1; rivers, 26, 80; salt production, 309n.13; as scientific center, 4–5, 10; sexual tension in, 57, 60, 96, 314n.49; social life, 96–97; sociology, 41–42, 47–61; soil erosion/exhaustion, 32; studies of, 16–17; sugar production in, 64–65; sumptuary laws in, 314n.55; territory, 24; urban civilization, 3, 75; as venue for research, 117, 296; wealth, 63–64. *See also* Cap François; Diseases; Les Cayes; Maréchaussée; Port-au-Prince; Slavery